ELECTRON HOLOGRAPHY

North-Holland
Delta Series

ELSEVIER
Amsterdam – Lausanne – New York – Oxford – Shannon – Tokyo

Electron Holography

Proceedings of the International Workshop on
Electron Holography, Holiday Inn World's Fair,
Knoxville, Tennessee, USA, August 29 – 31, 1994

Edited by

A. Tonomura

Exploratory Research for Advanced Technology (ERATO)
Research Development Corporation of Japan (JRDC)
and Advanced Research Laboratory, Hitachi, Ltd., Saitama, Japan

L.F. Allard

Oak Ridge National Laboratory
Oak Ridge, Tennessee, USA

G. Pozzi

Department of Physics, University of Bologna
Bologna, Italy

D.C. Joy

EM Facility, University of Tennessee
Knoxville, Tennessee, USA

Y.A. Ono

Exploratory Research for Advanced Technology (ERATO)
Research Development Corporation of Japan (JRDC)
and Advanced Research Laboratory, Hitachi, Ltd., Saitama, Japan

1995

ELSEVIER
Amsterdam – Lausanne – New York – Oxford – Shannon – Tokyo

ELSEVIER SCIENCE B.V.
Sara Burgerhartstraat 25
P.O. Box 211, 1000 AE Amsterdam
The Netherlands

ISBN: 0-444-82051-5

© 1995 ELSEVIER SCIENCE B.V. All rights reserved.

No part of this publication may be reproduced, stored in a retrieval system or transmitted, in any form or by any means, electronic, mechanical, photocopying, recording or otherwise, without the prior written permission of the publisher, Elsevier Science B.V., Copyright & Permissions Department, P.O. Box 521, 1000 AM Amsterdam, The Netherlands.

Special regulations for readers in the U.S.A. - This publication has been registered with the Copyright Clearance Center Inc. (CCC), 222 Rosewood Drive, Danvers, MA 01923. Information can be obtained from the CCC about conditions under which photocopies of parts of this publication may be made in the U.S.A. All other copyright questions, including photocopying outside of the U.S.A., should be referred to the copyright owner, Elsevier Science B.V., unless otherwise specified.

No responsibility is assumed by the publisher for any injury and/or damage to persons or property as a matter of products liability, negligence or otherwise, or from any use or operation of any methods, products, instructions or ideas contained in the material herein.

This book is printed on acid-free paper.

Printed in The Netherlands.

PREFACE

The International Workshop on Electron Holography was held on August 29-31, 1994 at the Holiday Inn World's Fair, Knoxville, Tennessee, USA. This workshop was the first international meeting on electron holography and almost all of the world's electron holography experts on three continents convened to discuss the state-of-the-art developments in this field.

Electron holography is a very powerful technique which uses the phase information in the electron image to study fundamental physics of matter at the micrometer to nanometer scale, or even to Ångstrom scale. A few years ago, successful examples of electron holography included confirmation of Aharonov-Bohm (AB) effect and direct visualization of magnetic fields not only outside but also inside of magnetic materials. Since then, fields of application of electron holography have expanded from solid state physics to materials science and the biological sciences, and new techniques have been invented including real-time holography and computed tomography of the reconstructed phase image. Therefore we thought it appropriate and timely to hold an international gathering of electron holography experts to discuss recent developments and future prospects.

The workshop was co-organized by the Research Development Corporation of Japan (JRDC) and Oak Ridge National Laboratory (ORNL) and was held, primarily, to celebrate the completion of two major research projects, the Tonomura Electron Wavefront Project under the Exploratory Research for Advanced Technology (ERATO) program of JRDC, and the ORNL electron holography project funded by the Laboratory Directed Research and Development Program. Because of the close connections these two projects have had with both the University of Bologna, Italy, through Professor G. Pozzi, a long-time collaborator with Dr. Tonomura, and the University of Tennessee (UT), through Professor D.C. Joy, an ORNL/UT Distinguished Scientist, these two institutions also co-organized the meeting. Major funding for the workshop was contributed by JRDC and ORNL.

The meeting comprised 29 invited oral presentations and an additional dozen poster presentations. Time was allowed for discussion of the talks and posters, and some interesting debates ensued which were exciting and suggestive for future work. These presentations covered a wide area of application fields and new techniques; magnetic fluxes (vortices) in superconductors, magnetic domains, magnetic and electric microfields, and p-n junctions, catalysts, biological filaments, fine particles, and fullerenes for materials applications. Talks were also presented on digital recording and processing, real-time electron holography, computed tomography, and STEM holography.

In this proceedings, we have edited the manuscripts of presentations and classified them into four main parts: Introduction to Electron Holography, Holographic

Interference Electron Microscopy - Techniques, Holographic Interference Electron Microscopy - Applications, and New Types of Electron Holography. In addition, D.C. Joy's concluding remarks explain the atmosphere of the workshop. We hope that the proceedings will play a role of not only a good introductory book for newcomers in this field but also a reference book of new techniques and applications for experts.

We would like to thank Mr. T.A. Nolan and Dr. E. Völkl of Oak Ridge National Laboratory, Dr. B.G. Frost of the University of Tennessee, and Dr. T. Tanji of ERATO, JRDC for their help in editing the proceedings.

February 1995

A. Tonomura
L.F. Allard
G. Pozzi
D. C. Joy
Y. A. Ono

International Workshop on Electron Holography

COMMITTEES

Organizing and Program Committee:

A. Tonomura, ERATO, JRDC and Advanced Research Laboratory, Hitachi, Ltd.
L.F. Allard, Oak Ridge National Laboratory
T.A. Nolan, Oak Ridge National Laboratory
G. Pozzi, University of Bologna
D.C. Joy, University of Tennessee
Y.A. Ono, ERATO, JRDC and Advanced Research Laboratory, Hitachi, Ltd.

TABLE OF CONTENTS

Preface	v
Photo	vii
Committees	viii

INTRODUCTION TO ELECTRON HOLOGRAPHY

Progress in holographic interference electron microscopy
   A. Tonomura   1

Electron holography: State and experimental steps towards 0.1 nm with the CM30-Special Tübingen
   H. Lichte   11

HOLOGRAPHIC INTERFERENCE ELECTRON MICROSCOPY - TECHNIQUES-

Digital recording and processing of electron off-axis holography
   G. Ade   33

Observation of atomic surface potential by electron holography
   T. Tanji and K. Ishizuka   45

Phase-shifting techniques in electron holography
   Q. Ru   55

Holographic reconstruction methods
   M. Lehmann and H. Lichte   69

Real-time electron holography using a liquid-crystal panel
   J. Chen, T. Hirayama, G. Lai, T. Tanji, K. Ishizuka and A. Tonomura   81

Electron holographic computed tomography
   G. Lai, T. Hirayama, A. Fukuhara, K. Ishizuka, T. Tanji and A. Tonomura   93

Practical electron holography: Applications of advanced hologram processing techniques to materials science problems
   E. Völkl, L.F. Allard and B. Frost   103

Retrieval of atomic displacements from reconstructed electron waves
as an ill-posed inverse problem
    K. Scheerschmidt and F. Knoll   117

HOLOGRAPHIC INTERFERENCE ELECTRON MICROSCOPY - APPLICATIONS -

Interference- and Lorentz-image simulations of vortices
in superconductors
    G. Pozzi, J. E. Bonevich, K. Harada, H. Kasai,
    T. Matsuda, T. Yoshida and A. Tonomura   125

Magnetic field observation of vortices in superconductors
by electron holography
    J. E. Bonevich, K. Harada, H. Kasai, T. Matsuda,
    T. Yoshida, G. Pozzi and A. Tonomura   135

Holographic studies on magnetic phenomena in small regions
    T. Hirayama, J. Chen, Q. Ru, K. Ishizuka, T. Tanji and
    A. Tonomura   145

Electron holography of magnetic and electric microfields
    G. Matteucci, M. Muccini and D. Cavalcoli   159

Holography of electrostatic fields
    B.G. Frost, L.F. Allard, E. Völkl and D.C. Joy   169

Quantitative applications of off-axis electron holography
    D.J. Smith, W.J. de Ruijter, M. Gajdardziska-Josifovska,
    M.R. McCartney and J.K. Weiss   181

Electron holography of p-n junctions
    M.R. McCartney, B. Frost, R. Hull, M.R. Scheinfein,
    D.J. Smith and E. Voelkl   189

Electron holography of heterogeneous catalysts
    A.K. Datye, D.S. Kalakkad, E. Völkl and L.F. Allard   199

Electron holography in materials science
    X. Lin, V. Ravikumar, R.P. Rodrigues, N. Wilcox and
    V. P. Dravid   209

Electron holography applied to the study of fullerene materials
L.F. Allard, E. Völkl, S. Subramoney and R.S. Ruoff   219

Transmission electron holography of polymer microstructure
M. Libera, J. Ott and Y.C. Wang   231

Electron holographic observation of thin biological filaments
K. Aoyama, G. Lai and Q. Ru   239

Fraunhofer in-line electron holography of small weak-phase objects
T. Matsumoto, T. Tanji and A. Tonomura   249

NEW TYPES OF ELECTRON HOLOGRAPHY

State of the art of low-energy electron holography
H.-W. Fink, H.J. Kreuzer and H. Schmid   257

On the reconstruction of low voltage point projection holograms
J.C.H. Spence, X. Zhang and W. Qian   267

Coherent electron diffraction and holography
J.W. Steeds, P.A. Midgley, P. Spellward and R. Vincent   277

Modeling of convergent beam interferometry
R.A. Herring and G. Pozzi   287

Interpreting the reconstructed object waves
D. Van Dyck and M. Op de Beeck   297

Focal series wave function reconstruction in HRTEM
M. Op de Beeck, D. Van Dyck and W. Coene   307

High-resolution tilted single-sideband holography
K. Ishizuka   317

STEM holography of magnetic materials
M. Mankos, P. de Haan, V. Kambersky, G. Matteucci, M.R. McCartney, Z. Yang, M.R. Scheinfein and J.M. Cowley   329

Amplitude-division electron holography
Q. Ru   343

CLOSING ADDRESS

Concluding remarks
   D.C. Joy   355

List of participants   357

Author index   367

Progress in Holographic Interference Electron Microscopy

Akira Tonomura

Advanced Research Laboratory, Hitachi, Ltd.
Tonomura Electron Wavefront Project, ERATO, JRDC

Hatoyama, Saitama 350-03, Japan

Abstract
   Electron interferometry, which reveals microscopic objects and fields utilizing its extremely short wavelength, has recently progressed significantly with the development of a "coherent" field-emission electron beam and electron-holographic image-reconstruction techniques.

1. INTRODUCTION

   The phase shift of an electron wave can be measured from an interference pattern between the wave of an object to be observed and a reference wave. The first electron interference pattern was observed by Boersch [1] in 1940: The interference fringes are called "Fresnel fringes", and are produced by interference between a cylindrical wave scattered from a sample edge and a plane wave passing through away from the edge.
   An interferometer for phase measurement was developed by Marton [2] in 1952, which was a Mach-Zehnder type interferometer using Bragg reflections off of three sheets of single-crystalline film. The idea was derived from the interference fringes obtained in the preceding year by Uyeda et al. [3], which was formed by transmitted and Bragg-reflected beams from two sheets of single-crystalline film. This interferometer was not practical for use as an interferometer for electrons, but is now used for X-rays and neutron beams [4].
   The electron interferometer now widely used is an electron biprism devised by Möllenstedt and Düker [5] in 1955. Extremely interesting, though not many, experiments were carried out until the early 1970s, first at Tübingen University [6], and CNRS Toulouse [7], and later at the Technical University of Berlin [8], Tohoku University [9], Bologna University [10], PTB Braunschweig [11], Hitachi [12] and elsewhere. These experiments investigated the inner and contact potentials [6,7], quantized magnetic fluxes [8] and magnetic fields [10, 12], which are well summarized in a review article by Missiroli et al. [13].
   These electron-interference experiments have been carried out using a transmission electron microscope equipped with an electron biprism. The advent of a "coherent" field-emission electron beam [14] has greatly expanded the feasibility of electron-interference experiments: The interference patterns

have become directly observable on the fluorescent screen and the total number of interference fringes has increased from 300 to more than 3000.

This development has, furthermore, made it possible to utilize electron holography [15] to open new possibilities in electron interferometry. For example, phase contour maps have become obtainable and the measurement precision in the phase has increased to 1/100 of the wavelength [16]. The historical developments of electron holography are omitted in this paper, but are given elsewhere [17]. This paper presents recent developments in electron interferometry.

2. ELECTRON HOLOGRAPHY

Electron holography is a two-step imaging method: A hologram is formed by the interference between an object wave and a reference wave using an electron biprism, and then an image is reconstructed optically by illuminating a laser beam onto the hologram. Once electron wavefronts are transformed into optical wavefronts, versatile optical techniques can be applied to do what has been impossible in electron optics.

Optical reconstruction with laser light is simple, but must be done off-line as a result of the time involved in developing the film. Therefore, on-line reconstruction techniques are being developed using computers and optical devices. An image can be numerically reconstructed from a hologram recorded on a charge-coupled device (CCD) attached to an electron microscope. By doing this, an amplitude image, phase image, and interference image can be displayed. These images can be obtained in a fairly short time, depending on the performance of the computer used, but not yet in real-time.

This speed problem is now being addressed with the development of real-time methods, for example, using a liquid-crystal panel as a phase hologram [18]. The image signal of the hologram detected with a TV camera attached to the electron microscope is transferred to the liquid-crystal panel, where the intensity distribution is transformed into a phase shifting function for an illuminating light beam. The time resolution of 1/30 s is limited by the scanning rate of the TV system. Dynamic phenomena can therefore be observed in real time with this method. This method has been used to observe the dynamics of magnetic domain walls [19].

The liquid-crystal panel can also be used to obtain arbitrarily defocused images or corrected images from a single hologram by placing the panel at the back focal plane of an image-forming lens in the optical reconstruction stage. This is because a two-dimensional phase-shifter having an arbitrary phase distribution function can be obtained by applying the appropriate electric signal to the liquid-crystal panel. Therefore, the liquid-crystal panel acts as a phase-shifter for spatial filtering [20], compensating for the lens aberrations caused by the electron lens, thereby producing images under arbitrary focusing conditions.

Information about an object, which cannot be obtained from a single hologram, can be obtained from multiple holograms. An example is a phase-shifting method of producing interference micrographs with higher precision. In this method, many electron holograms of an object are recorded on video tape in such a way that the carrier fringes are displaced little by little, for example, by slightly changing the angle of the incident electron beam to the object while the in-focus image of the object is kept still. Then, the phase distribution of the transmitted electron beam is calculated numerically from these holograms [21]. The measurement precision in the phase distribution increased from $2\pi/4$ in the case of a single hologram to less than $2\pi/100$ when 200 holograms were used [22].

When electron holograms of an object are formed at different incident angles within range of ±60°, three-dimensional information about the object can be obtained by numerical calculation [23]. Furthermore, when electron holograms are formed at various angles tilting around two axes, the distribution of vector fields such as magnetic fields can be determined [24].

3. PRINCIPLE BEHIND ELECTRON INTERFEROMETRY

When a parallel electron beam is incident to electromagnetic fields, the electron beam is phase-shifted. The phase shift $\Delta S / \hbar$ is calculated from the Schrödinger equation as,

$$\Delta S / \hbar = (1/\hbar) \oint (mv - eA) \cdot ds$$
$$= (1/\hbar) \oint \left(\sqrt{2meV} - e\mathbf{t} \cdot \mathbf{A}\right) ds, \qquad (1)$$

where integration is carried out along a closed path connecting two electron trajectories and $\mathbf{t}$ is the unit tangent vector of the electron trajectory. This equation shows that electromagnetic potentials $(\mathbf{A}, V)$ can be detected by measuring the phase shift in an electron beam, although what we detect is not the electromagnetic potentials themselves but their integrals along the electron-beam trajectory.

4. APPLICATIONS

4.1. Specimen thickness measurement

An electron beam is accelerated by the inner potential when it enters a specimen. The specimen is regarded as a space within which the electrostatic potential is different from that in the vacuum. An electron beam transmitted through the specimen thus undergoes a phase shift depending on the specimen material and thicknes as can be seen from Equation (1). When a specimen is made of a uniform material, a contour map of the electron phase distribution indicates its thickness contours. An example, an interferogram reconstructed from an incoherent hologram of fine gold particles [25], is shown in Fig. 1. The carrier fringes of this hologram were lattice fringes of a single-crystalline thin film, which could be formed even with an "incoherent" incident electron beam. Either a transmitted beam or a Bragg-reflected beam is modulated by an object located a small distance from the film to form a hologram. The thickness distribution can be quantitatively measured from the interferogram, because one fringe spacing corresponds to a thickness change of 330 Å. The particles here are shown to take the shape of a truncated pyramid with a triangular base

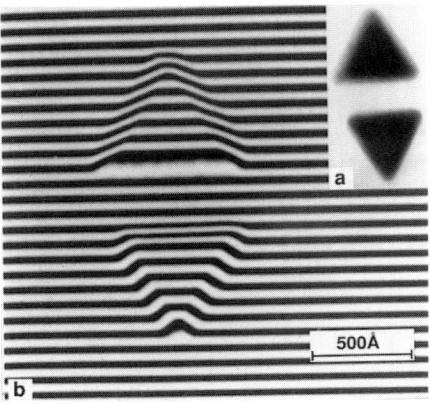

Fig. 1. Interference micrograph of fine gold particles reconstructed from an incoherent hologram: (a) Electron micrograph and (b) Interference micrograph.

4.2. Magnetic-field observation

For a magnetic object, the phase difference between two electron beams passing through it is given by

$$\Delta S / \hbar = -(e/\hbar) \oint A \cdot ds = -(e/\hbar) \int B \cdot dS. \tag{2}$$

It is concluded from this equation [26] that an interference micrograph can be interpreted in a straightforward way, as follows:

(1) Contour fringes in the interference micrograph indicate magnetic lines of force, because the phase difference $\Delta S / \hbar$ vanishes between two beams passing through arbitrary points along a magnetic line.
(2) A magnetic flux of h/e flows between two adjacent contour fringes.

An example of this is shown in Fig. 2. The object here is a smoke particle prepared by gas evaporation in an inert-gas atmosphere. No contrast can be seen in electron micrograph (a), which represents the electron intensity distribution. However, circular contour fringes appear in contour map (b).
Since the phase distribution is amplified two times, these contour lines indicate magnetic lines of force in $h/2e$ units. It can be seen at a glance how magnetic lines of force rotate in such a fine particle.

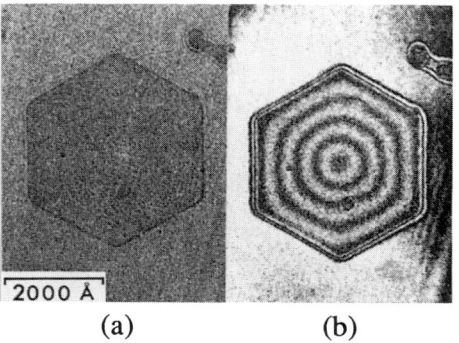

Fig. 2. Interference micrograph of a fine particle: (a) Electron micrograph and (b) contour map (phase amplification: ×2).

The diameter of this particle is around 3000 Å. For smaller or anisotropic particles, magnetization is not closed inside, but the particle is uniformly magnetized. An example of a barium-ferrite particle [27] is shown in Fig. 3. Here, magnetic fields leak from the upper, north, pole of the particle and are then sucked up at the south pole. The particle clearly has a single magnetic domain.

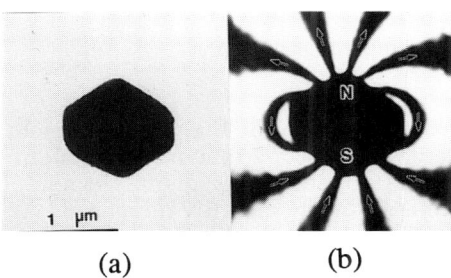

Fig 3. Interference micrograph of a barium-ferrite particle:
(a) Electron micrograph and (b) Interference micrograph.

4.3. Observation of flux lines in superconductors

Flux lines in superconductors were observed quantitatively by interference microscopy [28, 29] and Lorentz microscopy [30, 31] with a 350-kV holography electron microscope [32].

The experimental arrangement is shown in Fig. 4(a). A thin-film sample, set on a low-temperature stage, was tilted 45° to an incident electron beam so that the electron beam could be influenced by the flux-line magnetic fields. An external magnetic field of up to 150 gauss was applied horizontally. An example of a flux-line array in single-crystalline Nb thin film [28] is shown in Fig. 4(b). In this interference micrograph, projected magnetic lines of force can be observed. These lines become dense in the localized regions indicated by circles in the photograph, which correspond to individual flux lines.

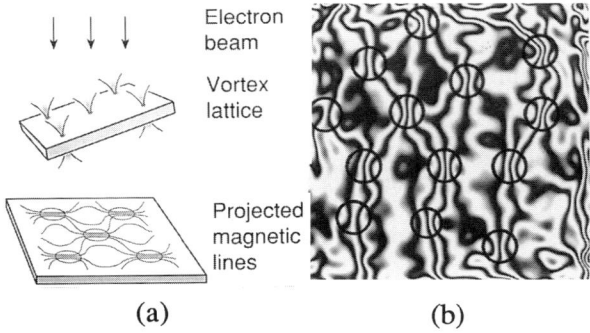

Fig 4. Interference micrograph of flux lines in Nb film at $T = 4.5$ K, $B = 20$ gauss (Phase amplification: ×16):(a) Schematic and (b) interference micrograph.

Although interference microscopy has high spatial resolution and produces quantitative results, Lorentz microscopy is more convenient for observing the dynamic behavior of flux lines. In this experiment, the sample was first cooled to 4.5 K and the magnetic field B was gradually increased. As B increased, flux lines suddenly began to penetrate the film at around $B = 30$ gauss, and the number of flux lines increased as B was increased further. Their dynamic behavior was quite interesting. At first, only a few flux lines appeared here and there in a 15×10-μm^2 field of view. They oscillated around their own pinning centers and occasionally hopped from one center to another. These movements continued for a few minutes until arriving at an equilibrium state.

An equilibrium Lorentz micrograph at $B = 100$ gauss [30] is shown in Fig. 5. The film has a fairly uniform thickness in the region shown, but is bent along the black curves, called bend contours, due to Bragg reflections from the atomic plane brought to a favorable angle by bending. Each spot showing a black and white contrast is an image of a single flux line. This contrast reversed, as expected, when the applied magnetic field was reversed. The polarity of a vortex can be read from the line dividing the black and white parts of the spots. Because the black part is on the same side of all the spots, the polarities of all the flux lines seen in the region are the same. At low B, i.e., 30 - 50 gauss, the flux lines are too sparse to form a lattice, even in equilibrium. At B = 100 gauss, the flux-line density is so high that they cannot form anything but a hexagonal lattice.

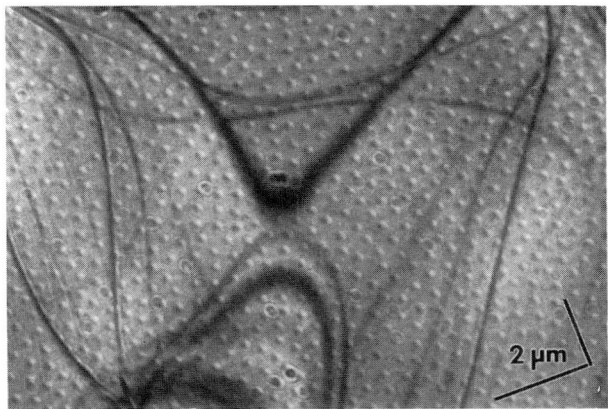

Fig 5. Lorentz micrograph of a two-dimensional array of flux lines in superconducting Nb film at T= 4.5 K, B = 100 gauss.

A high-T_C superconductor has been investigated by Lorentz microscopy [31]. High-T_C superconductors have been difficult to use practically, because the critical current vanishes at high temperatures and under high magnetic fields, even when the temperature is well below the critical temperature T_C. This phenomenon most probably arises from the behavior of flux lines, but has not yet been proven concretely. Some researchers believe that these flux lines melt like molecules in a liquid, and, as a result, it is difficult to fix flux lines at some pinning sites [33]. Evidence for flux-line melting was provided by a Bitter BSCCO figure in which the flux-line image was blurred even at 15 K and 20 gauss (T_C = 85 K) [34]. Accordingly, the maximum temperature for practical use would not be T_C but rather the melting temperature T_m. Others, however, disagree with this, and attribute this phenomenon to weak-pinning effects.

The flux lines were dynamically observed to test whether flux lines begin to move under such conditions. The observation was made under a fixed magnetic field B , while increasing the sample temperature from 4.5 K to above T_C. In theLorentz micrograph at T = 4.5 K and B = 20 gauss shown in Fig. 6 (a), the flux lines are distributed at random. When the temperature was raised stepwise by a few K, the flux lines moved. After a few minutes, flux lines arrived at an equilibrium state and became still. They did not melt even at 20 K. The flux-line configuration changed between 40 K and 50 K. Flux lines form a regular lattice (c) above this transition region. The flux-line lattice persisted at higher temperatures though the image contrast gradually decreased and then disappeared above 77 K.

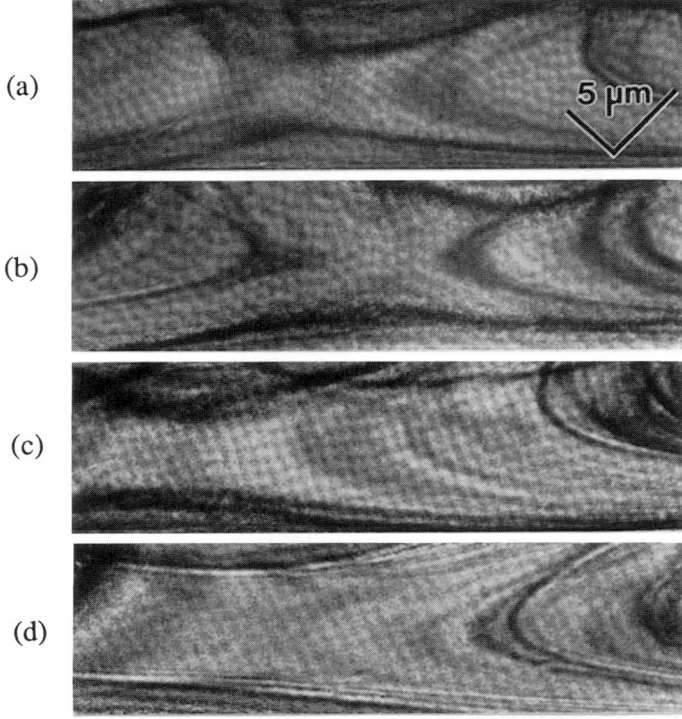

Fig. 6. Lorentz micrographs of BSCCO (2212) film: (a) $T = 4.5$ K, (b) $T = 20$ K, (c) $T = 56$ K and (d) $T = 68$ K.

5. CONCLUSIONS

Electron interferometry has a long history spanning more than 40 years. With the development of a coherent electron beam and electron holography presented in this paper, the phase distribution of an electron beam has become measurable to within $2\pi/100$, which has opened the way to observing microscopic objects and fields. Various applications will be developed to explore the microscopic world using this technique.

REFERENCES

1. H. Boersch, Naturwissenschaften 28 (1940) 711.
2. L. Marton, Phys. Rev. 85 (1952) 1957.
3. T. Mitsuishi, H. Nagasaki and R. Uyeda, Proc. Jpn. Acad. 27 (1951) 86.
4. For example, see A. G. Klein and S. A. Werner, Rep. Prog. Phys. 46 (1983) 259.
5. G. Möllenstedt and H. Düker, Naturwissenschften 42 (1955) 41.
6. G. Möllenstedt and H. Keller, Z. Phys. 148 (1957) 34.
7. J. Faget and C. Fert, Cah. Phys. 83 (1957) 285.
8. H. Boersch and B. Lischke, Z. Phys. 237 (1970) 449.
9. T. Hibi and K. Yada, *Principles and Techniques of Electron Microscopy*, ed. M. A. Hayat, Vol. 6 (Van Nostrand, New York, 1976) p. 312.
10. G. Pozzi and G. F. Missiroli, J. Microsc. 18 (1973) 103.
11. K. Hanszen, *Advances in Electron and Electron Physics,* ed. J. Marton, Vol. 59 (Academic Press, New York, 1982) p.1 .
12. A. Tonomura, Jpn. J. Appl. Phys. 11 (1972) 493.
13. G. F. Missiroli, G. Pozzi and U. Valdre, J. Phys. E14 (1981) 649.
14. A. Tonomura, T. Matsuda, J. Endo, H. Todokoro and T. Komoda, J. Electron Microscopy 28 (1979) 1.
15. D. Gabor, R. Soc. London A 197 (1949) 454.
16. A. Tonomura, T. Matsuda, T. Kawasaki, J. Endo, S. Yano and H. Yamada, Phys. Rev. Lett. 54 (1985) 60.
17. A. Tonomura, *Electron Holography*, Springer Series in Optical Sciences, Vol. 70 (Springer, Heidelberg, 1993).
18. J. Chen, T. Hirayama, G. M. Lai, T. Tanji, K. Ishizuka and A. Tonomura, Opt. Lett. 18 (1993) 1877.
19. T. Hirayama, J. Chen, T. Tanji and A. Tonomura, Ultramicroscopy 54 (1994) 9.
20. J. Chen, G. M. Lai, K. Ishizuka and A. Tonomura, Appl. Opt. 33 (1994) 1187.
21. Q. Ru, J. Endo, T. Tanji and A. Tonomura, Appl. Phys. Lett. 59 (1991) 2372.
22. Q. Ru, G. Lai, K. Aoyama, J. Endo and A. Tonomura, to be published in Ultramicroscopy.
23. G. Lai, T. Hirayama, K. Ishizuka, T. Tanji and A. Tonomura, Appl. Opt. 33 (1994) 829.
24. G. Lai, T. Hirayama, K. Ishizuka, T. Tanji and A. Tonomura, J. Appl. Phys. 75 (1994) 4593.
25. Q. Ru, N. Osakabe, J. Endo A. Tonomura, Ultramicroscopy 53 (1994) 1.
26. A. Tonomura, T. Matsuda, J. Endo, T. Arii and K. Mihama, Phys. Rev. Lett, 44 (1980) 1430.
27. T. Hirayama, Q. Ru, T. Tanji and A. Tonomura, Appl. Phys. Lett. 63 (1993) 418.
28. J. E. Bonevich, K. Harada, T. Matsuda, H. Kasai, T. Yoshida, G. Pozzi and A. Tonomura, Phys. Rev. Lett. 70 (1993) 2952.
29. J. Bonevich, K. Harada, H. Kasai, T. Matsuda, T. Yoshida, G. Pozzi and A. Tonomura, Phys. Rev. B 49 (1994) 6800.
30. K. Harada, T. Matsuda, J. E. Bonevich, M. Igarashi, S. Kondo, G. Pozzi, U. Kawabe and A. Tonomura, Nature 360 (1992) 51.

31. K. Harada. T. Matsuda, H. Kasai, J. E. Bonevich, T. Yoshida, U. Kawabe and A. Tonomura, Phys. Rev. Lett. 70 (1993) 3371.
32. T. Kawasaki, T. Matsuda, J. Endo and A. Tonomura, Jpn. J. Appl. Phys. 29 (1990) L 508.
33. D. J. Bishop, P. L. Gammel, D. A. Huse and C. A. Murray, Science 255 (1992) 165.
34. R. N. Kleiman, P. L. Gammel, L. F. Schneemeyer, J. V. Waszczak and D. J. Bishop, Phys. Rev. Lett. 62 (1989) 2231.

Electron Holography: State and experimental steps towards 0.1nm with the CM30-Special Tübingen

Hannes Lichte, Institut für Angewandte Physik

Universität Tübingen, Auf der Morgenstelle 12, D72076 Tübingen, Germany

Abstract

Gabor's initial idea of electron holography to overcome the resolution limit is coming close to realization: For the first time, the point resolution of medium voltage electron microscopes of about 0.17nm has been improved to about 0.13nm by means of off-axis electron holography using the Möllenstedt electron biprism. Additionally, full use of wave optical analysis of the object exit wave can be made by means of numerical wave optical processing.

1. Introduction

Gabor proposed electron holography [1] to overcome the resolution limit in electron microscopy due to spherical aberration which - since the famous work of Scherzer [2] - was known to him to be unavoidable with the usual electron lenses. The effect of spherical aberration is to distort the electron wave function by the wave aberration function. Actually, this distortion does not mean any loss of information about the object structure; instead it is a mere disorder of amplitude and phase in the image plane. The restriction of resolution in conventional electron microscopy comes about because only the intensity (i.e. the squared amplitude) and not the phase of the electron image wave is recorded in a micrograph. Gabor showed theroretically that, if the electron wave is recorded completely by amplitude and phase, the distortion of the electron wave due to spherical aberration can be corrected a-posteriori, by reordering amplitude and phase. Later, Gabor et al. [3] noted that additional benefit can be expected from holography in that phase contrast techniques become available which, while well-developed in light optical microscopy, are not applicable in electron microscopy up to now.

2. The importance of phases and their application in holography

Observing the intensity i.e. the electron current distribution on the screen of the electron microscope, one is tempted to ignore the importance of the phases for the imaging process because they never show up directly. Therefore, in the following the role of phases for the imaging process is illustrated.

A wave spreads in space according to the wave equation. Given a wave function on a surface, e.g. at the exit face of an object, it can be determined in each point in space by means of the Kirchhoff integral; in the vicinity of the object it can be approximated by the Fresnel integral ("Fresnel diffraction"). At a large distance, the approximation is called Fraunhofer diffraction, mathematically given by a Fourier transform. Generally, the resulting waves are given by amplitude and phase, and the procedures can be inverted if both are known.

For illustration, assume a simple hypothetical object modulating the amplitude of the electron wave according to the object function

$$o(x) = 1 + a\cos(2\pi R_0 x) \tag{1}$$

with the spatial frequency R_0; for simplicity, a is assumed small.

The modulated wave spreads in space according to the wave equation. With distance z from the specimen, in Fresnel approximation the wave can be written as

$$\psi(x,z) = 1 + \exp(i\chi(z)) \cdot a\cos(2\pi R_0 x) \tag{2}$$

$$= 1 + \cos(\chi(z)) \cdot a\cos(2\pi R_0 x) + i\sin(\chi(z)) \cdot a\cos(2\pi R_0 x)$$

The function $\chi(z)$ reads as

$$\chi(z) = \pi R_0^2 / k \cdot z \tag{3}$$

with $k=1/\lambda$ the wavenumber. For example, with 300keV electrons (k=508/nm) and R_0=5/nm, χ changes by $\pi/2$ for each z=10nm.

Evidently, at distance z from the object, the wave ψ can be written as

$$\psi(x,z) = 1 + A(x,z) + i\Phi(x,z) \tag{4}$$

i.e. is modulated both in amplitude by

$$A(x,z) = \cos(\chi(z)) \cdot a\cos(2\pi R_0 x) \tag{5}$$

and in phase by

$$\Phi(x,z) = \sin(\chi(z)) \cdot a\cos(2\pi R_0 x). \tag{6}$$

With increasing distance z, the information about the object structure oscillates between amplitude and phase; for $\chi(z) = 0, \pi \ldots \mod 2\pi$, it is found only in the amplitude A, for $\chi(z) = \pi/2, 3\pi/2 \ldots \mod 2\pi$ only in the phase Φ, at the other values it is complementarily found in either of them.

Similarly, in the case of a weak phase grating

$$o(x) = 1 + i\varphi\cos(2\pi R_0 x), \quad (\varphi << 2\pi) \tag{7}$$

the object information is found in the amplitude $A(x,z) = -\sin(\chi(z)) \cdot \varphi \cos(2\pi R_0 x)$ for $\chi(z) = \pi/2, 3\pi/2 \ldots \mod 2\pi$, and in the phase $\Phi(x,z) = \cos(\chi(z)) \cdot \varphi \cos(2\pi R_0 x)$ for $\chi(z) = 0, \pi \ldots \mod 2\pi$.

Recording the intensity yields

$$I(x,z) = \psi(x,z)\psi^*(x,z)$$
$$= 1 + 2A(x,z) \qquad (8)$$

if one neglects terms of higher order in A and Φ. Consequently, the fraction of object information present in the phase Φ is lost for further evaluation from a micrograph; in the best case, all object information is in the recorded amplitude, in the worst case, no object information is recorded at all and the object structure is completely lost. However, these cases only occur with pure amplitude and pure phase objects, generally the object information is always distributed over A and Φ.

These problems do not occur, if the wave $\psi(x,z)$ is holographically recorded, i.e. with $A(x,z)$ and $\Phi(x,z)$: then, by inversion of the propagation along the z-direction, the wave can be back-propagated into the object plane ("reconstruction of the object wave"), and the object wave can be analyzed without loss of information. The back-propagation can then be performed by means of any medium which obeys the wave equation. For example, Gabor proposed to take the hologram with electrons to make use of the high resolution offered by the short electron wavelength and of the strong interaction of electrons with atomic structures, and to reconstruct the object wave with light because light optics offers the advantage of a much better handling and a much higher-developed wave optical technology. Today, the reconstruction is usually performed with a computer programmed with the laws of wave optics.

At a large distance from the object, say in the order of ten milimeters, a Fraunhofer diffraction pattern arises, i.e. a wave which is given by the Fourier transform of the object function

$$s(R) = \int o(x) \exp(i2\pi R x) dx. \qquad (9)$$

In the case of an amplitude and phase grating

$$o(x) = 1 + a\cos(2\pi R_0 x + \varepsilon) + i\varphi\cos(2\pi R_0 x + \varepsilon) \qquad (10)$$

it reads

$$s(R) = \delta(R) + t\exp(i\vartheta)\{\exp(i\varepsilon)\delta(R - R_0) + \exp(-i\varepsilon)\delta(R + R_0)\} \qquad (11)$$

with $t = 0.5\sqrt{a^2 + \varphi^2}$ and $\vartheta = \arctan(\varphi/a)$. Each of the off-axis reflections is given by an amplitude and by a phase: the amplitude t gives a measure for the total strength

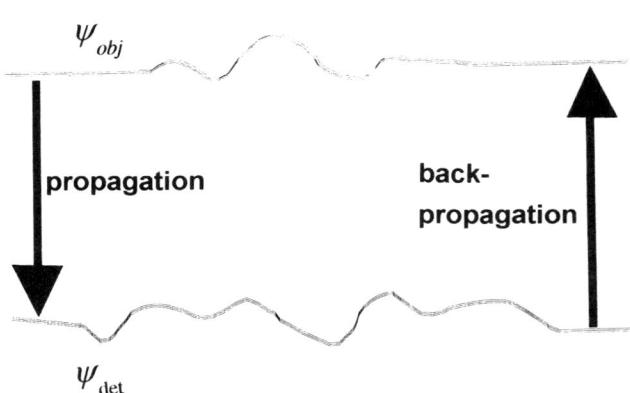

Figure 1. A wave propagating from the object plane to the detector plane can likewise be back-propagated according to the Kirchhoff diffraction integral, if the detection process recovers amplitude and phase. This led Gabor to the invention of holography.

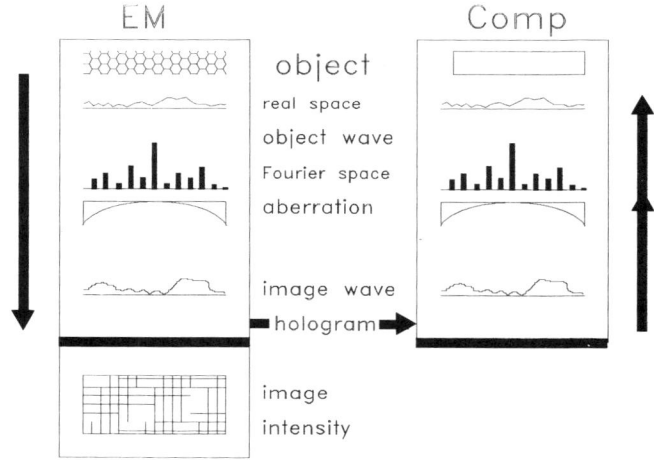

Figure 2. Scheme of Image plane holography.

In the electron microscope, instead of the intensity, the complex electron wave is recorded by means of a hologram and transferred to a computer; there, it can be back-propagated to Fourier-plane or object plane under correction of aberrations.

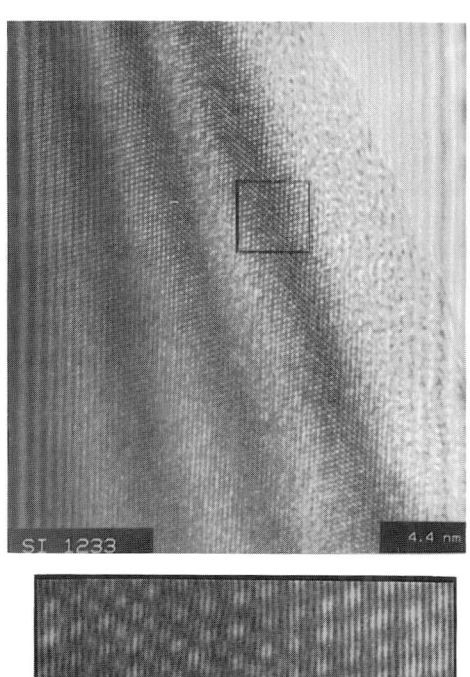
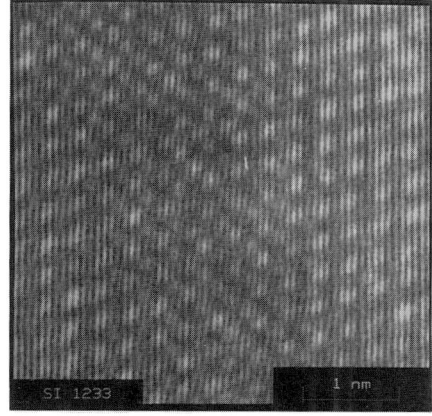

Figure 3. Hologram of a wedge-shaped Si crystal in <110> orientation taken at 100kV.
At high magnification (bottom), the hologram fringes show strong phase shifts due to the atomic columns and dynamic effects at the extinction thicknesses.

of the modulation by a or φ, whereas the phase tells one to which extent it is an amplitude or a phase grating, and exhibits the lateral position ε of the grating.

Again, recording the intensity $s(R)s^*(R)$ the phases ϑ and ε are lost, hence one cannot tell anything, either about the nature of the grating - amplitude, phase or both - or about its position. One can only discern how strong the spatial frequency R_0 contributes. If, by any means, one can record the Fraunhofer wave holographically, the Fourier transform can be completely inverted, again delivering the complete object wave.

These examples show that, if the electron wave is holographically recorded in one plane, it can be determined in any plane, e.g. in the object plane or in the Fraunhofer diffraction plane (fig.1). This led Gabor to the idea of lensless imaging thus avoiding the aberrations of an objective lens. In light optical holography, using a laser as a very powerful coherent source, lensless imaging proved to be very successfull. In electron optics, however, this technique was comparably poor in resolution because of the poor coherence of electron sources. Recently, Fink et al. [4] with the help of a highly coherent monoatomic field emission electron source started a very promising approach to lensless imaging at atomic dimensions.

High resolution holography with usual electron sources, e.g. a conventional field emission gun, can be performed, if the hologram is taken close to the image plane in an electron microscope: Then the lateral coherence limits the field of view but does not limit the resolution. It is true that the image wave recorded in the hologram is affected e.g. by the spherical aberration of the objective lens, but this can be taken into account under reconstruction of the object wave (fig.2).

3. Scheme of image plane off-axis holography.

The scheme of image plane holography is the following [5]: In the image plane of the electron microscope, a coherent plane reference wave is superimposed on the image wave by means of the Möllenstedt electron biprism [6]. The resulting interference pattern is a hologram in the sense of Gabor, because both amplitude and phase of the image wave are encoded in the appearance of the interference fringes (fig.3). For reconstruction, today the hologram is fed to a computer which unveils the image wave in the form of an array of complex data. Programmed with the laws of wave optics, the computer establishes a perfect wave optical microscope: starting from the given image wave, the wave can be back-propagated into any plane of interest, e.g. into the object plane or into the plane of Fraunhofer diffraction (fig.4). Consequently, all kinds of wave optics needed to extract and to display quantitatively the complex electron wave can very flexibly be performed. There are no technical restrictions to be faced in the electron microscope, e.g. with the application of a phase plate in the back focal plane of the objective lens. Since the

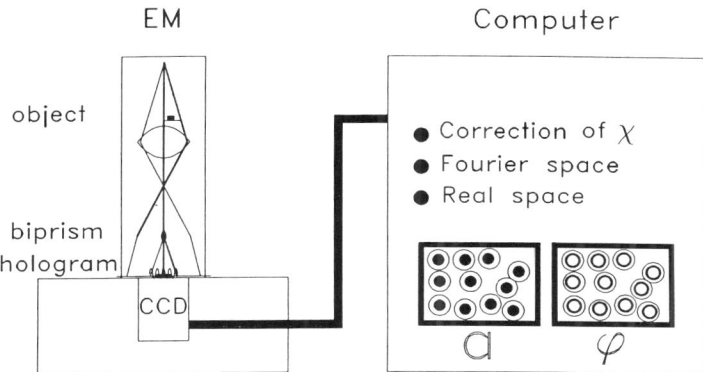

Figure 4. Realization of off-axis electron holography.

The hologram taken by means of the Möllenstedt electron biprism in the image plane is recorded with a CCD camera and fed to the computer for back-propagation, i.e. correction of aberrations and wave optical analysis in Fourier space and real space.

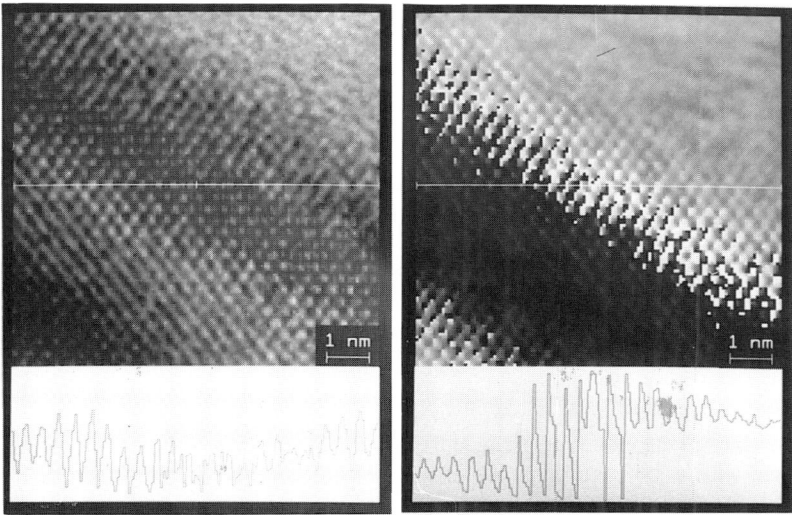

Figure 5. Amplitude (left) and phase of the object wave of the Si crystal. Across the extinction thickness, the amplitude ahows half spacings while jumps of about $\pi/2$ are found in the phase image [22].

hologram reveals all properties of the object which modulate the object wave at the moment of hologram recording, the information gained by the different reconstruction procedures, in real space or in Fourier space, can unambiguously be related with the structure of the object exit wave.

After solving the problems involved with taking electron holograms at atomic dimensions with a 100kV-Philips-EM420 electron microscope equipped with a field emission gun [7], some effort has been put into the development of the numerical reconstruction [8] and the realization of the wave optical procedures sketched above. Meanwhile, the state of the method can be summarized as follows [9 -11]

1. Correction of aberration.

In principle, all aberrations contributing to the wave aberration of the objective lens can be corrected. For the correction, a phase plate has to be generated numerically which is conjugate to the one in the microscope, and has to be applied on the Fourier spectrum of the reconstructed image wave. At the beginning, the most essential ones are spherical aberration of 3rd order, astigmatism of 2nd order, and defocus; with improved resolution, one has to think also about spherical aberration and astigmatism of higher order, and about the influence of beam tilt (axial coma). Meanwhile the correction of spherical aberration of 3rd order, astigmatism of 2nd order, defocus and beam tilt has been shown experimentally; three-fold astigmatism is in progress.

2. Quantitative evaluation of amplitude and phase of the electron wave (fig.5).

After obtaining an inverse Fourier transform of the corrected Fourier spectrum, the object exit wave can be displayed by amplitude and phase uniquely in the sense that there is no crosstalk between the amplitude and the phase structure as in conventional images. Since amplitude and phase are given quantitatively, there is no need for additional contrast-producing procedures. Compared with a conventional micrograph, the wave is free from artifacts in that it represents linear and zero-loss information.

3. Analysis of Fourier space

The Fourier transform of the complex object wave corresponds to the electron diffraction pattern found in the back focal plane of the electron objective lens. In comparison with a diffractogram determined from a conventional micrograph, there are some very essential advantages. The main advantage is that the phases in the Fourier spectrum are available; there is no "phase problem" with holography. The phase information may be utilized for the analysis of scattering phases, or of the phase distribution due to dynamical scattering with characteristic phase jumps at the extinction thicknesses of crystals; likewise, it may be used for the measurement of thickness and tilt of crystals (fig.6). In addition, the complex diffraction pattern

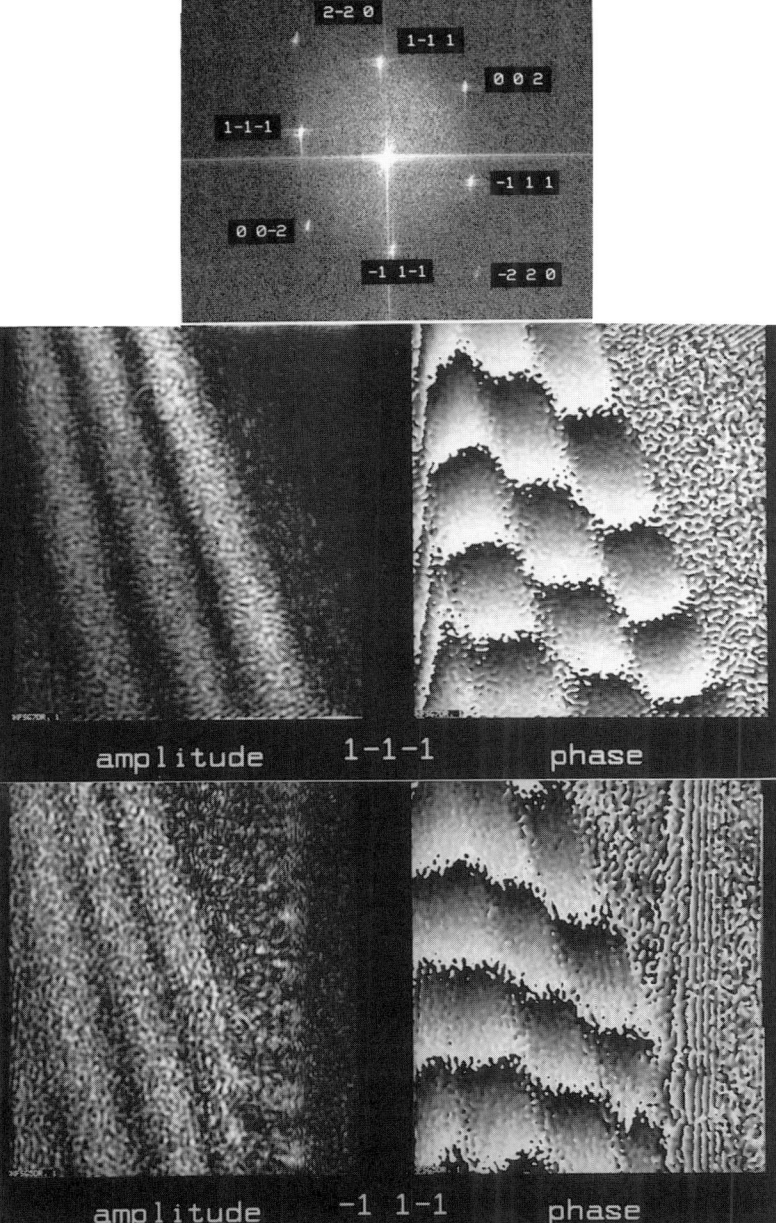

Figure 6. Amplitude (left) and phase distribution reconstructed from single reflections give additional information about thickness and tilt distribution of the Si crystal.

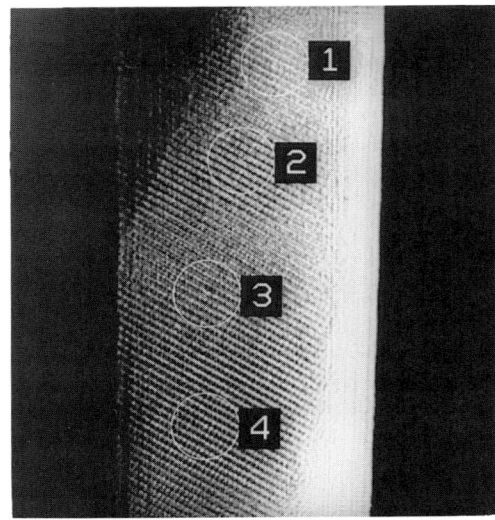
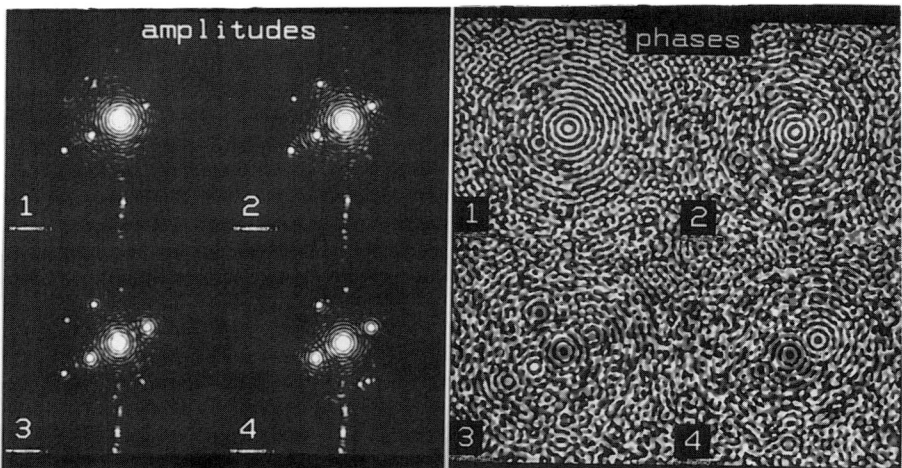

Figure 7. Nanodiffraction.

At the different positions of the numerical SA aperture (∅ 4nm), strong differences show up in the respective Fourier spectra indicating a considerable twist along the Nb_2O_5 crystal.

reveals asymmetries in the intensities of opposite reflections, e.g. in the case of crystal mistilt; surprisingly, often a hologram shows crystal tilt in spite of the fact that the crystal was carefully aligned using the diffraction pattern in the microscope. The reason may either be that in the microscope a much larger area of the object is used for alignment, so a mistilt averages out, or the fact already described by Smith et.al [12] that, when switching from diffraction to imaging mode, the beam tilt changes. In addition, apertures can be applied in the Fourier plane in order to see specific properties of the object wave by "Selective Imaging": The most simple ones are imaging with single reflections in bright field or dark field. Lattice images at an arbitrary combination of diffracted waves can be produced, superlattice or sublattice reflections can be isolated to form the corresponding image wave, etc.

4. Nanodiffraction

Selected area diffraction is an essential tool in electron microscopy for the analysis of variations of crystal tilt or thickness over the field of view, of defects, precipitations, or grains in the case of polycrystalline material. However, the analyzed area cannot be smaller than about 100nm in diameter, because the selected area (SA) aperture can not be made smaller. In addition, due to the spherical aberration and to insufficient accuracy in axial positioning of the SA aperture, for very small SA apertures there would be an error in that reflections not belonging to the selected area would show up in the diffraction pattern.

Nanodiffraction, i.e. diffraction at areas only a few nanometers wide, is easily possible with the object wave reconstructed from the hologram. Numerically generated apertures of arbitrary shape can be placed on the object wave and the corresponding complex diffraction pattern is found after Fourier transform. Since spherical aberration is corrected and positioning in z-direction is very accurately in focus, the above mentioned problems do not occur; apertures as small as 1 nm have successfully been applied. In most cases, shifting the aperture over the field of view, one finds a significant change in the nanodiffraction pattern due to changes in thickness, tilt or composition of the crystals, especially when viewing the thin objects needed for high resolution work (fig.7).

These holographic techniques allow a more thorough analysis of the object exit wave than conventional microscopy. However, it is not yet clear whether the collected information is in principle sufficient to determine uniquely the underlying object structure, particularly in a direction along the electron beam. Therefore, calculations for modelling the object structure and simulating the object exit wave are still needed for the interpretation. Nevertheless, thanks to the improved data about amplitudes and phases, and about crystal tilt and thickness, the modelled structures will be more accurate and the interpretation more reliable with holography.

4. Requirements for holography

For an assessment of the limits of holography one has to consider the main components involved in the whole procedure: the microscope, the holographic technique and the numerical reconstruction.

4.1 Information transfer of the microscope

The image wave to be recorded in the hologram contains only the information that is transferred by the objective lens coherently, i.e. up to the information limit set by the incoherent aberrations. The chromatic aberration must be considered, which, given by the energy spread of the electron radiation and the coefficient of chromatic aberration, cannot be influenced by the operator of the microscope. Next, the envelope function due to the illumination aperture has to be considered which is given by

$$E_{ill}(R) = \exp\{-(\pi \vartheta_c k \cdot grad \chi(R))^2\} \tag{12}$$

damping the transfer of the spatial frequency R if the object is illuminated with the wavenumber $k = 1/\lambda$ at the illumination aperture ϑ_c.

Since the wave aberration

$$\chi(R) = 2\pi k \cdot \left\{ 0.25 C_S \left(\frac{R}{k}\right)^4 - 0.5 Dz \left(\frac{R}{k}\right)^2 \right\} \tag{13}$$

depends not only on the spherical aberration Cs but also on defocus Dz, the gradient hence the damping can be minimized over a broad range of spatial frequencies. To achieve this, the optimum defocus for holography given by

$$Dz_{opt} = 0.75 C_S \left(\frac{R_{max}}{k}\right)^2 \tag{14}$$

should be selected if one intends to record the spatial frequencies up to R_{max} [13]. It is worth noting also that for other reasons, like noise and correctability of spherical aberration, in holography it is most favourable to optimize the gradient of the wave aberration, whereas in conventional microscopy the wave aberration is optimized with respect to image contrast, usually at Scherzer focus. For the performance of holography, image contrast in the microscope is of minor importance because, at the end, ideal phase and amplitude contrast are achieved by means of the reconstruction procedure.

4.2 Properties of the hologram

Superposition of a plane reference wave with unity modulus on the image wave

$$b(x,y) = A(x,y) \cdot \exp(i\Phi(x,y)) \tag{15}$$

results in a hologram with the intensity distribution

$$I_{hol}(x,y) = 1 + A^2(x,y) + 2V \cdot A(x,y)\cos(2\pi R_c x + \Phi(x,y)). \tag{16}$$

Besides the intensities A²(x,y) and 1, one finds the holographic term sampling the image wave at the carrier frequency R_C. The holographic term is easily separated from the remainder in Fourier space if the carrier frequency fulfills $R_c \geq nR_{max}$ where R_{max} is the maximum spatial frequency present in the image wave. In the special case of a pure phase modulation of the image wave, i.e. A(x,y)≡1, n=1 is sufficient; in the general case of arbitrarily strong A(x,y), n≥3 has to be selected [5].

At first glance, the width w of the hologram does not seem to play a major role other than to limit the reconstructable field of view of the wave. However, one must keep in mind that the image wave represents the object wave convoluted with the point spread function which, neglecting the incoherent envelope functions, is the Fourier transform of the transfer function exp(iχ(R)). Each object point is smeared out in the image wave according to the point spread function; in this sense, the later correction of aberrations means the re-concentration of the spread information into the respective point. This is only possible if the point spread area is completely collected in the hologram for all object points to be reconstructed. A more detailed analysis [14] shows that the hologram must be at least as wide as four times the diameter of the point spread function, if one cannot build up the object structure from a smaller area by periodic continuation.

The holographic term is damped by the contrast of the fringes $V = V_{coh} \cdot V_{inel} \cdot V_{inst}$ with $0 \leq V \leq 1$. $V_{inel} = \sqrt{1 - P_{inel}}$ is given by the probability P_{inel} that an electron suffers at least one inelastic interaction with an energy transfer of more than about $4. \cdot 10^{-15} eV$ [15]; the inelastic electrons contribute only to the background; hence, if all electrons have undergone an inelastic interaction, a hologram cannot be recorded. Conversely, one can say that the hologram and hence the reconstructed wave represents only zero-loss-information.

V_{coh} describes the damping of fringe contrast contributed from the degree of coherence between the image wave and the reference wave. Since the coherence length of the electrons is greater than 100,000 wavelengths, longitudinal coherence is always sufficient, and we only need to care about lateral coherence. Then, following the van Cittert-Zernike theorem, V_{coh} is given by the modulus of the Fourier transform of the intensity distribution of the electron source. In the case of a Gaussian electron source it turns out [16] that the mean current density j_{coh} in a hologram of contrast V_{coh} and width w is related to the axial brightness B of the electron gun by means of

$$j_{coh} = -\frac{2B \cdot \ln(V_{coh})}{k^2 \cdot \pi w^2}. \tag{17}$$

Usually it is increased by approximately up to one order of magnitude by employing elliptical illumination [9].

During exposure time t a certain number of electrons per unit area is collected in the hologram; insufficient brightness can in principle be compensated by longer exposure times. However, at the high carrier spatial frequencies Rc needed for high resolution work, residual instabilities reduce the contrast V by the factor V_{inst}, as the exposure time is increased. In fact, V_{inst} seems to be the most severe limiting factor for the value of V, that can be reached in today's high resolution holograms.

4.3 Recording the hologram

Formerly, the holograms were recorded on photographic film. However, as pointed out by W.D. Rau [17], CCD cameras are increasingly applied in holography because, concerning the essential figures of merit such as linearity, dynamic range, speed of processing, etc., they are far superior. However, with respect to the available number N_{pix} of pixels along the field of view, there are severe limitations.

The number of pixels needed for processing the wave is related to the performance of the electron microscope and the maximum spatial frequency R_{max} present in the image wave. It can be found in the following way: For a given wave aberration $\chi(R)$, i.e. spherical aberration and defocus, the diameter PSF of the point spread function can be estimated as $PSF = grad\chi(R)_{max} / \pi$, if the incoherent envelope functions are neglected; $grad\chi(R)_{max}$ means the maximum absolute value of $grad\chi(R)$ in the interval $[0, R_{max}]$. PSF increases with the maximum spatial frequency R_{max} present in the image wave. From minimum width w=4*PSF and fringe spacing $R_c = 3 R_{max}$ of the hologram, one can compute the minimum number of fringes which have to be recorded. Since each fringe has to be sampled at four pixels [18], the necessary number of pixels steeply increases with R_{max}. Again optimum focus $Dz_{opt} = 0.75 C_s (R_{max}/k)^2$ is most favourable, since the PSF takes the smallest possible value ("circle of least confusion") for given R_{max}, hence fewer pixels are sufficient. It can be summarized that, if the hologram is taken in a microscope with the Scherzer resolution R_{sch} and is recorded by means of a camera with $N_{pix} \cdot N_{pix}$ pixels, after correction of aberrations the resolution

$$R_{max} = 0.3 \cdot N_{pix}^{1/4} \cdot R_{sch} \tag{18}$$

can be obtained, from the point of view of image processing.

4.4 Noise

Noise in the hologram and in the reconstructed wave arises mainly due to shot noise. Together with the often poor contrast V of the hologram fringes it hampers the detection of weak modulations of the reconstructed wave. For example, between two adjacent picture elements of the wave a phase difference can only be discerned at a Signal/Noise ratio (SNR) if larger than

$$\delta\varphi = \frac{SNR}{V} \sqrt{\frac{2}{N_{el}}} \tag{19}$$

with N_{el} the number of electrons collected per picture element [19].

Taking into account the Modulation Transfer Function MTF of the CCD camera, the contrast reads

$$V = V_{coh} \cdot V_{inel} \cdot V_{inst} \cdot MTF \tag{20}$$

N_{el} can be determined from the data of the hologram by

$$N_{el} = j_{coh} \cdot \kappa \cdot DQE \cdot d^2 \cdot t \tag{21}$$

with κ the ellipticity of illumination, d^2 the area of the picture element, DQE the Detection Quantum Efficiency of the CCD camera, t the exposure time, and with the coherent current density

$$j_{coh} = -\frac{2B \cdot \ln(V_{coh})}{k^2 \cdot \pi w^2} . \tag{22}$$

Assuming optimum focus, the width of the point spread function is given by

$$PSF = 0.5 C_s \left(\frac{R_{max}}{k}\right)^3, \tag{23}$$

and with the width of the hologram $w = 4 \cdot PSF$, and setting $d=1/R_{max}$, one finds

$$\delta\varphi = \frac{2 \cdot SNR \cdot C_s \frac{R_{max}^4}{k^2}}{V_{coh} \cdot V_{inel} \cdot V_{inst} \cdot MTF} \sqrt{\frac{\pi}{-B \cdot \ln(V_{coh}) \cdot \kappa \cdot DQE \cdot t}} \tag{24}$$

plotted in fig.8.

For a given exposure time, V_{coh} can be optimized by selecting $V_{coh} = \exp(-0.5) \approx 0.61$. It is interesting to note that Cs, though it is corrected at the end, has a strong effect on noise in the reconstructed wave.

5. Experimental approach to 0.1nm-resolution: The CM30-Special Tübingen microscope.

To the authors knowledge, the best resolution obtained so far by holographic correction of aberrations with a 100kV electron microscope is 0.2nm, recently achieved in our group [20]. However, if one evaluates the most important parameters, it becomes evident that a resolution limit of 0.1nm cannot be reached with today's 100keV electron microscopes: The information limit is about 0.17nm; furthermore, the hologram would have to be approximately 120nm wide and the pixel capacity of the image processing system would have to be larger than N_{pix}=14000. In addition, because of the high noise level, only phase shifts larger than $2\pi/6$ could be detected while for single gold atoms $2\pi/15$ is needed. The situation looks much more favourable with a 300keV microscope. At optimum holography focus, an information limit close to 0.1nm seems within reach. The width of the point spread function can be estimated as 5nm at Rmax=10/nm; hence the width of the hologram must be 20nm and the needed pixel number is less than 2000. The noise level can be estimated to allow to detect phases as small as $2\pi/50$, sufficient for light atoms like oxygen.

In 1989, the Philips company agreed to design and build a special 300 kV holography microscope with the following specifications:

- High brightness electron gun with $B \geq 5 \cdot 10^8 A / cm^2 / sr$ at 300kV

- Information limit of 0.1nm at optimum holography focus

- Cs ≤1.2mm

- Magnification of the objective lens higher than 50

- Magnification on the photographic plate higher than 2Mio, at the level of a CCD camera close to 4Mio

- High mechanical and electrical stability

- low AC-stray field level at the microscope column from the power supplies.

The microscope was installed in summer 1992 in a specially prepared laboratory. Special care was taken to minimize AC-stray fields and acoustic noise; together with the high voltage supply the microscope is installed on a platform suspended on air cushions to isolate it from mechanical vibrations. A 1k*1k CCD-camera (Photometrics) is installed; the image processing is done with a Tietz computer. The

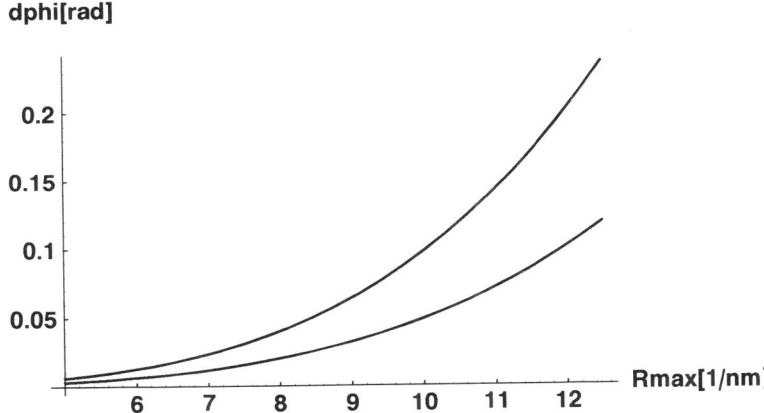

Figure 8. Minimum phase shift detectable from a hologram taken in a 300kV microscope (Cs=1.2mm (top), Cs=.6mm (bottom); exposure time 10sec; fringe contrast 20%). At a resolution of Rmax=10/nm, the phase shift due to a gold atom and an oxygen atom amounts to about 0.5rad and 0.1rad, respectively.

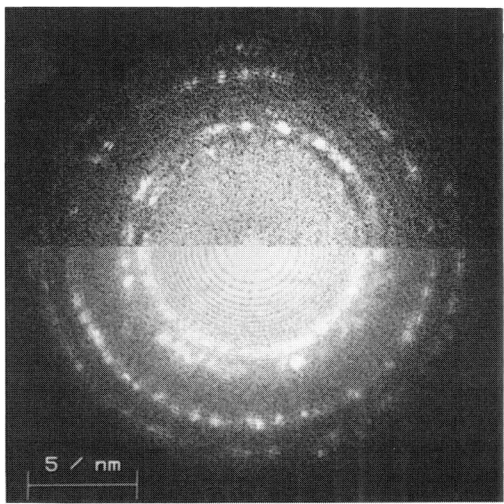

Figure 9. Diffractograms taken at the CM30-Special Tübingen microscope show the improvement of information limit close to 0.1nm at 420nm underfocus (bottom), as compared with Scherzer focus (top). Note the fine oscillations due to the PCTF in the outermost reflections.

performance of this image processing system, in particular the facilities of on-line reconstruction, are described in more details by W.-D. Rau [17].

Biprism

A biprism holder is inserted instead of the regular SA-holder. The biprism constant σ_{bp}, which relates the filament voltage U_f to the fringe spacing at the object s_{obj} by means of $s_{obj} \cdot U_f = \sigma_{bp}$ was determined as 25 V·nm [21]. Consequently, for s_{obj}=0.05nm one needs a filament voltage of 500V; fringes with a spacing of 0.031nm have been recorded at U_f=800V.

Information limit 0.1nm

The usual technique of diffractometry was applied for the determination of the information limit using the Tietz-image processing system. However, two problems showed up: First, the diffraction efficiency of the usual amorphous carbon or germanium foil was so poor that the obtained data were limited by the specimen rather than by the microscope. With a carbon foil covered with a submonolayer of W, we finally succeeded in getting diffracted information out to 0.1nm. Second, at the beginning the foils were too thick in that the diffractogram showed diffracted intensity but not the expected dark and bright rings due to the phase contrast transfer function. Finally, using areas thinner than about 3nm, we found the dark and bright rings reaching out to 0.1nm, indicating unambigously that the linear information transfer is reaching that far at the optimum holography focus (fig.9).

Brightness 5 10^8 A/(cm²sr)

The axial brightness can easily be measured interferometrically. At a strong excitation of the condensor lens an interference pattern is produced on the final screen with a good contrast and low current density. Then the excitation of the condensor lens is reduced causing the current density to increase, and the contrast of the fringes drops below the threshold of 5%. Measuring at this point the current density and the width w of the hologram, the brightness B can be calculated from the relation

$$j_{coh} = -\frac{2B \cdot \ln(V_{coh})}{k^2 \cdot \pi w^2} \tag{25}$$

It turned out that the brightness of the gun meets the specification of B=5 10^8 A/(cm²sr) at 300kV.

Figure 10. Object wave of Si in <110> orientation reconstructed from a 300kV hologram. In both amplitude (left) and phase, the dumbbells with a spacing of 0.136nm can be discerned. top: holographic reconstruction, bottom: corresponding results simulated with the EMS program package. More recent results can be found in more detail in [23].

Holograms

Presently, holograms are usually taken at a fringe spacing of 0.05nm and a width of 20-30nm, which, after correction of aberration, permit to reconstruct information down to a resolution close to 0.13 nm. From a hologram of Si taken in <110> orientation, the dumbbells can be distinguished both in the reconstructed amplitude and phase (fig.10). The very careful correction of aberrations needed at this resolution can more easily be accomplished by means of the new facilities presented by M. Lehmann during the workshop [24].

Nevertheless, the contrast of the hologram fringes of about 20% is not yet satisfactory, as seen from the noise of the reconstructed images, as compared to the 30% contrast usual at our 100kV microscope. The main reasons presumably are some tiny residual disturbances of unknown origin which, by means of V_{inst}, have a strong influence on the quality of the holograms. These are presently carefully analyzed.

Acknowledgements

The discussion with Dr. Gottfried Möllenstedt, Dr. Karl-Heinz Herrmann and Dr. Friedrich Lenz, as well as the cooperation of Alexander Harscher, Peter Kessler, Günter Lang, Michael Lehmann, Dimitrios Malamidis, Alexander Orchowski, and Dr. Wolf-Dieter Rau have substantially contributed to the achieved state of high resolution holography in Tübingen. Our work on electron holography is supported by the Körber Stiftung, Volkswagen-Stiftung, Deutsche Forschungsgemeinschaft and the European Community.

References

[1] D. Gabor, A new microscopic principle, Nature 161 (1948), 777

[2] O. Scherzer, Über einige Fehler von Elektronenlinsen, Z. Physik 101(1936), 593

[3] D. Gabor, G.W. Stroke, D. Brumm, A. Funkhouser and A. Labeyrie, Reconstruction of phase objects by holography, Nature 208(1965),1159

[4] H.-W. Fink, H. Schmid, H.J. Kreuzer and A. Wierzbicki, Atomic Resolution in Lensless Low-Energy Electron Holography, Phys.Rev.Lett. 67(1991), 1543

[5] H. Wahl, Bildebenenholographie mit Elektronen, Thesis Tübingen, 1975

[6] G. Möllenstedt and H. Düker, Beobachtungen und Messungen an Biprisma-Interferenzen mit Elektronenwellen, Z. Physik 145(1956), 377

[7] H. Lichte, Electron holography approaching atomic resolution, Ultramicroscopy 20(1986), 293

[8] F.J. Franke, K.-H. Herrmann and H. Lichte, Numerical reconstruction of the electron object wave from an electron hologram including the correction of aberrations, Scanning Microscopy Suppl. 2(1988), 59

[9] H. Lichte, Electron Image Plane Off-axis Electron Holography of Atomic structures, Advances in Optical and Electron Microscopy, 12(1991),25

[10] Q. Fu, H. Lichte and E. Völkl, Correction of Aberrations of an Electron Microscope by Means of Electron Holography, Phys. Rev. Lett. 67(1991), 2319

[11] H. Lichte, E. Völkl and K. Scheerschmidt, Electron Holography II. First Steps of high resolution electron holography into materials science, Ultramicroscopy 47(1992), 231

[12] D.J. Smith, W.O. Saxton, M.A. O'Keefe, G.J. Woods and W.M. Stobbs, The importance of beam alignment and crystal tilt in high resolution electron microscopy, Ultramicroscopy 11(1983), 263

[13] H. Lichte, Optimum focus for taking electron holograms, Ultramicroscopy 38(1991), 13

[14] H. Lichte, Parameters for high-resolution electron holography, Ultramicroscopy 51(1993)15

[15] A. Harscher, F. Lenz and H. Lichte, Electron Holography provides zero-loss images, X. Europ. Cong. on Electron Microscopy, 1992, Granada, in: Last Minute Brochure p.35

[16] H. Lichte, P. Kessler, F. Lenz and W.-D. Rau, 0.1nm Information Limit with the CM30FEG-Special Tübingen, Ultramicroscopy 52(1993),575

[17] W.-D. Rau, On-line Reconstruction of Electron Holograms, MSA Bulletin 24(1994),459

[18] F. Lenz and E. Völkl, Stochastic Limitations to Phase and Contrast Determination in Electron Holography, Proc. XII. Int. Cong. for Electron Microscopy, Seattle, 1990, vol.I, p.228

[19] H. Lichte, K.-H. Herrmann and F. Lenz, Electron noise in off-axis image plane holography, Optik 77(1987), 135

[20] A. Harscher, G. Lang and H. Lichte, Interpretable resolution of 0.2nm at 100kV using electron holography, submitted to Ultramicroscopy

[21] H. Lichte and D. Malamidis, Electron holography at 300kV, X. Europ. Cong. on Electron Microscopy 1992, Granada, Last Minute Brochure p.3

[22] H. Lichte, A. Orchowski and H. Lichte, Electron Holography for Analysis of a wedge shaped Silicon crystal, X. Europ. Congr. on Electron Microscopy 1992, Granada; Last Minute Brochure, p.31

[23] A. Orchowski, W.-D. Rau and H. Lichte, Electron Holography surmounts Resolution Limit of Electron Microscopy; submitted to Phys.Rev.Letters.

[24] M. Lehmann and H. Lichte, Holographic Reconstruction Methods, this conference.

Digital recording and processing in electron off-axis holography

G. Ade

Physikalisch-Technische Bundesanstalt, Bundesallee 100,
38116 Braunschweig, Germany

Abstract
Electron holography provides a powerful tool for making a quantitative determination of the complex object function of interest. However, its realization, especially at high resolution, is restricted by several parameters. The major difficulty seems to lie in determining the aberration coefficients with sufficient accuracy.

1. INTRODUCTION

Holography was suggested by Gabor [1] to compensate for aberrations and so to enhance the resolution of the electron microscopes. As a scientific discipline, holography has had its ups and downs, but throughout these fluctuations, steady progress was being made and important applications were being discovered. In its early history, the technique could not be put to practical use, because of the lack of *coherent* beams. The invention of the laser over 30 years ago allowed *optical* holography to come into full bloom, but *electron* holography had to wait for the development of field-emission guns. The *use* of such guns to obtain high-quality holograms and the *employment* of digital image-processing systems for reconstruction purposes may be considered the main factors in pushing the technique towards high-resolution applications, cf. e.g. [2, 3].

Holography is normally understood as a two-step imaging technique: The amplitude and phase distributions of the object wave are first recorded in an electron hologram by superposing the object wave with a reference wave, and then reconstructed by optical or digital means. Holographic reconstructions are now increasingly performed digitally. This is because the light optical reconstruction is beset by many difficulties, such as the nonlinearity of the photographic process, photographic noise, and imperfection of the optical elements used. Furthermore, digital techniques are extremely flexible with regard to arbitrary manipulations, such as contrast enhancement, signal mixing, and filtering. A full review of digital techniques with an extended list of references has been given in [4].

2. PRINCIPLE

Several methods have been employed in electron holography, but the main stream is image-plane *off-axis* holography using a Möllenstedt Biprism [5]. In this method, the reference wave is somewhat tilted relative to the image-forming wave, and the hologram is recorded in a slightly defocused image plane of the object.

2.1. Hologram recording

As schematically shown in Fig. 1, the object is illuminated by a coherent electron beam. The part of this beam covering the area on the other side of the optical axis is used as a reference. The object and the reference beams are brought to interference in an intermediate image plane by means of an electrostatic *biprism*. The resulting fringe pattern is then recorded as a *hologram* in the final image plane either photographically or digitally.

For the digital recording, high-grade CCD camera systems *(slow-scan cameras)* are now commercially available which have pixel sizes equivalent to those of photographic plates and a dynamic range which exceeds that of the plates by a factor of about ten.

2.2. Digital reconstruction

The original object wave can be retrieved from the hologram by performing the so-called reconstruction step. For this purpose, the digitized hologram is fed into an image-processing system, and the reconstruction procedure is performed according to the following scheme (see Fig.2):

• The digitized hologram is Fourier transformed to obtain the corresponding spectrum. This spectrum usually contains an *unwanted* autocorrelation (or intermodulation) part and *two* sidebands which carry the desired information on the object.

- One of these first-order sidebands is selected for reconstruction, and a *digital filter* is employed to correct aberrations.

- The filtered sideband is then inverse Fourier transformed to obtain the complex object function of interest which can be displayed as *amplitude* or *phase* image.

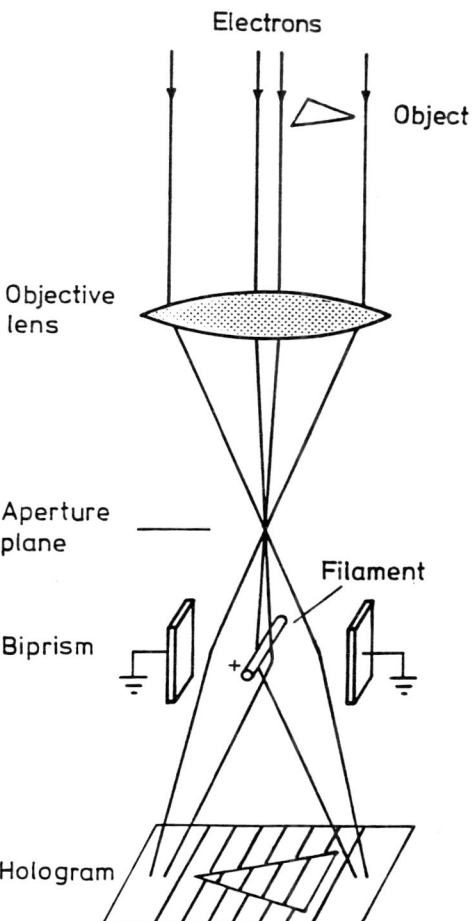

Figure 1. Schematic arrangement for recording an electron off-axis hologram [4].

3. REQUIREMENTS AND LIMITATIONS

Although electron off-axis holography is a straightforward method for obtaining the original object function, its realization, especially at high resolution, is restricted by several parameters. The most important ones are:

- Carrier frequency
- Number of fringes and available pixels
- Partial coherence
- Noise and instabilities of the electron microscope.

Problems connected with the photographic recording of the hologram *(i.e. DQE, Modulation transfer function, and nonlinearity effects)* will not be considered here. A brief discussion of these problems can be found in [4].

3.1. Carrier frequency

To avoid the occurrence of spurious structures and a limitation of resolution in the reconstructed image as a result of an *overlap* between the sidebands and the autocorrelation or intermodulation spectrum (see Fig. 3), the carrier frequency R_c of the hologram *(i.e. the reciprocal of the fringe distance)* must be at least *three* times larger than the maximum spatial frequency R_m to be reconstructed.

Thus to achieve the highest resolutions, electron holograms with extremely narrow fringes are required, but because of the limited coherence and instabilities of the electron microscope, the fringe quality becomes inferior when the fringes are made narrower. Consequently, the *noise level* increases in the reconstructed images. As already reported in [6], however, this problem can be conveniently avoided by eliminating the intermodulation spectrum by the digital subtraction of *two* phase-shifted holograms. The spatial resolution can then be extended up to the actual value of the carrier frequency R_c.

3.2. Number of fringes and available pixels

The procedure of correcting aberrations does not only require an *exact knowledge* of the aberration coefficients [7, 8] but also *correct sampling* of the wave aberration. This implies that a *minimum* number of fringes are contained in the hologram and a *minimum* number of pixels are available in the image-processing system used.

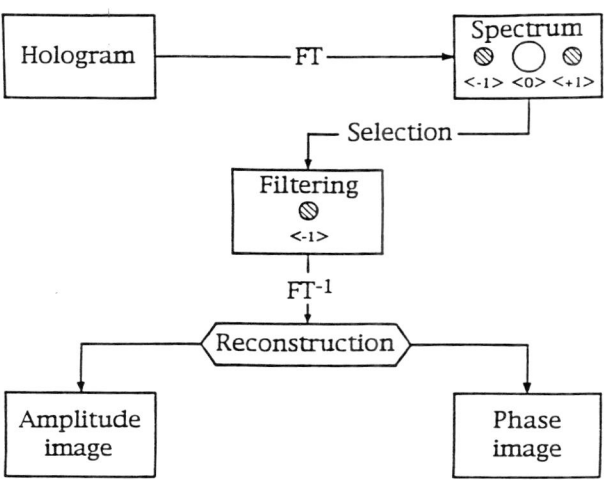

Figure 2. Schematic diagram of the digital reconstruction of the object function from an off-axis hologram [6].

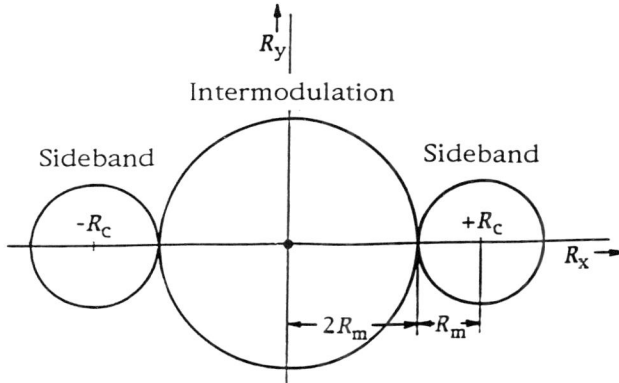

Figure 3. Fourier spectrum of an off-axis hologram (shown schematically). To avoid an overlap between the sidebands and the intermodulation part of the spectrum, the carrier frequency R_c of the hologram must be at least 3 times larger than the maximum spatial frequency R_m to be reconstructed from the hologram.

Following the *Rayleigh criterion*, the phase uncertainty in a sampling interval δR in the Fourier space should not exceed $\pi/2$ [9, 4]. This means that δR must satisfy the conditon

$$\delta R \leq 1/4 \; |\text{grad} W(R)|_{\max}, \tag{1}$$

where $W(R)$ is the reduced wave aberration and R denotes the spatial frequencies of the object. In the simplest case, the function $W(R)$ can be written as

$$W(R) = -0.5 \Delta z \lambda R^2 + 0.25 C_s \lambda^3 R^4, \tag{2}$$

where Δz represents the defocus and C_s is the spherical aberration coefficient of the objective lens. Since W depends on Δz, the sampling interval δR is generally a function of defocus. At optimum focus [9], i.e. when

$$\Delta z_{\text{opt}} = 0.75 \, C_s (\lambda R_m)^2, \tag{3}$$

where R_m is maximum spatial frequency, the condition (1) for δR can be rewritten as

$$\delta R \leq 1/C_s (\lambda R_m)^3. \tag{4}$$

Using the relation

$$\delta R = R_c / N_f \tag{5}$$

and noting that the carrier frequency R_c must be at least three times larger than R_m, we conclude that a minimum number

$$N_f \geq 3 \, C_s \lambda^3 R_m^4 \tag{6}$$

of hologram fringes are needed. Since a sampling rate of at least *four* pixels per fringe is required [10], the number of pixels needed to reach a specific value of R_m

is given by

$$N_p = 4N_f \geq 12 C_s \lambda^3 R_m^4. \tag{7}$$

The maximum spatial frequency R_m is then determined by the number N_p of available pixels according to

$$R_m \leq [N_p / 12 C_s \lambda^3]^{1/4}. \tag{8}$$

Example: If we insert the numerical values C_s = 1.2 mm and λ = 2.5 pm (corresponding to an electron energy of 200 keV) in Eqs. (7) and (8), we find that
- the minimum number of pixels needed for the achievement of a resolution of 0.1 nm (R_m = 10 nm^{-1}) is 2250.
- the highest resolution achievable with 512 pixels per line is 0.145 nm.

Note that the number of pixels (7) increases by a factor of 4 if the hologram to be processed is recorded at zero defocus instead of optimum focus. The maximum reconstructable spatial frequency (8) then decreases by a factor of $1/\sqrt{2}$.

3.3. Partial coherence

The effect of *chromatic* partial coherence can be described by an envelope function

$$G_c(R) = \exp[-(\pi \delta_e \lambda R^2/2)^2] \tag{9}$$

with

$$\delta_e = C_c / 2 (\ln 2)^{1/2} \, \delta E / E \tag{10}$$

that is mainly determined by the *relative energy width* $\delta E/E$ of the electron beam and the *chromatic* aberration coefficient C_c of the objective lens [4]. This envelope determines the information limit of the electron microscope, and hence the maximum spatial frequency R_m that can be reconstructed from the hologram. Spatial frequencies R damped below the noise level (dashed line in Fig. 4) cannot generally be reconstructed.

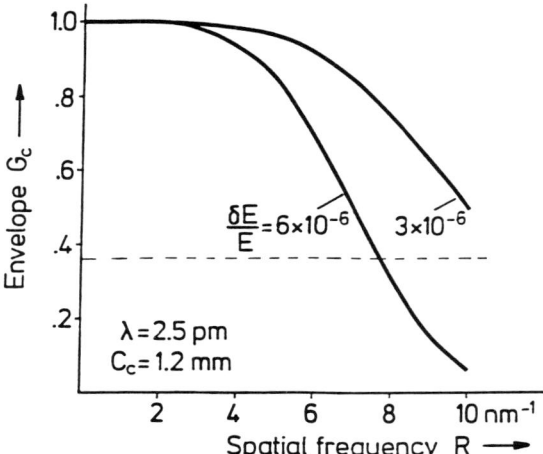

Figure 4. Envelope G_c as a function of the spatial frequency R in the case of *chromatic* partial coherence with a Gausssian distribution. The parameter is the relative energy width of the electron beam.

Note that the information limit is not only determined by the envelope due to *chromatic partial coherence* but also by the attenuation effect due to the *finite* illumination aperture *(spatial partial coherence)*. However, since this effect strongly depends on the defocus, it can be minimized by choosing optimum-focus conditions [9] in the recording step of the hologram.

As discussed in [4], the finite illumination aperture does not only attenuate the object spectrum but also the *contrast* of the hologram fringes. The contrast κ is found to depend on the half width φ_e of the illumination aperture according to

$$\kappa(D) = exp[-(\pi\varphi_e D/\lambda)^2], \tag{11}$$

where the hologram width D is given by

$$D = \lambda b f^2 / [a(a+b)\varepsilon_c] - 2f r_F/a. \tag{12}$$

In this equation, the parameter a represents the distance between the biprism filament and the back focal plane of the objective lens, b is the distance between the biprism and the intermediate image plane (cf., e.g., Fig. 8 in [4]). Furthermore, the parameters f, r_F, and ε_c represent the focal length of the objective lens, the radius of the biprism filament, and the hologram fringe distance, respectively.

To obtain high fringe contrast, the value of κ must be close to unity. This implies the use of *small* illumination apertures ($\leq 10^{-6}$) and *low* current densities in the hologram plane. On the other hand, to operate with small exposure times, the current density should be *high*. These requirements can be fulfilled if a field emission gun is used and *astigmatic illumination* is realized.

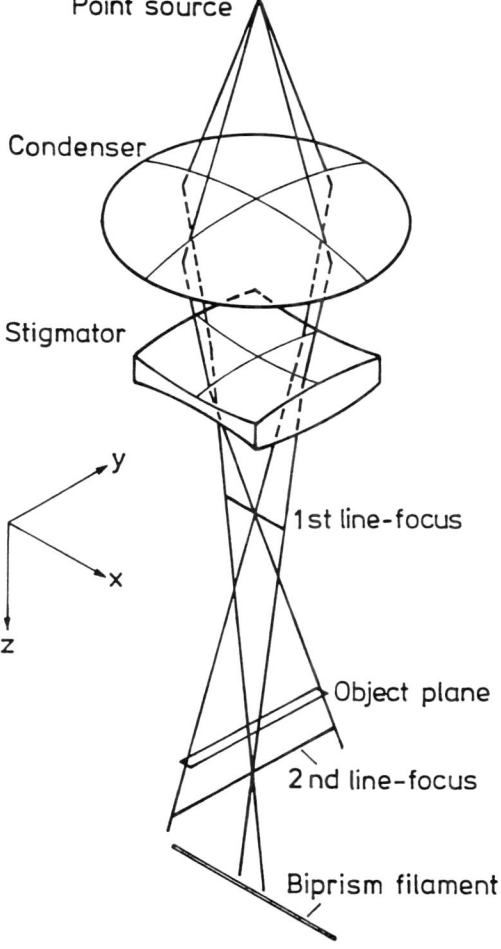

Figure 5. Object illumination by a coherent line-shaped focus obtained by *astigmatic* imaging of the point-like source of a field emission gun [12, 4]. This line focus is adjusted parallel to the biprism filament. The second focus extending in a perpendicular direction is located at a short distance from the object. Because of the elliptical shape of the illumination aperture, the current density is very high in this case.

This kind of illumination can be obtained by overexciting the condenser stigmator to such a degree that *two* line foci with a small separation are produced: one *in front* of the object, and the other *behind* it (see Fig. 5). In this case, the illumination aperture is elliptical in shape and the corresponding envelope function is somewhat complicated in form. This function is shown in Fig. 6 in the form of projected cross sections.

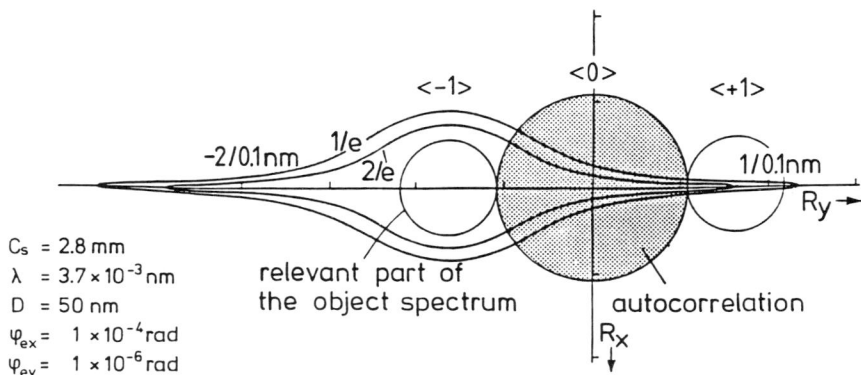

Figure 6. Two-dimensional display of the normalized envelope function in the case of *spatial* partial coherence and astigmatic illumination [13, 4]. It is assumed that the sideband <-1> is used for reconstruction.

It follows from this result that
• the relevant part of the sideband used for reconstruction remains almost unaffected by the envelope.
• the transfer properties of the reconstructed image are determined by the illumination aperture *perpendicular* to the hologram fringes.
• the illumination aperture *parallel* to the fringes can be made at least 100x larger than that in the orthogonal direction.

Consequently, the exposure time can be made about 100x shorter than in the case of rotationally symmetric illumination.

3.4. Noise and instabilities of the electron microscope

The accuracy in the determination of the amplitude and phase distributions of the object wave is found to be limited by the *number of electrons* collected in a given area and by the *contrast* of the hologram fringes [10, 11].

Generally, the *number of electrons* depends on the width of the illumination aperture and the brightness of the electron gun used [4]. As already discussed, the *fringe contrast* decreases with increasing illumination aperture. Furthermore, an additional reduction of contrast results from the *instabilities* of the electron microscope. This decrease of contrast can, in principle, be compensated by proper demagnification of the electron source, but this leads to a reduction of the number of electrons and, as a result, to severe noise problems.

4. CONCLUSIONS AND FUTURE PROSPECTS

The *digital* reconstruction of electron holograms provides a powerful tool for making a quantitative determination of the object function of interest. The *ability* to correct aberrations makes it possible to attain ultra-high resolutions of less than 0.1 nm. Such an ability is, of course, indispensable for new technologies aimed at investigating materials in the atomic range. There are, however, some obstacles to be removed. These are:
- The number of hologram fringes
- On-line reconstruction
- Aberration coefficients

The problem of generating holograms with a *sufficient* number of fine fringes to enable high resolutions to be achieved can be overcome by employing modern electron microscopes with field-emission guns and taking measures against instabilities.

The rapid progress in CCD slow-scan technology and the continuing increase in computing power will surely allow digital reconstruction to be performed *on-line* in the very near future.

The major *difficulty* in reaching the highest resolutions seems to lie in the determination of the aberration coefficients with sufficient accuracy [14]. However, there are positive signs in this direction and a lot of progress [15, 16] has already been made.

REFERENCES

[1] D. Gabor, Proc. Roy. Soc. (London) A 197 (1949) 454.

[2] T. Tanji, K. Urata, and K. Ishizuka in: Proc. 49th Annual EMSA Meeting, San Jose, 1991, p. 672.

[3] H. Lichte, E. Völkl, and K. Scheerschmidt, Ultramicroscopy 47 (1992) 231.

[4] G. Ade in: Advances in Electronics and Electron Physics, Vol. 89, in press.

[5] G. Möllenstedt and H. Düker, Z. Physik 145 (1956) 377.

[6] G. Ade and R. Lauer, Electron Microscopy 1994, Proc 13th Int. Congress, Paris, Vol. 1, p. 327; see also: Beitr. elektronenmikrosk. Direktabb. Oberfl. 26 (1994), in press.

[7] H. Lichte, Ultramicroscopy 47 (1992) 223.

[8] K. Ishizuka, T. Tanji, A. Tonomura, T. Ohno, and Y. Murayama, Ultramicroscopy 53 (1994) 361.

[9] H. Lichte, Ultramicroscopy 38 (1991) 13.

[10] F. Lenz and E. Völkl, Electron Microscopy 1990, Proc. 12th Int. Congress, Seattle, Vol. 1, p. 228.

[11] F. Lenz, Optik 79 (1988) 13.

[12] R. Lauer, Electron Microscopy 1982, Proc. 10th Int. Congr., Hamburg, Vol. 1, p. 427.

[13] G.Ade, R. Lauer, and K.-J. Hanßen, Ber. APh-25, Phys.-Techn. Bundesanstalt, Braunschweig, 1985, p. 45.

[14] H. Lichte, Ultramicroscopy 51 (1993) 15.

[15] W.M.J. Coene and T.J.J. Denteneer, Ultramicropcopy 38 (1991) 225.

[16] A.J. Koster and A.F. de Jong, Ultramicropcopy 38 (1991) 235.

Observation of atomic surface potential by electron holography

Takayoshi Tanji and Kazuo Ishizuka

Tonomura Electron Wave Front Project
ERATO, Research Development Corporation of Japan (JRDC)
P.O. BOX 5, Hatoyama Saitama, 350-03 Japan

The utility of electron holography for observing crystal surfaces in a profile imaging mode is shown by visualizing the phase of an object wave and correcting aberrations. A new method for estimating the amount of aberration is proposed, and its effectiveness is confirmed by using a carbon thin film. The technique is applied to the observation of a MgO (001) surface and reveals clear atomic structure at the surface.

1. INTRODUCTION

On the surface of a crystal, the electrostatic potential is not rigidly restricted but extends into a vacuum. Accurately measuring this surface potential profile and determining the atomic alignment will offer a lot of information about the physical and chemical properties on the surface. Such properties are closely related to fundamental mechanisms of crystal growth, catalysis, etc. Direct microscopical observations of surfaces are very useful in such investigations. Today, typical techniques for this purpose are scanning probe microscopy (SPM), reflection electron microscopy (REM), and transmission electron microscopy (TEM). TEM observation with a glancing angle provides the information on crystal surface structures in both real and reciprocal spaces with high spatial resolution [1]. However, observation in real space using this technique (profile imaging [1,2]) is not easy, because the optimum condition for observing fine structures is far from the in-focus condition, so strong Fresnel diffraction at the crystal edge hides the surface and disturbs the image contrast.

In the present paper, electron holography was employed to obtain clear surface profile images free from both the aberration and the defocusing effect. Electron holography can separately extrapolate both the phase- and the amplitude-information of the object electron wave. We show that the phase image is independent of the Fresnel diffraction, although the defocusing effect remains. We also show the necessity of aberration correction. A new method is proposed for estimating the amount of aberration which is the most important point in the process. This method is applied to the surface profile image.

2. EXPERIMENTAL DETAILS

Specimens of MgO fine particles were prepared by burning a magnesium ribbon in dry air and collecting the smoke on a copper grid covered with a holey carbon film. The plastic film used to prepare the holey carbon film had been dissolved. An alcohol droplet was put on the holey carbon film and removed to fix stably the particles by the surface tension.

The MgO small cubic crystal with edges of 100 ~ 300 nm was oriented so that an electron beam was incident along the [110] direction of the crystal. With this orientation, the (001) surface is parallel to the incident beam and the specimen has a 90° wedge shape, as shown in Figure 1. Thus, Mg and O atoms were aligned independently along the viewing direction. An electron biprism wire was positioned in the vacuum parallel to the edge of the crystal near the image plane.

The holograms were recorded on Kodak 4489 electron microscope films using a HF-2000 field emission TEM instrument. The wave function of the object wave was reconstructed and restored numerically. Holograms (negative) were digitized by CCD video cameras into 512x512 or 1024x1024 pixels, and processed by personal computer (an IBM PC/AT clone), or a SUN work station. The main software was a combination of SEMPER and some original routines added for holography. The film transparency was transferred to electron density in preference to all other processes. The results of the reproduced amplitude and phase were represented in 256 gray levels converted from their normalized values and angles.

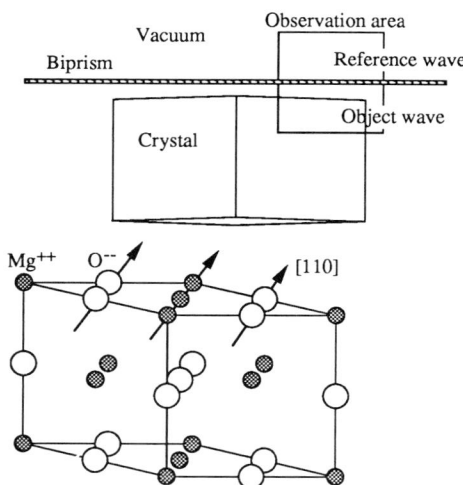

Figure 1. MgO crystal oriented along the [110] direction aligning each kind of atom. An electron hologram is made by overlapping the vacuum area (reference wave) and the specimen area (object wave) with a biprism.

3. OBSERVATION OF PHASE IMAGE

In order to investigate the effect of the Fresnel fringes upon the wave diffracted at a crystal edge, the amplitude and the phase of the wave function were calculated by using an ideal phase object. A profile image of a MgO (001) surface was observed with the phase of electrons.

3.1. Calculation of Fresnel fringes

The phase distribution of an electron wave diffracted by a semi-infinite flat edge was simulated, as shown in Figure 2 [3]. The phase change in the specimen was assumed to be $\pi/6$ and the observation at 40 nm under-focus (Gabor focus [4]) for 200 keV electrons, as illustrated in Figure 2a. A strong oscillation appearing in the intensity distribution in Figure 2b disappears in the phase distribution in Figure 2c. Significantly, however, the difference in the phase between the inside and outside of the crystal does not appear as a sharp edge but as a broad slope. Therefore, we may use the phase image to observe the surface without Fresnel fringes, but we have to remove the effects of both aberrations and defocusing to discuss the potential distribution quantitatively.

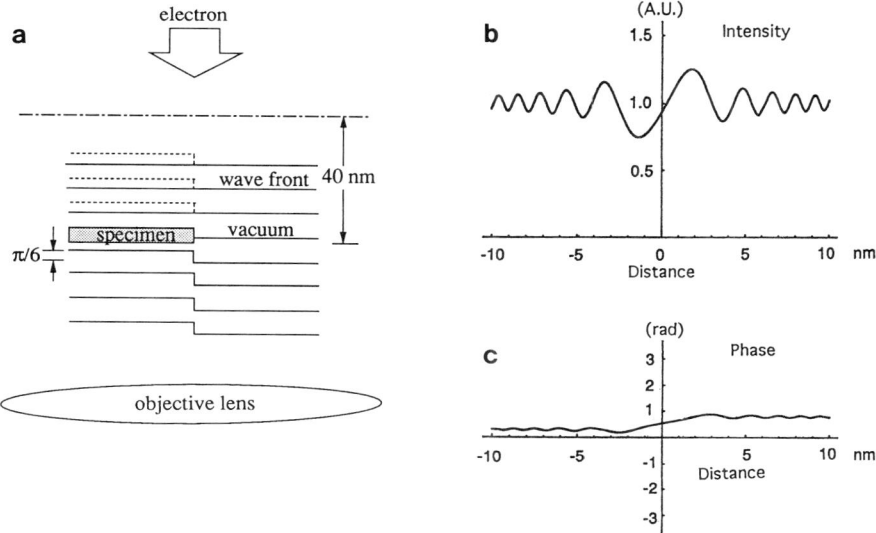

Figure 2. Simulation of Fresnel fringes made by a half infinite phase plate assuming the phase shift of π/6 and defocusing of -40 nm (under focus) (a). The intensity distribution around the edge of the plate shows a strong oscillation (b) and the phase distribution blurs the edge without the oscillation (c).

3.2. Experimental Results [5]

An amplitude image and phase image reconstructed from the same hologram of the MgO (001) surface are shown in Figure 3. The strong Fresnel fringe disturbing the observation of the specimen edge in amplitude image (a) has disappeared in phase image (b), and a clear surface is revealed there. These two digital processed images have exactly the same coordinates. The position of the true surface is more obvious in Figure 3b than in Figure 3a. However, a weak and gradual phase shift is recognized over 1 nm width.

Higher-resolution holography of this surface is shown in Figure 4. From the original hologram in Figure 4a, which consists of 0.03 nm interference fringes, the surface atomic structure was reconstructed with some focus adjustment as shown in Figure 4b. The surface potential extends into a vacuum, depending on the kind of ions facing the vacuum. The defocusing effect remains here, as shown in the previous section.

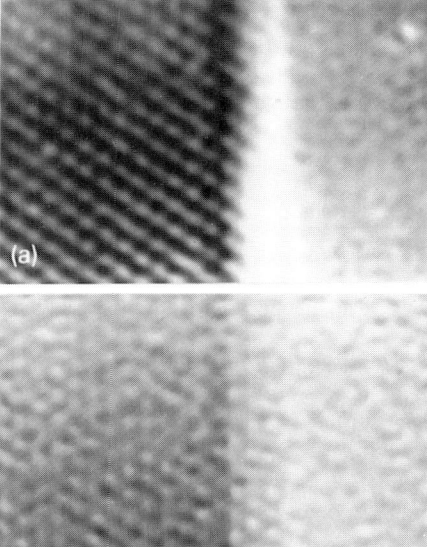

Figure 3. Fresnel fringe appearing at the surface in the amplitude image (a) disappears in the phase image (b).

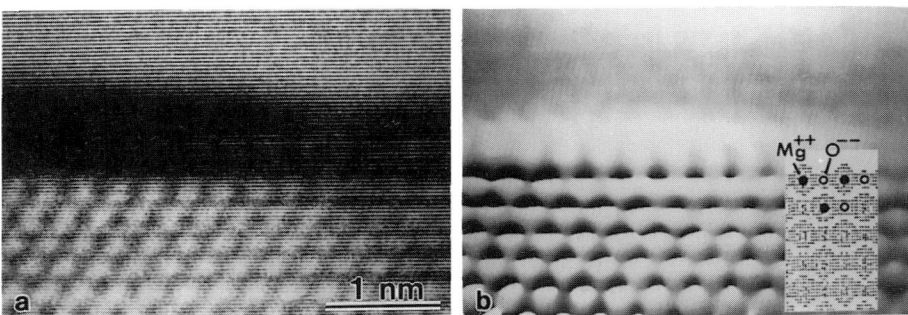

Figure 4. The potential distribution at the MgO (001) surface hidden by the Fresnel fringe in the original hologram (a) is shown clearly in the reconstructed phase image (b).

4. ABERRATION CORRECTION

4.1. Principle of correction

In conventional transmission electron microscopy, the Fourier spectrum, $\Psi(u)$, of the electron wave in an image plane ($\psi(r)$) is characterized by a transfer function, $H(u)$, with the relation to the Fourier spectrum, $O(u)$, of the wave at the exit surface of a specimen, ($o(r)$), as

$$\Psi(u)=O(u)H(u), \tag{1}$$

$$H(u)=A(u)\exp\{\pi i(0.5Cs\lambda |u|^4+df\lambda|u|^2)\}, \tag{2}$$

$$A(u)= \begin{array}{ll} 1, & \text{if } |u| < u_0 \\ 0, & \text{else,} \end{array} \tag{3}$$

where $A(u)$ denotes the objective aperture; λ is the electron wavelength; Cs, the spherical aberration coefficient of the objective lens; and df, the amount of defocusing.

In off-axis electron holography, the wave, $\psi(r)$, is superimposed on a reference plane wave by using an electron biprism with an angle of a_0. The intensity of the recorded hologram is

$$i(r) = |\psi(r)\exp\{\pi i u_0 r\}+\exp\{-\pi i u_0 r\}|^2$$

$$= 1+|\psi(r)|^2+\psi(r)\exp\{2\pi i u_0 r\}+\psi^*(r)\exp\{-2\pi i u_0 r\}, \tag{4}$$

where $|u_0|=a_0/\lambda$ denotes the inclination introduced by the electron biprism and * shows a complex conjugate. The Fourier transform of equation (4) gives

$$I(u)=FT[i(r)]$$

$$=\delta(u)+FT[|\psi(r)|^2]+\Psi(u)\delta(u-u_0)+\Psi^*(-u)\delta(u+u_0). \tag{5}$$

Selecting the third term in equation (5) and correcting the origin to $|u|=0$ in real space [6], we obtain a spectrum similar to equation (1). If we know the values of Cs and df, the original wave function is reconstructed as

$$o(r) = FT^{-1}[O(u)]$$

$$= FT^{-1}[\Psi(u)/H(u)] \qquad (6)$$

where FT^{-1} denotes an inverse Fourier transform.

The correction was evaluated as follows: In high resolution electron microscopy, many specimens may be considered to be phase objects. Therefore, *if the restoration of the wave front in equation (6) is performed perfectly, the contrast of the reconstructed amplitude, $|o(r)|$ should be constant*. The wave function, o(r), may have a small amplitude component, and an example was shown in another paper [8]. This assumption is a looser restriction than the weak phase object approximation.

4.2. Amorphous thin film [7,8]

Figure 5a shows a hologram (positive) of a thin amorphous carbon film, whose simply reconstructed intensity distribution is given in Figure 5b. The background contrast of the hologram, which corresponds to an ordinary electron microscope image, agrees well with the image in Figure 5b, except that the reconstructed image looks sharper than the background of the hologram. This is because of the energy-filtering effect of holography. That is, only the electrons without energy loss in the specimen can interfere with the reference wave. Figure 5c shows the phase distribution of the restored wave using the aberration parameters estimated as described below.

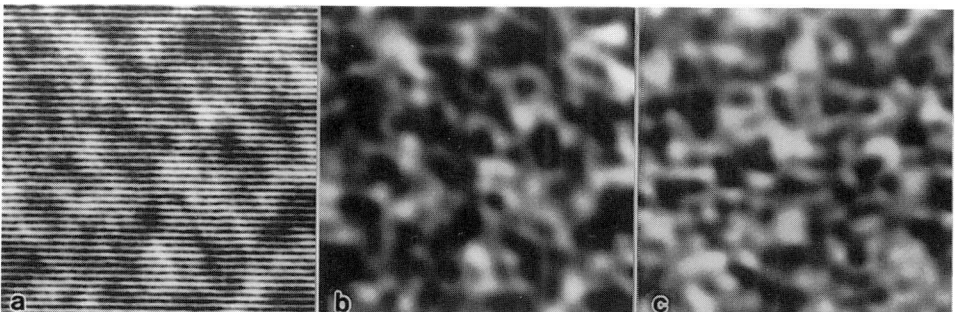

Figure 5. Aberration correction of a thin carbon film. (a) the original hologram; (b) the intensity reconstructed simply from (a); and (c) the phase distribution restored with Cs=1.9 mm and df=-328 nm.

In order to estimate the amount of aberration affecting the hologram, the standard deviation of the amplitude or intensity distribution of the processed image was monitored for various Cs and df in the restoration function. Figure 6 plots the normalized standard deviation of the amplitude images restored from the hologram in Figure 5a, for various correction values of df from 100 to 500 nm, assuming that Cs=1.7 mm. The standard deviations were calculated from 256x256 pixels at the center of the restored images and normalized by the mean values. They show a clear minimum at df=-320 nm (under focus). The deviation of the amplitude simply

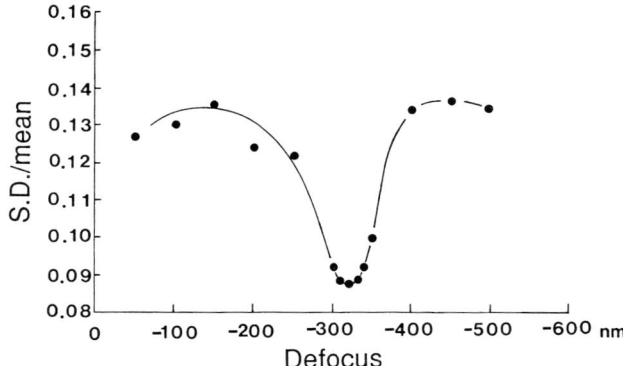

Figure 6. Normalized standard deviations of the amplitude images of carbon thin film processed with various defocus values for Cs=1.7 mm show only one minimum.

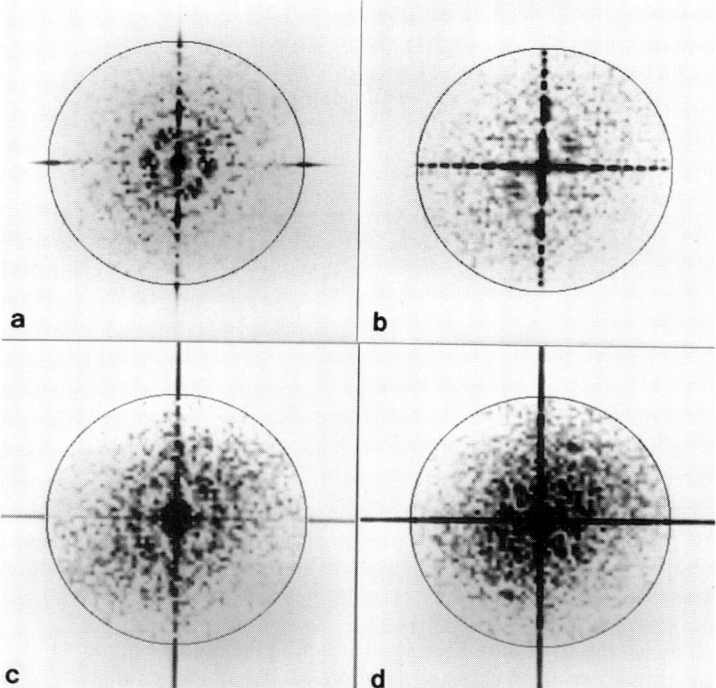

Figure 7. Fourier transforms of amplitude images before (a) and after (b) the aberration correction for the wave reconstructed from Figure 5, and those of phase images before (c) and after (d).

reconstructed is 0.13, while the minimum value in Figure 6 is 0.085. Similarly calculating the normalized standard deviations for various values of Cs shows the smallest point at Cs=1.9 mm and df=-328 nm, which were used in the processing of Figure 5c. Figure 7 shows four diffractograms of the amplitude and phase images before (a) and (c) and after (b) and (d) the aberration correction. Diffractograms before the aberration correction have rings whose contrast are the reverse of each other, because the pattern in Figure 7a shows the imaginary part of the transfer function, and that in Figure 7c, the real part. These rings disappear after the correction, demonstrating that the aberration correction has been carried out successfully.

4.3. MgO (001) surface [3]

In the case of crystalline specimens, the standard deviations of intensity plotted against the parameters (Cs and df) may have many minima, from which the best condition for restoration has to be selected. The method which we propose for choosing the best condition is to monitor simultaneously the amplitude or intensity distribution patterns of the restored images while monitoring their contrast.

In the observation of the crystal surface in the profile imaging mode, Fresnel fringes on the edge are monitored. Figure 8a shows an original electron hologram. The specimen had a 90° wedge shape, and the thickness of the specimen was between about 5 nm (left side) and 13 nm (right side). The hologram consists of an interference area of 10 nm wide with interference

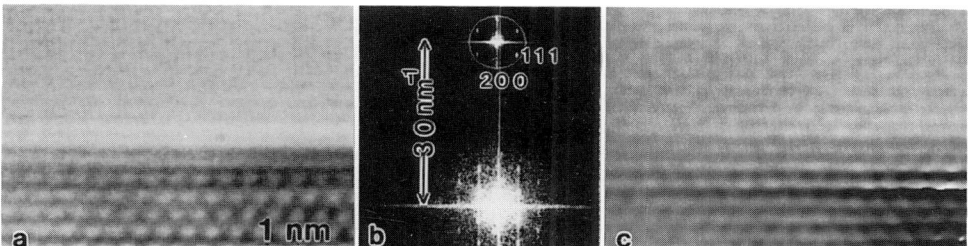

Figure 8. Electron hologram (positive) of MgO (001) surface (a), the Fourier transform (b) and the phase image (c) simply reconstructed with the spectrum encircled in (b).

fringes of 0.033 nm period as shown in its Fourier transform Figure 8b. In the hologram background, which is an ordinary TEM image, strong Fresnel fringes disturb the observation of the surface structure. The phase distribution, which was simply reconstructed with the side-band encircled in the Fourier spectrum in Figure 8b, is shown in Figure 8c. The wave front reconstructed here is still distorted due to the spherical aberration of the objective lens and the out-of focusing from the Gaussian plane.

The standard deviations of the restored wave amplitude are plotted in Figure 9 for various correction values of df from -100 to -200 nm, assuming that Cs =1.7 mm. These standard deviations were calculated over an area of 128x128 pixels at the center of the frame and were normalized with mean values. They show many minima with distances from 20 to 25 nm. These distances correspond to the period with which the transfer function of the electron microscope has a large value for 111 reflections. By comparing the images processed with the defocus values near each minimum of the curve in Figure 9, we found that the minimum point df = -165 nm is the best condition for restoration. Three examples of amplitude distributions processed at two minimum points and one maximum point are presented beside the plot in Figure 9. The image processed with df = -165 nm shows the lowest contrast and no Fresnel

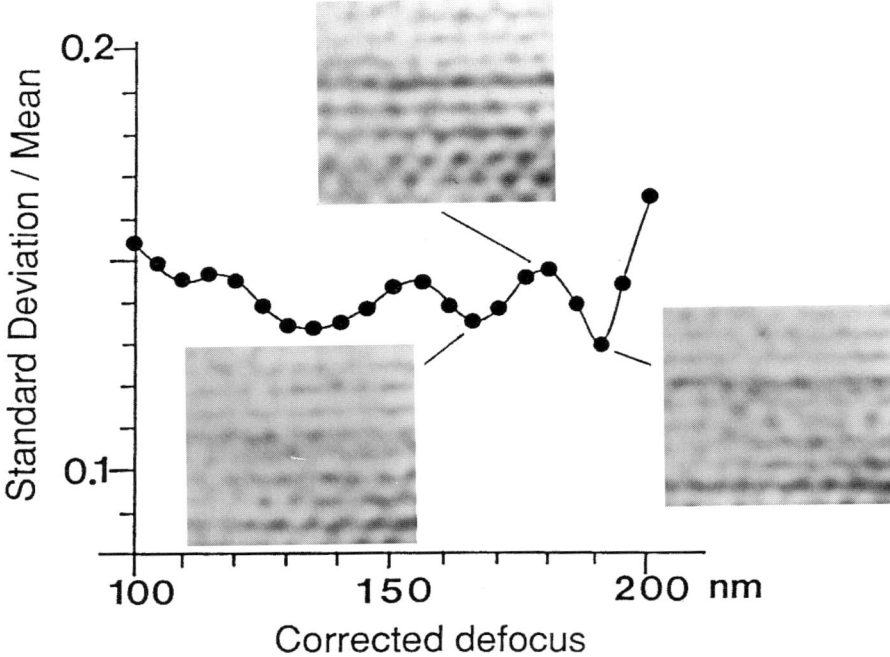

Figure 9. Standard deviations have many minima for various correction values of df. Insets are three examples of amplitude images processed with df = -190 nm, -179 nm and -165 nm.

fringe. Periodic contrast appearing in the vacuum area is a ghost induced by truncating elongated diffractions due to a sharp end at the crystal surface during the processing. The phase distribution of the wave freed from the aberration effects reveals the extension of the surface potential with a same period as the (110) lattice planes, as shown in Figure 10a. This result supports the previous preliminary finding in Section 3.2 in which the Gabor focus was utilized in recording the hologram and no aberration correction was applied. However, the length of the extension is shorter in Figure 10a than that in Figure 4b. The dark area represents the phase delayed by the crystal potential, which is attractive to incident electrons. Therefore, cations (Mg^{++}) appear dark and anions (O^-) appear bright, as indicated in the simulated phase image Figure 10b for t = 10 nm, Cs = 0, and df = 0. The surface atomic structure in the model is assumed to be a simple truncation of the bulk structure. The spacing between the first and the second atomic layer in Figure 10a looks 20~25% wider than the inside one, while the simulated image Figure 10b shows only an expansion of about 5%.

5. DISCUSSION

The abrupt change in phase shown in Figure 4b is due to uncertainty of the phase at the point of very weak amplitude. The phase determined by the arctangent of the ratio of the imaginary part to the real part of the wave function at each point is uncertain if the amplitude is

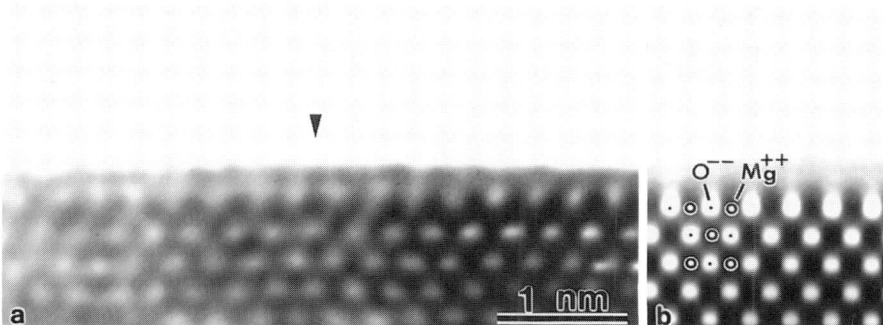

Figure 10. Phase image of MgO (001) surface restored with values Cs = 1.7 mm and df = -165 nm (a) coincides with a simulated one for t = 10 nm (indicated by an arrow in (a)) and Cs = df = 0 (b). Open circles show Mg^{++} and black dots O^-.

nearly zero. This condition is often observed in thick regions of the specimen. The expansion in Figure 10a seems to be too large to be considered an electro-optical effect, even if a little of effect remains. Relaxation of surface atoms have been reported in some metals, and that on the gold (111) surface has been observed by transmission electron microscopy [9]. However, very little is known about surface relaxation on MgO crystals, except for a work with low energy electron diffraction [10]. A detailed discussion requires more accurate computer simulation. The effect of charging up, which often occurs in the observation of fine particles [11], may have to be considered, too.

6. CONCLUSIONS

It was shown that phase images obtained by electron holography are free from Fresnel diffraction. Therefore, the image obtained under proper focus conditions can present a clear profile image of the surface. However, the defocusing effect remains in such images because the proper focus is far from the Gauss focus. Therefore, aberration correction is indispensable for quantitative analysis. Estimating the amount of aberration, the last important problem to be solved in the aberration correction, was surmounted by monitoring the standard deviations of the restored amplitude. Minima and maxima appearing in the standard deviation curves of crystalline specimens were distinguished by paying attention to the pattern of the amplitude image processed under each condition. The atomic potential distribution on the (001) surface of MgO was observed to extend into a vacuum, depending on the kind of ions facing the vacuum. The aberration correction made the extension length shorter and expands the distance between the first and the second atomic layers. More detailed investigation is required to confirm if this expansion represents a surface relaxation.

ACKNOWLEDGEMENTS

The authors are very grateful to Mr. Takahisa Ohno for his assistance in the image processing through this work. We also thank Mr. Takao Matsumoto for his help in the calculation in Section 3.1.

REFERENCES

1. T.Tanji and J.M. Cowley, Ultramicroscopy, 17 (1985) 287.
2. L.D. Marks and D.J. Smith, 303 (1983) 316.
3. T.Tanji, T.Ohno, K.Ishizuka and A.Tonomura, J. Electron Microsc., 43 (1994) in press
4. H.Lichte, Ultramicroscopy, 38 (1991) 13.
5. T.Tanji, K.Urata, K.Ishizuka, Q.Ru and A.Tonomura, Ultramicroscopy, 49 (1993) 259.
6. K.Urata, K.Ishizuka, T.Tanji and A.Tonomura, J. Electron Microsc., 42 (1993) 88.
7. K.Ishizuka, T.Tanji, A.Tonomura, T.Ohno and Y.Murayama, Ultramicroscopy, 53 (1994) 361.
8. T.Tanji and K.Ishizuka, Microsc. Soc. of America Bulletin, 24(1994) 494.
9. L.D.Marks, Surface Science, 139 (1984) 281.
10. Y.Murata and S.Murakami, In: P.J.Dobson, J.B.Pendry and C.J.Humphreys (eds.), Electron diffraction 1927-1977, Inst Phys, Bristol, (1978) 218.
11. T.Tanji, H.Masaoka, J.Ito, K.Yada and J.M.Cowley, Ultramicroscopy, 27 (1989) 223.

Electron Holography
A. Tonomura, L.F. Allard, G. Pozzi, D.C. Joy and Y.A. Ono (Editors)
© 1995 Elsevier Science B.V. All rights reserved.

Phase-shifting techniques in electron holography

Q. Ru

Tonomura Electron Wavefront Project, ERATO
Research Development Corporation of Japan (JRDC)
P. O. Box 5, Hatoyama, Saitama 350-03, Japan

A phase-shifting method is developed and described for reconstructing off-axis wavefront-division electron holograms obtained with an electron biprism. The ability of high sensitivity, high precision and high resolution of phase measurements is shown with theoretical analysis and demonstrated with experimental results. A modified method is also described for reconstructing amplitude-division holograms obtained with a crystal beam splitter.

1. INTRODUCTION

Since the successful development of the field-emission electron gun and electron biprism, lots of kinds of interesting specimens have been observed and investigated by electron holography [1]. Recently, two topics of electron holography are getting interesting and studied actively. One is how to achieve high resolution images, such as atomic structures at Å-order resolution.[2-4] The other is how to observe and measure very weak phase objects such as biological specimens and weak electric/magnetic fields [5-8].

Since electron holography has a high potential for correcting the aberrations of electron lenses, many high resolution studies by electron holography have been reported. For increasing the resolution of reconstructed images, making narrow fringes is most emphasized and studied actively. This is because that the widely used Fourier transform reconstruction methods, including the numerical [9,10] and optical [11] methods based on Fourier theory, need sufficiently narrow fringes compared with the resolution to be observed. A rough estimation of necessary spacing of fringes is less than 1/3 of the resolution to be observed. On the other hand, due to the finite coherence of electron source, the quality of fringes is inferior when fringe are made narrower. The incompatibility of fringe quality and fringe spacing is one of the most difficult problems for high resolution observations.

For weak-phase measurement, the sensitivity and precision of phase reconstruction is most important. Although both numerical and optical reconstruction methods have been successful to detect $2\pi/50$ of phase change [12], the sensitivity is still limited at this level. The limitation arises mainly from quantum noise of electrons and from the filtering process in the Fourier domain of hologram. In principle, any finite object or hologram produces an infinitely broad spectrum distribution, so that the center band and the side bands in the Fourier domain can never be separated perfectly for each other, and therefore errors can never be avoided.

In order to achieve both high resolution and precision measurements without making narrow fringes, we have developed a phase-shifting method. The so-called phase-shifting technique proposed first by Bruning [13] for optical wavefront measurement is based on use of number of holograms with properly different initial phases. It has been shown that the initial phase of a wavefront-division hologram obtained with a biprism can be shifted by tilting the incident beam [14-16], and that of an amplitude-division hologram obtained with a crystal beam splitter can be shifted by shifting the crystal [17]. The object wave can therefore be calculated from a series of holograms of different initial phases, independent of the fringe spacing. Because summation operations are included in the calculations, the random quantum noise of electrons and shot noise of detectors can be greatly reduced to achieve high precision measurements.

2. PRINCIPLE OF PHASE-SHIFTING HOLOGRAPHY

Phase-shifting electron holography consists of three processes: (i) tilting the incident electron beam to obtain a series of holograms of different initial phases, (ii) measuring the initial phase of each hologram by calculating the fringe shift of each hologram, and (iii) calculating the object wave from whole holograms and their initial phase values.

2.1 Shifting initial phase

Phase-shifting electron holography is based on tilting the incident electron beam to obtain a series of holograms with different initial phases. The electron optics and instruments used to do this are shown in Figure 1. An in-focus image of the object to be measured is first formed, and then a hologram of the object is produced by an electron biprism. After the illuminating, imaging and interfering electron optics conditions have been adjusted, the current in the beam-tilt coil is increased (or decreased) manually, or automatically by linearly raising the voltage of a battery which is connected to the beam-tilt coil.

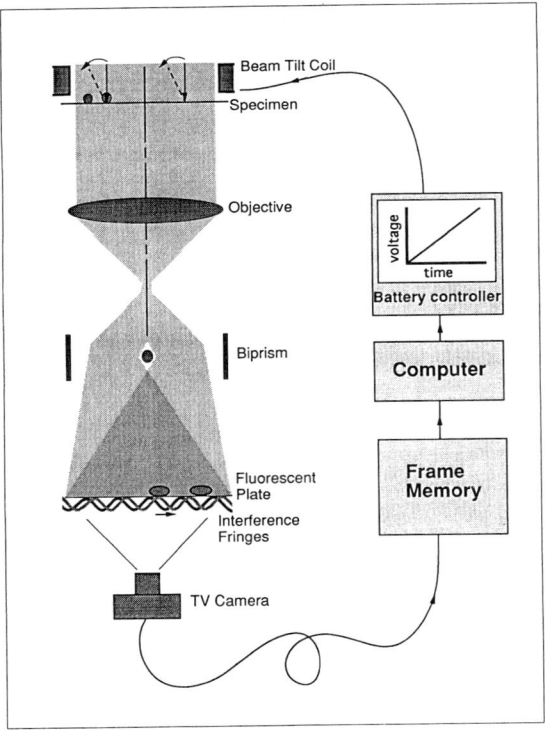

Figure 1. Electron optics and instruments for shifting the initial phase of a hologram.

Most electron microscopes have two beam-tilt coils: one tilts the incident beam in the x direction and one tilts it in the y direction. If the fringes of the hologram are parallel to the x direction, the initial phase is only sensitive to the y-direction beam tilt, and consequently fringes parallel to the y direction are only sensitive to the x-direction beam tilt. The incident beam is tilted continuously until the initial phase is shifted by 2π (or more), which corresponds to one fringe shift (or more), while the initial phase-shifting holograms are recorded in series by a TV camera that detects the holograms frame by frame, with a time resolution of 1/30 second for an NTSC signal format. The hologram data for each frame is immediately transferred to a frame memory and then to the computer. If the electron microscope is equipped with a slow-scan CCD camera, the initial phase may be shifted step by step.

The relation between the initial phase and the angle of the incident beam can be derived as the following. the path difference S between the two interfering beams

$$S = W \tan \theta \approx W\theta, \qquad (1)$$

introduces an initial phase ϕ_0 of

$$\phi_0 = (2\pi/\lambda)S$$
$$= (2\pi/\lambda)W\theta, \qquad (2)$$

where W denotes the width of the interference area, λ the wavelength of the incident electron beam, and θ the incident angle, which is usually so small that the approximation $\tan\theta = \theta$ is always well satisfied. The initial phase must shift by at least 2π. The necessary beam tilt θ_0 for a 2π-initial phase shift can be derived as

$$\theta_0 = \lambda/W. \qquad (3)$$

For 200-kV electrons ($\lambda = 0.0025$ nm) and an interference area 250 nm wide for example, the beam tilt must be at least 10^{-5} radians.

It has been pointed out that shifting the biprism in a direction orthogonal to the biprism wire can also introduce a shift of the initial phase.

2.2 Measuring initial phase

It is obvious from a series of electron holograms with different initial phases in Figure 2 that

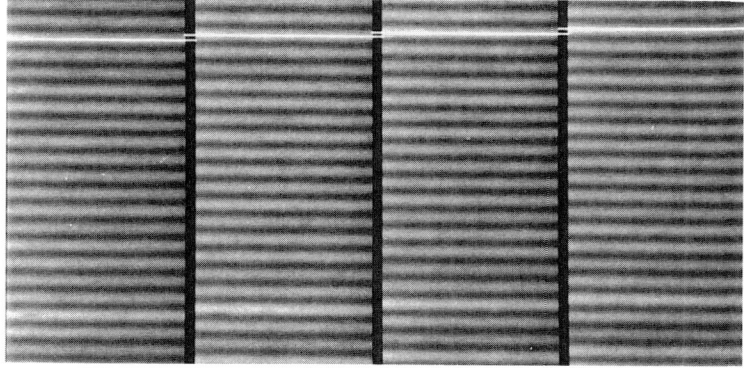

Figure 2. Holograms with different initial phases shifted by tilting the incident beam.

a shift in the initial phase of an off-axis hologram causes a fringe shift. If the initial phase is shifted by 2π, all the fringes in the hologram shift by one fringe. The initial phase shift can thus be equivalently calculated by measuring the fringe shift.

An accurate way to measure the fringe shift is to Fourier transform the hologram and then measure the phase in the complex Fourier spectrum. There are always three bands, a center band and two side bands, appearing in the Fourier spectrum of an off-axis hologram. A typical complex Fourier spectrum is shown in Figure 3a. The real and imaginary components given by

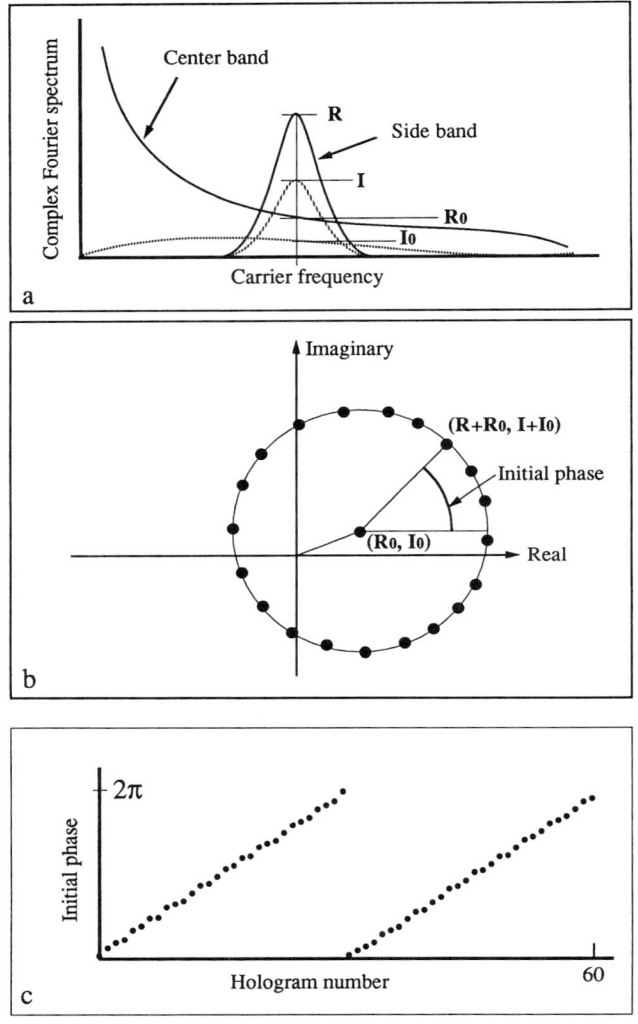

Figure 3. (a) Typical profile of a complex Fourier spectrum where the real and imaginary parts are shown by solid and broken lines, respectively. (b) Complex plane illustrating the circular trajectory of the spectrum data at the carrier frequency. (c) Calculated initial phase data for a series of holograms.

the center band are denoted by R_0 and I_0 and those given by the side band are denoted by R and I. Since shifting the initial phase of a hologram does not change the complex spectrum of the center band nor the absolute amplitude of the side band, R_0, I_0 and $(R^2+I^2)^{1/2}$ are invariable during initial phase shifting. A series of the real-imaginary pair values obtained from a series of holograms with different initial phases therefore forms a circular trajectory, with its radius equal to the amplitude $(R^2+I^2)^{1/2}$ of the complex spectrum of the side band and its center located at point (R_0, I_0). The spectrum series data are shown in Figure 3b as black dots. The values of R_0 and I_0 are determined by fitting the data trajectory with a circle, and the components R and I given by the side band are determined by subtracting R_0 and I_0 from the measured spectrum data. The initial phase of each hologram is determined by calculating the arc tangent of the ratio I/R. A series of calculated initial phases is plotted in Figure 3c, where up to a 4π- shift in the initial phase (i.e., two-fringe shift) occurred.

2.3 Reconstructing object wave

The relation between the obtained hologram intensity data I, the initial phase data ϕ_0 and the object wave to be reconstructed can be expressed by

$$I(x,y,n) = a(x,y) + b(x,y)\cos\{2\pi x/T_x + 2\pi y/T_y + \phi(x,y) + \phi_0(n)\}, \qquad (4)$$

where $a(x,y)$ is the background contrast proportional to the microscopic intensity image of the object, $b(x,y)$ is the fringe contrast proportional to the amplitude distribution of the object wave, $\phi(x,y)$ is the phase image, T_x and T_y are the fringe spacing in the x and y directions and n is an integer numbering the time-sequence holograms. For convenience, Equation 4 is rewritten in complex form as

$$I(n) = C_1 + C_2 \exp\{+i\phi_0(n)\} + C_3 \exp\{-i\phi_0(n)\}, \qquad (5)$$

where

$C_1 = a(x,y),$
$C_2 = (1/2)b(x,y)\exp\{+i2\pi x/T_x + i2\pi x/T_y + i\phi(x,y)\},$
$C_3 = (1/2)b(x,y)\exp\{-i2\pi x/T_x - i2\pi x/T_y - i\phi(x,y)\}.$

Equation 5 can be expressed in matrix form as

$$[1 \quad \exp\{+i\phi_0(n)\} \quad \exp\{-i\phi_0(n)\}] \begin{bmatrix} C_1 \\ C_2 \\ C_3 \end{bmatrix} = I(n). \qquad (6)$$

Multiplying both sides of the Equation by 1, $\exp\{-i\phi_0(n)\}$, and $\exp\{+i\phi_0(n)\}$ alternatively, and then summing both sides of the Equation over n leads to a 3×3 matrix expression:

$$\begin{bmatrix} N & \sum_{n=1}^{N}\exp\{+i\phi_0(n)\} & \sum_{n=1}^{N}\exp\{-i\phi_0(n)\} \\ \sum_{n=1}^{N}\exp\{-i\phi_0(n)\} & N & \sum_{n=1}^{N}\exp\{-2i\phi_0(n)\} \\ \sum_{n=1}^{N}\exp\{+i\phi_0(n)\} & \sum_{n=1}^{N}\exp\{+2i\phi_0(n)\} & N \end{bmatrix} \begin{bmatrix} C_1 \\ C_2 \\ C_3 \end{bmatrix} = \begin{bmatrix} \sum_{n=1}^{N}I(n) \\ \sum_{n=1}^{N}I(n)\exp\{-i\phi_0(n)\} \\ \sum_{n=1}^{N}I(n)\exp\{+i\phi_0(n)\} \end{bmatrix} \quad (7)$$

The 3x3 matrix in Equation 7 is called the characteristic matrix. C_1, C_2, and C_3 can be determined by computing the inverse of the characteristic matrix:

$$\begin{bmatrix} C_1 \\ C_2 \\ C_3 \end{bmatrix} = \begin{bmatrix} N & \sum_{n=1}^{N}\exp\{+i\phi_0(n)\} & \sum_{n=1}^{N}\exp\{-i\phi_0(n)\} \\ \sum_{n=1}^{N}\exp\{-i\phi_0(n)\} & N & \sum_{n=1}^{N}\exp\{-2i\phi_0(n)\} \\ \sum_{n=1}^{N}\exp\{+i\phi_0(n)\} & \sum_{n=1}^{N}\exp\{+2i\phi_0(n)\} & N \end{bmatrix}^{-1} \times \begin{bmatrix} \sum_{n=1}^{N}I(n) \\ \sum_{n=1}^{N}I(n)\exp\{-i\phi_0(n)\} \\ \sum_{n=1}^{N}I(n)\exp\{+i\phi_0(n)\} \end{bmatrix}. \quad (8)$$

In the special case where

$$\phi_0(n) = 2\pi n/N \text{ for } n=1,2....N, \qquad (N \geq 3) \qquad (9)$$

all the elements but the diagonal ones in the characteristic matrix become zero, and Equation 8 can be simplified to

$$\begin{bmatrix} C_1 \\ C_2 \\ C_3 \end{bmatrix} = (1/N) \begin{bmatrix} \sum_{n=1}^{N}I(n) \\ \sum_{n=1}^{N}I(n)\exp\{-i2\pi n/N\} \\ \sum_{n=1}^{N}I(n)\exp\{+i2\pi n/N\} \end{bmatrix}, \qquad (10)$$

which is equal to the formula derived from the original description of the conventional optical phase-shifting method.

After C_1, C_2, and C_3 are determined from Equation 8 or 10, the background contrast distribution $a(x,y)$, the fringe contrast distribution $b(x,y)$, and the phase distribution $\phi(x,y)$ can be obtained from Equation 5:

$$a(x,y) = C_1, \tag{11a}$$

$$b(x,y) = 2\sqrt{C_2 C_3}, \tag{11b}$$

$$\phi(x,y) = \tan^{-1}\left\{\frac{\text{Im}[C_2]}{\text{Re}[C_2]}\right\} - 2\pi x/T_x - 2\pi y/T_y$$

$$= \tan^{-1}\left\{\frac{\text{Im}[C_3]}{\text{Re}[C_3]}\right\} - 2\pi x/T_x - 2\pi y/T_y \tag{11c}$$

where Re and Im denote the real and imaginary parts of the complex value in brackets [].

Equation 10 is similar to the Fourier transform of the time-sequence intensity data $I(n)$. When $\phi_0(n)$ satisfies Equation 9, multiplying $I(n)$ by $\exp\{+i2\pi n/N\}$ and then summing the results is equivalent to the discrete Fourier transformation of $I(n)$ for $n=1,2,\ldots,N$. As schematically illustrated in Figure 4, the time-sequence intensity data $I(n)$ at any point (x,y) is a cosine function, with amplitude and phase proportional to those of the object wave and a bias value proportional to the background intensity, at the same (x,y) point. These three fundamental parameters of a cosine function are determined by the Fourier transform.

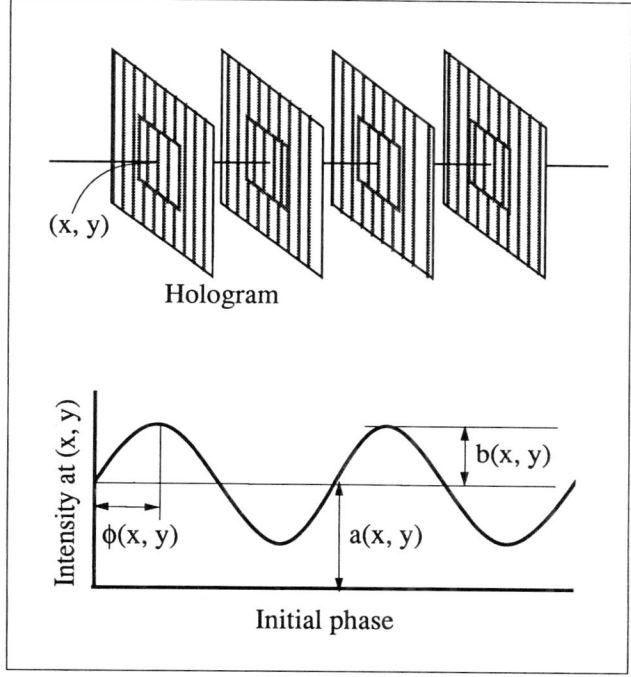

Figure 4. Relation between the object wave and the cosine curve obtained from a series of hologram intensities at point (x,y).

In an ideal case where no any noises exist in the fringes, 3 holograms are enough to reconstruct the object wave. In practice, however, quantum noise of electrons and shot noise of the detectors always exist. Because we have a large frame memory, we use as many holograms as possible to reduce the noises. In the following experiments, 100-300 holograms were used. Since the TV camera used in the experiments had the time resolution of 30 frames per second, total recording time was less than 10 seconds. The total recording time is reasonable and drift or radiation damage of the specimen can be neglected.

3. PERFORMANCE IN PHASE MEASUREMENT

3.1 Independence of fringe spacing

Since Equations 8 and 11 are derived without any assumption regarding the object size and fringe spacing, the spatial resolution of the reconstructed image is equal to that of the micrograph (if without spherical aberration correction), and is independent of the fringe spacing. This is the most important advantage of this method over the conventional Fourier transform reconstruction methods.

In most practical cases, holograms and objects have sharp edges. Because edges create a broad Fourier spectrum, infinitely broad in the strict sense, any finite filter used in the Fourier transform method produces an edge effect, which disturbs not only the edges themselves but also the image area away from the edges. The phase-shifting method, however, can reconstruct these sharp edges. Figure 5 shows the difference between the phase images reconstructed from an example hologram by using the Fourier transform method and phase-shifting method. In the

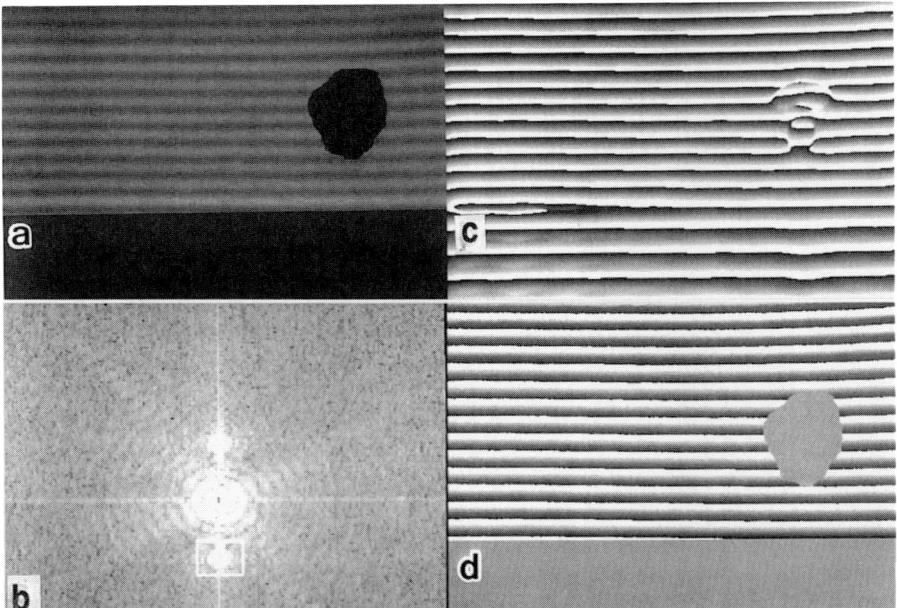

Figure 5. (a) Hologram including opaque objects and edges, (b) Fourier spectrum of the hologram, (c) phase image showing artifacts by Fourier transform method, and (d) correct phase image by phase-shifting method.

phase image obtained from the Fourier transform method, artificial phase distributions can be clearly seen in the areas where no fringes are recorded in the hologram. It is obvious from the Fourier spectrum shown in Figure 5b that the two side bands consist of long streaks overlapping with the center band. A side-band-pass filter, as marked in Figure 5b, produces an infinitely broad vibration in the image plane, i.e., phase distribution. Therefore, in spite of nothing in the opaque areas, artificial phase distributions are caused by the edge effect.

Application of the phase-shifting method to the observation of quantized magnetic fluxes (fluxons) leaking out from a superconducting film is shown in Figure 6, where 12 holograms were used for phase reconstruction. A Pb thin film is placed in the specimen plane, and then is cold down to 5 K so that superconducting state is generated and fluxons are trapped in pinning centers. By applying a uniform outer magnetic field to the thin film, the fluxons can be pulled out from the film to the outer vacuum space, and then detected by measuring the phase variation of the electron beam passed through the vacuum space.

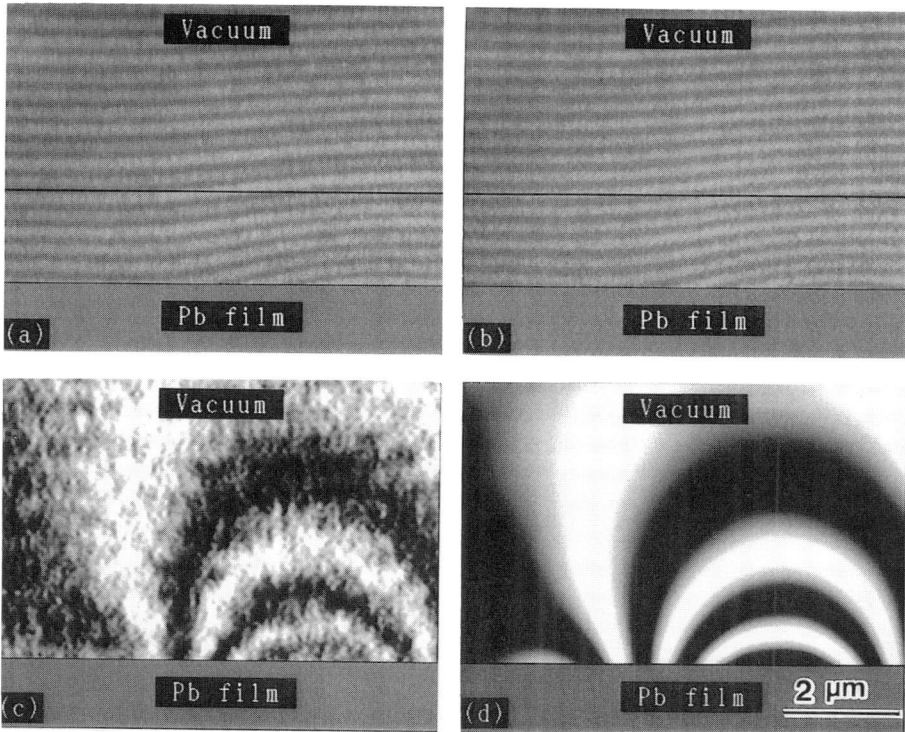

Figure 6. Two holograms (a) and (b) with different initial phases, reconstructed contour phase images (twice phase amplified) (c), and expected contour phase image (d).

An example of holographic observation of flagella filaments is shown in Figure 7. A series of holograms with 96 different initial phases, from 0 through 4π, were recorded and used in the phase computation. Although the filaments are as thin as the fringe width, the phase image has been reconstructed by the phase-shifting method.

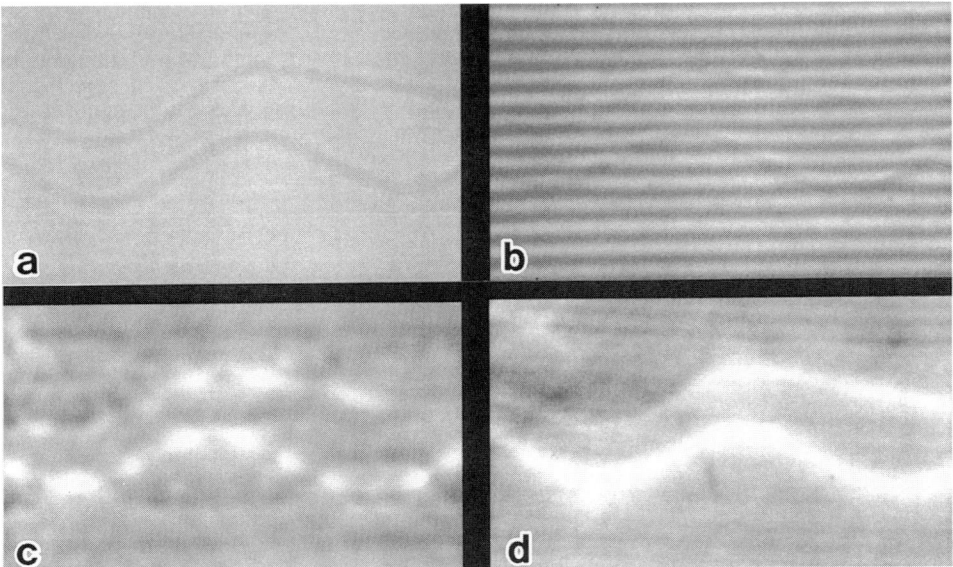

Figure 7. (a) Defocused TEM image of thin flagella filaments, (b) a hologram, (c) phase image obtained by Fourier transform method, and (d) phase image by phase-shifting method.

3.2 High precision performance

Another important advantage of the phase-shifting method is its ability to very accurately measure the phase of an object wave. Since all the elements in Equation 7 consist of a summing operation (i.e., averaging), such random noises as the quantum noise of electrons and the electric noise of detector arrays are significantly reduced by the time-domain summation. The more holograms are used, the higher the precision will become. This is the major reason phase-shifting is widely used in optical interferometry. Heterodyne optical interferometry based on analog analysis of a continuous intensity signal $I(t)$ (i.e., N is infinite) is as precise as $2\pi/1000$ in phase detection[19].

Since the object wave at any point is reconstructed from only the intensity data for that point, as mentioned in the previous section, any defects or contaminants are localized in their own areas and do not propagate to the area of interest. This is particularly useful in practical electron holography to reduce noise.

The experimental results obtained from fine measurement of atomic steps on a molybdenite crystal film are shown in Figure 8. The initial phase of a hologram was shifted continuously and linearly from 0 through 4π, and 192 frames of holograms were recorded by a TV camera, frame-by-frame in a frame buffer memory. The total time for recording these holograms was about 10 seconds. The phase difference over the step edge is $2\pi/25$, which corresponds to three atomic steps (1.86 nm height). The deviation of the background noise in the phase image is under $2\pi/100$; it means that the precision of the phase measurement is under $2\pi/100$.

Figure 8. Twenty-four phase-amplified contour image showing a three-atomic-height surface step on molybdenite crystal.

The precision allows us to observe more weak phase objects consisting of only light atoms, such as carbon atoms. Figure 9 shows the application results of observing single-shell carbon nanotubes.

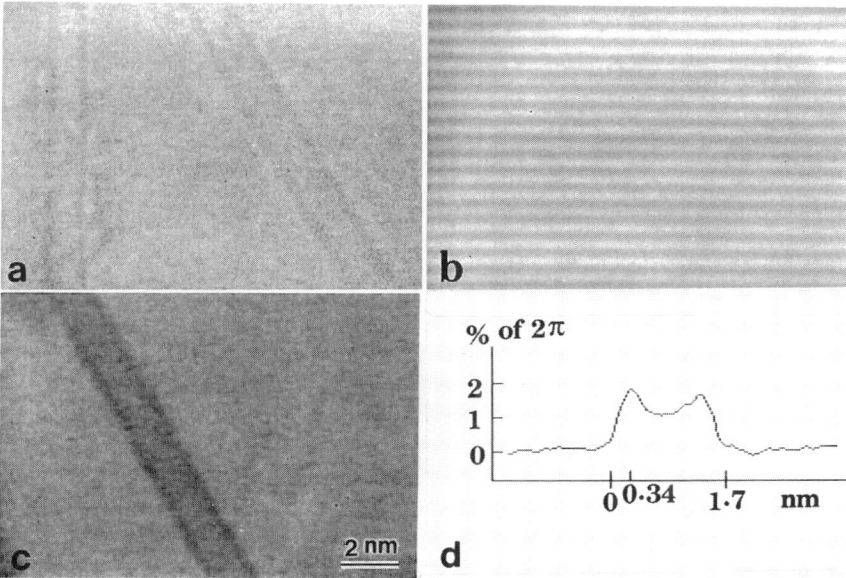

Figure 9. (a) Defocused micrograph of single-shell carbon nanotubes, (b) a hologram of the in-focus image, (c) phase image of the nanotube of 1.7 nm in diameter, and (d) sectional phase profile averaged along the tube axis.

It has been found by Iijima [20] that carbon nanotubes, particularly single-shell carbon nanotubes, can have interesting properties. Many electric and structural properties are still unknown. Electron holography is useful to investigate the projected potential distribution and to determine the 3-dimensional shape. However, the maximum phase shift caused by one carbon atom layer is as small as 1/200 of 2π which cannot be detected by the conventional methods. Here the present phase-shifting method is applied to observe the projected potential distribution of single-shell carbon nanotubes. Figure 9 shows the results. Although nothing can be observed from a hologram because of small phase change, the phase image of such a nanotube has been clearly obtained using 100 holograms of different initial phases. The measured phase profile well agrees with expectation. We found that the mean-inner potential of single carbon nanotubes is equal to 10 eV.

4. MODIFICATION FOR AMPLITUDE-DIVISION HOLOGRAPHY

Using a crystal film instead of an electron biprism to form interference fringes can greatly reduce the coherence requirement and facilitate hologram formation. Two modes, low-resolution [21-23] and high-resolution [24,25], have been developed. In the high-resolution mode, both the beam-splitting crystal film and the specimen film are placed in the normal specimen plane; in the low-resolution mode, the crystal film is placed in the normal specimen plane but the specimen to be investigated is placed in the select-area aperture plane under the objective lens of the microscope.

In both modes, the optical and digital Fourier transform methods were used for hologram reconstruction. Consequently, the resolution of the high-resolution mode cannot exceed the frequency of the lattice fringes of the beam-splitting crystal, and the resolution of the low-resolution mode cannot exceed 1/50-1/100 of the frequency because the lattice fringes in the specimen film have been enlarged 50-100 times by the objective lens.

In low-resolution mode [21,25], because the crystal film and the specimen film are in different holders, the initial phase can be easily achieved by slightly shifting the crystal film or by giving a drift to the crystal film. In high-resolution mode [24], however, because both the crystal and specimen films are in one holder, perfectly individual control of only the crystal film is impossible. In practical cases, two ways are available: 1) when the lattice fringes are much narrower than the specimen structure to be observed, simultaneously shifting both films is available because one-fringe shift of the specimen can be neglected; 2) the drifting effect of only the crystal film can be utilized.

The low-resolution mode was used to experimentally demonstrate the method. An externally charged micro tip hanging on a biprism wire (Figure 10a) was placed in the select-area aperture plane of an HF-2000 microscope. A gold crystal thin film was placed in the normal specimen plane. The 0.2-nm lattice fringes were used to form holograms of the electric field around the micro tip. Natural drift of the gold crystal film was achieved. A series of holograms were recorded during the 5 seconds while the crystal film drifted by one lattice fringe. Ninety-six holograms were used for the phase calculation.

The phase distribution close to the charged tip is shown in Figure 10b. The defocusing-correction process is not performed because the projected electric field varies so continuously and slowly that its phase distribution in the Fresnel plane was proportional to that in the in-focus plane.

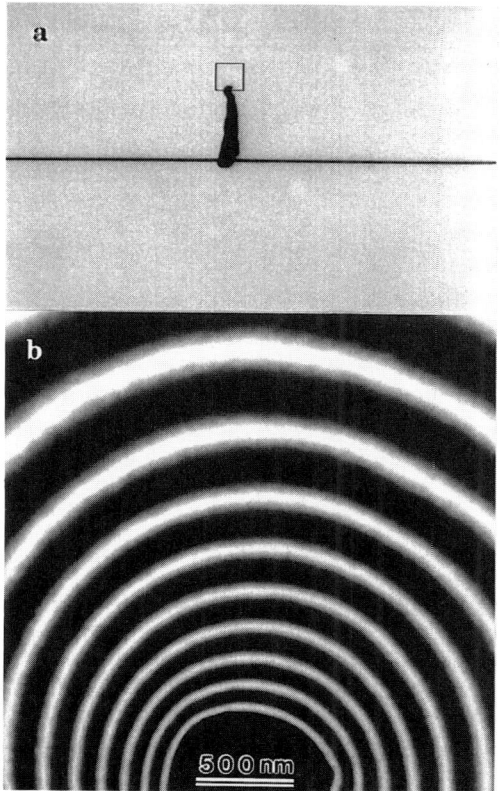

Figure 10. (a) An externally charged micro tip hanging on a biprism wire and (b) the measured contour phase image in the region as marked in (a).

5. CONCLUSIONS

We previously showed that tilting the incident electron beam or shifting the biprism can shift the initial phase of an electron hologram. A generalized phase-shifting method was described that reconstructs an object wave from a series of holograms having the same object wave but a different initial phase. The initial-phase measurement algorithm was improved to allow the initial phase of a hologram to be shifted more easily, even without external equipment. Theoretical expressions and experimental results show that the object wave can be reconstructed independently of the fringe spacing of the holograms. This is an important advantage over conventional Fourier-transform-based electron holography where fringes must be narrower than 1/3 of the spatial resolution to be reconstructed. The precision of the reconstructed object wave has also been improved to under $2\pi/100$.

6. ACKNOWLEDGMENT

I would like to thank the following colleagues for their collaboration in carrying out experiments: Dr. T. Tanji, Dr. J. Endo, Dr. G. Lai, Dr. K. Aoyama, Dr. N. Osakabe, and Dr. A. Tonomura. I would like also to thank Dr. S. Iijima for providing the single-shell nanotubes and for valuable discussions and suggestions.

7. REFERENCES

1. A. Tonomura, Electron Holography, Springer-Verlag, Berlin Heidelberg, 1993.
2. Q. Fu, H. Lichte, and E. Völkl, Phys. Rev. Lett., 67 (1991) 2319.
3. T. Tanji, K. Urata, K. Ishizuka, Q. Ru, and A. Tonomura, Ultramicroscopy, 49 (1993) 259.
4. T. Kawasaki, Q. Ru, T. Matsuda, Y. Bando, and A. Tonomura, Jpn. J. Appl. Phys. 30 (1991) 11830.
5. T. Matsumoto, T. Tanji, and A. Tonomura, Ultramicroscopy, 54 (1994) 317.
6. T. Hirayama, Q. Ru, T. Tanji, and A. Tonomura, Appl. Phys. Lett., 63 (1993) 418.
7. J.E. Bonevich, K. Harada, T. Matsuda, H. Kasai, T. Yoshida, G. Pozzi, and A. Tonomura, Phys. Rev. Lett., 70 (1993) 2952.
8. S. Frabboni, G. Matteucci, G. Pozzi, and M. Vanzi, Phys. Rev. Lett., 55 (1985) 2196.
9. M. Takeda and Q. Ru, Appl. Opt., 24 (1985) 3068.
10. Q. Ru, T. Matsuda, A. Fukuhara, and A. Tonomura, J. Opt. Soc. Am., A8 (1991) 1739.
11. J. Endo, T. Matsuda, and A. Tonomura, Jpn. J. Appl. Phys., 18 (1979) 2291.
12. A. Tonomura, T. Matsuda, T. Kawasaki, J. Endo, and N. Osakabe, Phys. Rev. Lett., 54 (1985) 60.
13. J.H. Bruning, Optical Shop Testing, Wiley-Interscience, New York, 1978, p.414.
14. Q. Ru, J. Endo, T. Tanji, and A. Tonomura, Appl. Phys. Lett., 59 (1991) 2372.
15. Q. Ru, J. Endo, T. Tanji, and A. Tonomura, Optik, 92 (1992) 51.
16. Q. Ru, G. Lai, K. Aoyama, J. Endo, and A. Tonomura, Ultramicroscopy, 55 (1994) 209.
17. Q. Ru, submitted.
18. G. Lai and T. Yatagai, J. Opt. Soc. Am., A8 (1991) 822.
19. G. Sommargren, Appl. Opt., 20 (1981) 610.
20. S. Iijima and T. Ichihashi, Nature, 363 (1993) 603.
21. G. Matteucci, G.F. Missiroli, and G. Pozzi, Ultramicroscopy, 6 (1981) 109.
22. G. Matteucci, G.F. Missiroli, and G. Pozzi, Ultramicroscopy, 8 (1982) 403.
23. G. Pozzi, Optik, 66 (1983) 91.
24. Q. Ru, N. Osakabe, J. Endo, and A. Tonomura, Ultramicroscopy, 53 (1994) 1.
25. Q. Ru, J. Appl. Phys., to be published in February 15, 1995 edition.

Holographic Reconstruction Methods

Michael Lehmann and Hannes Lichte

Institut für Angewandte Physik, Universität Tübingen,
Auf der Morgenstelle 12, D–72076 Tübingen, Germany

The complex image wave recorded in a single off–axis electron hologram must be reconstructed in order to perform the aberration correction process and to analyse the phase and amplitude images. Two methods of reconstruction are discussed here. The first one, a conventional reconstruction method using the rules of Fourier optics, is used for accurate determination of the image wave, in particular at high resolution. It is shown how some of the numerical artefacts arising due to discrete sampling may be removed. Also, a new program for the interactive correction of aberrations within one second processing time is introduced. The second method is an alternative reconstruction method which is used for a very fast preview of the image phase at the electron microscope. In the near future with an affordable computer system, the phase image will be achievable from a hologram at a refresh rate of 0.5 seconds.

1. INTRODUCTION

The development of modern slow–scan CCD cameras and fast computers available at a reasonable price offers a wide range of image processing possibilities. However, conventional micrographs do not contain the entire information of the image wave needed for wave optical processing, because the phase part of the wave is lost during recording. Consequently, wave optical image processing cannot be performed using a single conventional micrograph. Electron holography, however, provides the entire information of the image wave in a single off–axis electron hologram [1]. In order to retrieve this information, the wave has to be reconstructed.

In the early days of electron holography, holograms were recorded on a photoplate and reconstructed on an optical bench using a laser beam. Modern slow–scan CCD cameras and fast computers programmed with the rules of Fourier optics are much more advantageous because they offer a faster and more convenient way to reconstruct holograms [2,3]. However, new problems have arisen due to the discrete sampling of holograms. In the first part of this paper, we show our approach to get rid of such numerical artefacts. In the second part, we introduce an alternative method for the reconstruction which is specially designed for a very fast on–line display of the image phase.

2. CONVENTIONAL RECONSTRUCTION AND CORRECTION

We demonstrate the reconstruction of the image wave and the correction of the wave aberration using the example of a hologram of silicon nitride (Si_3N_4) which was oriented

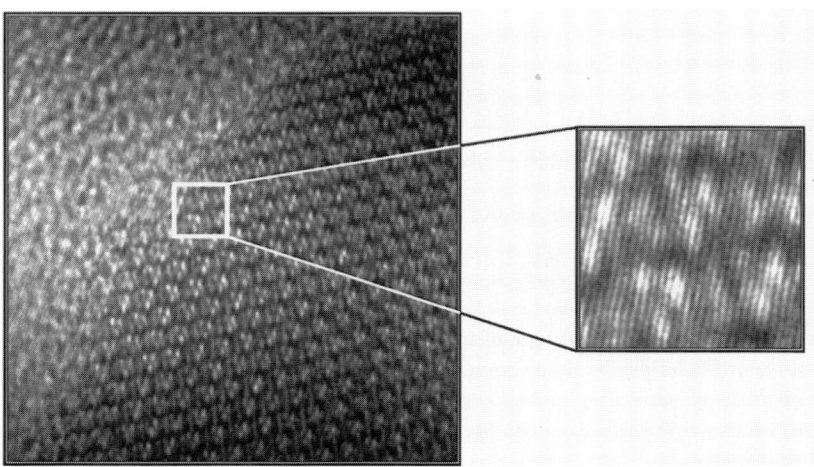

Figure 1. Hologram of silicon nitride (Si_3N_4) in $<001>$ direction; silicon atoms form a hexagonal structure. The magnified area shows the modulation of the interference fringes due to the image amplitude and phase.

with the beam along the $<001>$ zone axis (figure (1)). The hologram was recorded at the CM30 FEG–Special Tübingen electron microscope by means of a 1024 x 1024 slow–scan CCD camera.

In general, the intensity distribution of an off–axis electron hologram with the carrier frequency $\vec{q}_c = (u_c, v_c)$ can be written as

$$I(\vec{r}) = 1 + A^2(\vec{r}) + I_{inel}(\vec{r}) + 2\mu\, A(\vec{r}) \cos(2\pi \vec{q}_c \vec{r} + \Phi(\vec{r})) \quad , \qquad (1)$$

where μ denotes the contrast of the interference fringes and $I_{inel}(\vec{r})$ considers the intensity of inelastically scattered electrons which, due to their energy loss, cannot interfere with the reference wave. Amplitude and phase of the image wave $A(\vec{r})\, e^{i\Phi(\vec{r})}$ formed by the elastically scattered electrons are captured in the hologram as a contrast modulation and a bending of the interference fringes, respectively (figure (1)). The process of reconstructing this image wave consists of the steps outlined below.

2.1. Determination of Hologram Parameters

In a first step, the position of the sideband, which is part of the Fourier transform of the hologram, must be determined very accurately. This is severely hampered by the strong streaking of reflections resulting from leakage effect which shows up for spatial frequencies which are not exactly commensurate with the field of view in real space. Then, due to implicit periodic continuation, an abrupt discontinuity at the edges produces the streaking in Fourier space. For suppression of the leakage effect, a Hanning window

$$w_h(\vec{r}) = (1 - \cos(2\pi\, x/x_m))\,(1 - \cos(2\pi\, y/y_m)) \qquad (2)$$

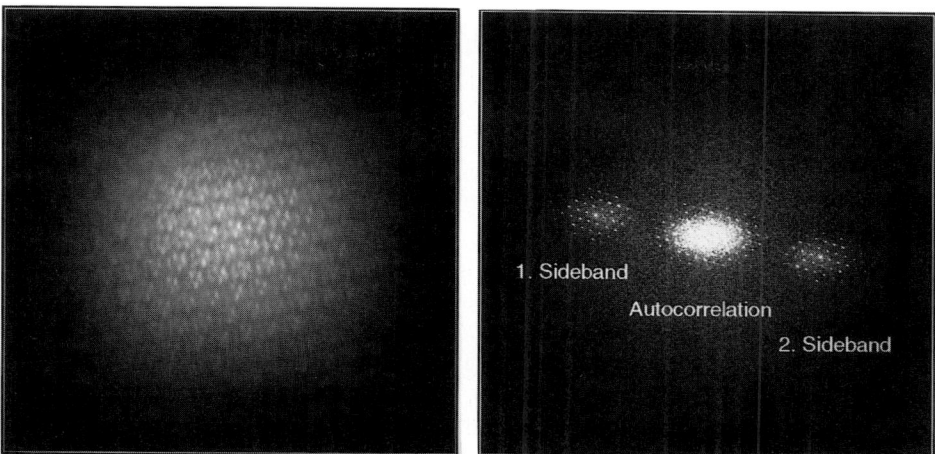

Figure 2. Left: The hologram is multiplied with a Hanning window. Right: Amplitude display of the Fourier transform of left image. Streaking is minimized due to the low intensity sidelobes of the Fourier transform of the Hanning window.

$(0 \leq x < x_m, 0 \leq y < y_m)$ is multiplied to the hologram. After a Fourier transformation, the complex Fourier spectrum of the hologram convoluted with Fourier transform of the Hanning window is obtained (figure (2)):

$$\begin{aligned}
\text{FT}\{I(\vec{r})\} &= \left(\text{FT}\{1 + A^2(\vec{r}) + I_{inel}(\vec{r})\}\right) \otimes \text{FT}\{w_h(\vec{r})\} && \text{center band} \\
&+ \left(\mu \, \text{FT}\{A(\vec{r})\, e^{i\Phi(\vec{r})}\} \otimes \delta(\vec{q} + \vec{q}_c)\right) \otimes \text{FT}\{w_h(\vec{r})\} && \text{1. sideband} \quad (3) \\
&+ \left(\mu \, \text{FT}\{A(\vec{r})\, e^{-i\Phi(\vec{r})}\} \otimes \delta(\vec{q} - \vec{q}_c)\right) \otimes \text{FT}\{w_h(\vec{r})\} && \text{2. sideband}
\end{aligned}$$

The center band mainly consists of the Fourier spectrum of the corresponding conventional image, whereas the two complex mutually conjugate sidebands represent the Fourier spectra of the complex image wave. The heavy streaking is suppressed because the Fourier transform of the Hanning window shows very low intensity sidelobes.

In order to find the exact position of a sideband i.e. the carrier frequency $\vec{q}_c$, an interpolation algorithm [4] which is based on the accurate knowledge of $\text{FT}\{w_h(\vec{r})\}$ can be applied. The procedure is the following: First, find the pixel $P_{f,g}$ with the largest amplitude of the sideband. Second, find the pixels with the next largest values in u– and v–direction adjacent to $P_{f,g}$. Then, the components of the carrier frequency $\vec{q}_c = (u_c, v_c)$ can be interpolated out of these three pixel amplitudes by means of

$$u_c = \frac{1}{x_m}\left[f - \alpha \frac{P_{f,g} - 2P_{f+\alpha,g}}{P_{f,g} + P_{f+\alpha,g}}\right] \quad , \quad v_c = \frac{1}{y_m}\left[g - \beta \frac{P_{f,g} - 2P_{f,g+\beta}}{P_{f,g} + P_{f,g+\beta}}\right] \quad (4)$$

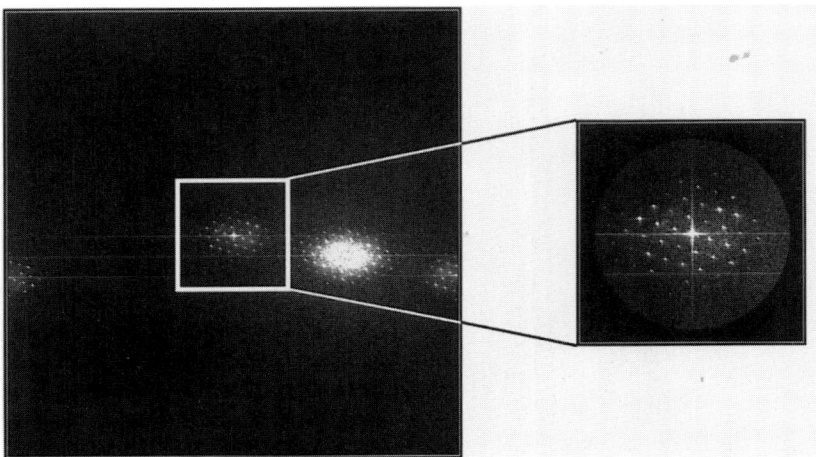

Figure 3. Left: The amplitude image of the Fourier transform shows the centered sideband. Right: A subimage and a circular aperture is taken in order to isolate the sideband from the autocorrelation and to save processing time.

The parameters α and β describe whether the next larger pixel value was found in positive coordinate direction (value: $+1$) or in negative direction (value: -1). By definition, in the center of reciprocal space $f, g = 0$ holds.

2.2. Reconstruction of the Image Wave

The second step performs the actual reconstruction of the image wave. In this case a Hanning window should not be applied because it heavily restricts the field of view due to the inaccuracy of numerical calculations. However, to minimize artefacts in the sideband due to heavy streaking of the center band, we subtract in the original hologram the average pixel value from each pixel.

Centering of the sideband at subpixel accuracy should not be done by interpolation because, according to our experience, heavy artefacts show up in the reconstructed wave. Instead, a phase wedge corresponding to the carrier frequency determined above is multiplied to the hologram in order to center the sideband about the origin of Fourier space. To achieve this goal, the real-valued hologram data are transferred to a complex data file such that each pixel value is represented by the "amplitude" part of the corresponding complex number. After applying the phase wedge, the "complex" intensity distribution can be written as

$$I'(\vec{r}) = \left(1 + A^2(\vec{r}) + I_{inel}(\vec{r}) + 2\mu\, A(\vec{r}) \cos(2\pi \vec{q}_c\, \vec{r} + \Phi(\vec{r}))\right) e^{2\pi i\, \vec{q}_c\, \vec{r}} \quad . \tag{5}$$

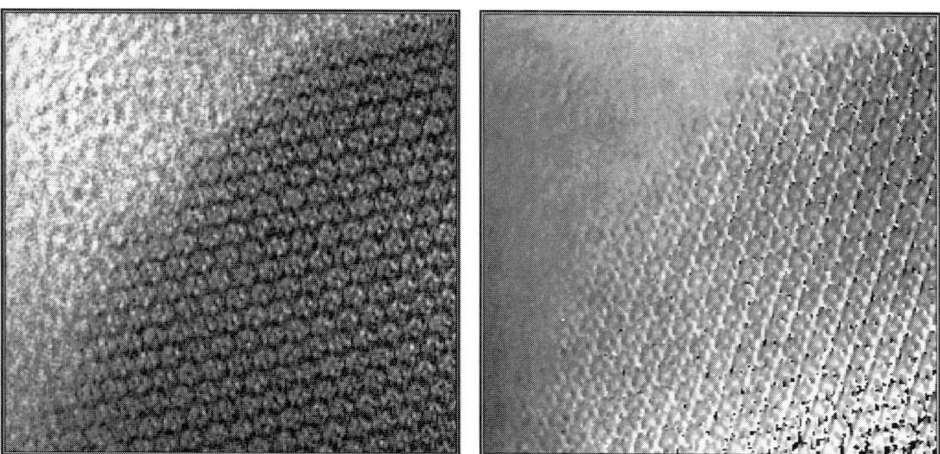

Figure 4. Image wave of the Si$_3$N$_4$ crystal. Left: amplitude. Right: phase.

Here, the finite field of view is not taken into account. The Fourier spectrum

$$\begin{aligned}\mathrm{FT}\{I'(\vec{r})\} &= \mathrm{FT}\{1 + A^2(\vec{r}) + I_{inel}(\vec{r})\} \otimes \delta(\vec{q} - \vec{q}_c) \quad \text{autocorrelation} \\ &+ \mu\,\mathrm{FT}\{A(\vec{r})\,e^{i\Phi(\vec{r})}\} \otimes \delta(\vec{q}) \quad \text{1. sideband} \\ &+ \mu\,\mathrm{FT}\{A(\vec{r})\,e^{-i\Phi(\vec{r})}\} \otimes \delta(\vec{q} - 2\vec{q}_c) \quad \text{2. sideband}\end{aligned} \qquad (6)$$

is obtained after a reverse Fourier transformation.

We find the first sideband is centered within a fraction of a pixel (figure (3) left), hence streaking caused by imperfect centering of the sideband is minimized. Thus, artefacts which appear when a circular aperture is applied in order to cut out the sideband will also be minimized. Additionally, we obtain a flat reconstructed phase image without the linear phase offset which otherwise occurs and which can easily cover the whole range of 2π and makes it impossible to display fine details in the phase. The combined process of multiplication with a phase wedge and subsequent Fourier transformation has turned out to be equivalent to the extended Fourier transform algorithm [5].

Since only the centered sideband is of interest for further processing, we take a subimage of 256 x 256 pixels to reduce the processing time of the following steps. In addition, a mask is applied to zero everything except the sideband (figure (3) right). This can be done with a numerical circular aperture, preferably with a soft edge, e.g. an exponential, which reduces artefacts in the reconstructed wave. A reverse Fourier transformation yields the image wave $\mu\,A(\vec{r})\,e^{i\Phi(\vec{r})}$ where amplitude $A(\vec{r})$ and phase $\Phi(\vec{r})$ can be displayed separately (figure (4)).

The reconstruction of the image wave from a hologram recorded with a 1024 x 1024 pixel slow–scan CCD camera can be performed at the electron microscope within 20 seconds using, for example, an on–line computer system from the Tietz company, which incorporates a 40 MFLOPS parallel signal processor for FFT calculations [6]. The reconstruction can be done automatically or can be interactively controlled by the operator, allowing on-line holographic investigations at the microscope.

2.3. Correction of Distortions

Distortions in the hologram can be caused by the projector lenses, by irregularities of the biprism, or by non–planar imaging conditions [6,7]. They produce parasitic bending of the hologram fringes which, by the reconstruction procedure, are interpreted as phase structures and which can give rise to very strong large area phase modulations in the image wave. Such distortions must be avoided because they hide the desired phase information, and prohibit the use of the full dynamics, in particular in the case of the weak phase shift occuring at high resolution. In order to correct these distortions, a second so-called reference hologram acquired with the object removed is taken immediately after the first one. In this case, we recommend first to reconstruct the reference hologram by the procedure described above, and afterwards the actual hologram with the same parameters. Then, both complex waves are divided by each other to obtain the distortion–free image wave.

With this procedure, if the position of the two holograms agrees very well, not only the distortion effect is corrected in the image phase but also the modulation of both amplitude and phase due to the Fresnel fringes of the biprism is removed.

2.4. Correction of Wave Aberrations

Amplitude and phase of the object wave are still mixed up in the image wave due to the wave aberration of the objective lens (e.g. figure (1) in [8]). In order to unscramble the information and to improve the interpretable resolution limit beyond the Scherzer resolution, a phaseplate (figure(5)) is generated which models the wave aberration $\chi(\vec{q})$ [9] including all possible contributions from coherent aberrations. So far, in the function $\chi(\vec{q})$ we have included defocus, spherical aberration, two–fold and three–fold astigmatism, and axial coma. Since axial coma is caused by a beam tilt, it is accommodated by a corresponding displacement of the phaseplate out of the center of reciprocal space [10]. The remaining contributions to the wave aberration $\chi(\vec{q})$ can be written with the wave number $k = 1/\lambda$ as

$$\chi(\vec{q}) = 2\pi k \Big\{ \tfrac{1}{4} C_s \left(\tfrac{q}{k}\right)^4 \quad \text{spherical aberration}$$
$$+ \tfrac{1}{2} D_z \left(\tfrac{q}{k}\right)^2 \quad \text{defocus} \quad (7)$$
$$+ \tfrac{1}{2} A_2 \left(\tfrac{q}{k}\right)^2 \cos(2(\alpha - \alpha_2)) \quad \text{two–fold astigmatism}$$
$$+ \tfrac{1}{3} A_3 \left(\tfrac{q}{k}\right)^3 \cos(3(\alpha - \alpha_3)) \Big\} \quad \text{three–fold astigmatism.}$$

The Fourier spectrum of the image wave is divided by the generated phaseplate (figure (5)). The aberration–free object wave (exit surface wave of the specimen) is obtained after a reverse Fourier transformation, if the true aberration parameters are known sufficiently well (figure (6)).

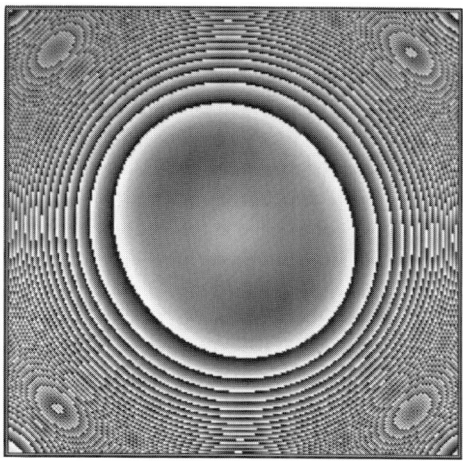

Figure 5. Phaseplate for $C_s = 1.2$ mm, $D_z = -45$ nm, $A_2 = 10$ nm, $\alpha_2 = 30°$

The application of a phaseplate always produces very nice looking images. The question then arises, which image really gives the object wave at highest point resolution? The aberration parameters effective in the hologram are not known at the required accuracy of e.g. a fraction of a percent needed for 0.1 nm resolution. In special cases, criteria can be found in the reconstructed wave; for example, the amplitude contrast can be minimized in the case of a thin amorphous area; the phase shift due to single atoms can be maximized, and so on. However, generally applicable criteria have not yet been found. Therefore, the comparison with simulated amplitude and phase images e.g. in areas of a perfect crystal may be very useful to optimize the different aberration parameters.

In any case, a very fast procedure is desired which allows the computer to simulate a microscope so that a series of images with different defocus, spherical aberration, astigmatism etc. can be produced by "turning the respective knobs". To achieve this, a new mouse–controlled program [11] has been developed for a DEC 3000 AXP 600 workstation (performance: 34 MFLOPS). Within one second, the wave is treated with an arbitrary set of aberration parameters; this includes the generation of the respective phaseplate, application to the Fourier spectrum, inverse Fourier transform, and the display of the amplitude and phase image. As an option, the power spectra of amplitude and phase can be displayed as well. This program not only helps the operator to find the correct aberration parameters e.g. by comparision with simulated images, but also helps to study tiny influences of the residual wave aberration $\chi(\vec{q})$ on amplitude and phase of the image wave in order to develop new criteria for correction.

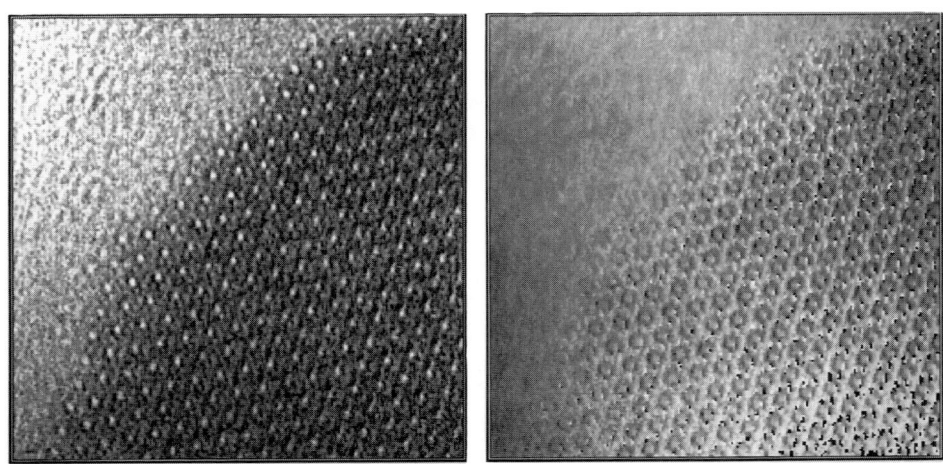

Figure 6. Object wave of the Si_3N_4 specimen. Left: object amplitude. Right: object phase. The hexagonal structure of the specimen can now be observed in the phase. However, the correction is not yet perfect as it can be seen in the reconstructed object amplitude.

2.5. Advanced Techniques

Since we have restored the whole complex wave, we can perform arbitrary numerical wave optical calculations for a comprehensive evaluation of the object structure which are not possible using a standard CTEM image. For example, nanodiffraction is a technique which permits the complex electron diffraction spectrum at the scale of a unit cell of a crystal to be analysed. So, changes in thickness, bending and misorientation of the crystal can be recognized on an atomic scale. Also, a focal series of images can be calculated for comparison with simulated images of crystals at varying focuses and thicknesses to assure that the correct aberration parameters were found and hence the correct object wave was reconstructed.

3. ALTERNATIVE RECONSTRUCTION

The conventional reconstruction of the image wave using Fourier optics needs at least two Fourier transformations and the operations to center and isolate one of the sidebands. As long as computers are still too slow, it is not possible to reconstruct the image phase in less than one second using this conventional approach. Therefore, Chen et al. [12] have taken another way. They have recently demonstrated that a real–time reconstruction of the image phase can be performed using a liquid crystal display (LCD) to transmit the hologram from the electron microscope to an optical bench and using the light optical setup as an optical Fourier computer.

Figure 7. Hologram of the MgO crystal. The amplitude of the image wave modulates the contrast of the fringes, whereas the image phase causes the bending of the fringes.

This technique has useful applications; nevertheless, numerical processing is more flexible and reproducible, and furthermore the data gained are more reliable. Therefore, the fast alternative numerical reconstruction method was established [13,14], which is based on a statistical method avoiding the two time–consuming Fourier transformations.

3.1. Reconstruction Algorithm

The mathematical description of the algorithm is very lengthy and is therefore omitted here; instead, the algorithm will be explained by the example of reconstruction of a hologram of a magnesium oxide crystal (MgO) (figure (7)).

We take a small subarea of the length of Z pixels of one line of the hologram. It is assumed that the intensity $1 + A^2(\vec{r}) + I_{inel}(\vec{r})$ and phase $\Phi(\vec{r})$ vary so slowly in this subarea that they can be treated as constant. Then, the intensity distribution (1) is fitted to this subarea using a least square fit. This yields two equations to calculate the pixel values of intensity and phase centered on the subarea. Subsequently, the subarea is shifted by one pixel and the fit is done again. Repeating this procedure over the whole hologram yields two images: the image phase and image intensity. Phase jumps, which may occur, are corrected by comparing with the previous calculated phase pixel value [7]. If the modulus of the difference between these two phases exceeds π, then the phase value of the new pixel is increased or diminished, respectively, by 2π, ensuring that the phase difference between two subsequent phase pixel values never exceeds π.

There are two parameters which have to be fed into the algorithm. One is the length of the subarea Z, which should be at least 1.5 periods of the interference fringes. The other is the carrier frequency $\vec{q}_c$, which is determined using two one–dimensional Fourier

Figure 8. Alternative reconstruction: Left: image intensity. Right: image phase

transforms of one line and one column of the hologram.

The reconstruction of intensity and phase of the MgO crystal are shown in figure (8). The phase values are spread over 4π but the phase jumps have been removed.

3.2. Speed and Application

For weak objects, the reconstructed image intensity corresponds reasonably well with the image amplitude. In this case, the alternative reconstruction method is a useful tool for preview imaging of amplitude and phase of the image wave in electron microscopy without loss of the large area phase contrast.

Besides its easy handling, the primary advantage of the alternative method compared to the conventional Fourier optics method is its superior speed. The reconstruction of intensity and phase of the image wave is performed within about 0.2 seconds for a 512 x 512 pixel hologram on a DEC 3000 AXP 600 workstation. So, a holographic electron microscope equipped with a slow–scan CCD camera with a read–out frequency of 2 MHz and a reasonable fast computer system makes it possible to record and display intensity and phase of the image wave within 0.5 seconds.

This opens a new window for TEM–holography: applications which cannot be performed in a CTEM, such as the investigation of electromagnetic or electrostatic fields, can then be performed by observing the phase images numerically reconstructed nearly in real-time.

4. CONCLUSION

Electron holography provides powerful possibilities to investigate objects at both low and high resolutions. The primary advantage of holography is that it provides highly flexible tools for manipulation and analysis of the waves by means of numerical wave optical processing. However, to make full use of the holographic technique, artefacts have to be avoided, which may easily arise e.g. due to discrete sampling of the hologram or due to application of incorrect aberration parameters for aberration correction. Furthermore, the time–consuming procedures has to be accelerated significantly; the goal is the interactive real–time analysis of the artefact–free object wave.

So far, in the Tübingen holography group three different approaches have been developed:

1. on-line reconstruction at the electron microscope by means of the usual Fourier algorithm.

2. most accurate correction of aberrations for selected holograms processed using a very fast off-line computer "like a microscope" by means of a dedicated program.

3. very fast alternative reconstruction method for real time analysis of phase structures.

5. ACKNOWLEDGEMENTS

The authors wish to thank Drs. F. Lenz, W.-D. Rau, E. Völkl, and L.F. Allard for helpful discussions, and A. Orchowski for providing the silicon nitride hologram. The financial support from Körber–Stiftung, Volkswagen–Stiftung, and Deutsche Forschungsgemeinschaft is gratefully acknowledged.

REFERENCES

1. H. Lichte, Ultramicroscopy 20 (1986) 293
2. F.J. Franke, K.-H. Herrmann, and H. Lichte, Scanning Microsc. Suppl. 2 (1988) 59
3. K. Ishizuka, Ultramicroscopy 52 (1993) 1
4. W.J. de Ruijter, M. Gajdardziska–Josifovska, M.R. McCartney, R. Sharma, D.J. Smith, and J.K. Weiss, Scanning Microsc. Suppl. 6 (1992) 347
5. E. Völkl and L.F. Allard, MSA Bulletin Vol.24, No.2 (1994) 466
6. W.D. Rau, H. Lichte, E. Völkl, and U. Weierstall, J. Comput. Assist. Microsc. 3 (1991) 51
7. W.J. de Ruijter and J.K. Weiss, Ultramicroscopy 50 (1993) 269
8. H. Lichte, Ultramicroscopy 38 (1991) 13
9. Q. Fu, H. Lichte, and E. Völkl, Phys. Rev. Lett. 67 (1991) 2319
10. F. Lenz, Proc. Internat. Symp. El. Optics, Beijing 1986, 122
11. M. Lehmann and H. Lichte, Proc. Internat. Conf. El. Microscopy ICEM13, Vol. 1 (1994) 293
12. J. Chen, T. Hirayama, G. Lai, T. Tanji, K. Ishizuka, and A. Tonomura, Optics Letters 18 (1993) 1887
13. M. Lehmann, F. Lenz, E. Völkl, and H. Lichte, Proc. 10th Eur. Congr. on Electron Microscopy EUREM92, Last–Minute Brochure, p.21
14. M. Lehmann, E. Völkl, and F. Lenz, Ultramicroscopy 54 (1994) 335

Real-time electron holography using a liquid-crystal panel

J. Chen, T. Hirayama, G. Lai, T. Tanji, K. Ishizuka, and A. Tonomura

Electron Wavefront Project, Research Development Corporation of Japan,
c/o Faculty of Engineering, Toyo University, Saitama 350 Japan.

Real-time electron holography has been realized by using a liquid-crystal panel as a rewritable electron hologram medium. In this technique, the electron off-axis hologram detected with a TV camera is transferred to a liquid-crystal panel (LC-panel) as a video signal. The LC-panel can function as a thin amplitude hologram or a thin phase hologram. A time-sequential interference micrograph is obtained at TV rate by superimposing a plane reference wave on the reconstructed object wave. Experimental results for observing a dynamic domain-wall motion in thin permalloy film are demonstrated. On the other hand, the LC-panel can also be used as a phase plate for compensating for the wave aberration. A new method for correcting the aberration of an electron-objective lens during the holographic reconstruction stage is developed. Experimental results of correcting the spherical aberration in an electron hologram of fine gold particle are shown.

1. INTRODUCTION

Electron holography [1] has been put to practical use with the development of an electron microscope with a field-emission gun. There are promising applications of electron holography. One, for example, is to attain the original objective of overcoming the resolution limitation of a transmission electron microscope (TEM). Another is to observe an electromagnetic field in a microscopic area directly by using the phase information provided by electron holography.

In electron-holographic interference microscopy [2], a magnetic field's distribution is visualized as contour-type interference fringes corresponding to magnetic lines of force. This technique has been demonstrated to be effective not only in fundamental physics [3], but also in many practical applications such as observation of quantized magnetic fluxes transmitted through a superconductor [4] and studies of the magnetic domain structure of magnet material [5]. The time-consuming photographic or digital processes involved in conventional electron-holographic interference microscopy, however, prevent this technique from being used to observe dynamic events. Electron holography is a two-step imaging method; an electron hologram is either recorded on photographic film and reconstructed by an optical reconstruction method [6] or it is detected with a TV camera and then reconstructed, by computer, by using a Fourier transform method [7, 8]. The time resolution in optical reconstruction is limited by the photographic processes, and in digital reconstruction it is limited by the numerical calculation processes. Efforts have been made to shorten the processing time. For instance, the dynamics of fluxons was studied using an off-line digital processing method, in which each frame of the electron hologram recorded on videotape was digitized and analyzed by the Fourier transform method [4]. But in addition to being time-consuming, that method cannot be used for on-line observation. Phase-shifting electron-wave interferometry [9] enables high-precision measurement; however, it requires three or more frames of interferogram for phase extraction. Video-rate observations can be made by exploiting the advantages of both the optical and digital reconstruction methods. An electron hologram can be detected with a TV camera at a video rate while the optical method is used for simultaneous reconstruction. A liquid-crystal panel (LC-panel), which has been used in optical holography [10] and in optical image-processing systems [11,12] can modulate the light intensity or phase. By using such a device, real-time electron holography was realized [13-16].

On the other hand, electron holography has the potential to overcome the TEM resolution limitations related to spherical aberration. In this technique, the aberration of the electron objective lens is corrected during the hologram reconstruction stage. The most important point is to appropriately cancel the wave function. The following two methods have been developed to compensate for aberration. Tonomura et al. [17] corrected for the spherical aberration in an electron lens by using optical lenses as a compensator. Lichte [18] used a numerical method in which the aberration is compensated by multiplying a conjugate wave aberration function to the Fourier spectrum of the reconstructed object wave in computer. Both methods have advantages and disadvantages. The former requires no digital processing for either Fourier transformations or aberration correction, thus enabling a high-speed processing. However, the adjustment of the aberration coefficients of the correction lenses to the condition in which the hologram was formed is rather difficult. On the other hand, Lichte's method has the advantage of high flexibility in processing distorted electron images. However, the numerical reconstruction processing for a large number of pixels is rather time-consuming. Moreover, the number of pixels that can be used for aberration correction is limited by the number of pixels covered by the side-band on the Fourier spectrum plane, which is only a small part of all the pixels used to sample the electron hologram [19]. A new method of compensating for aberration of the electron objective lens during the holographic reconstruction stage is developed by using the LC-panel [20, 21]. In this method, a LC-panel is employed as a computer-controlled phase plate on the Fourier plane of the optical reconstruction system. The method takes advantage of very fast optical parallel processing for both hologram reconstruction and aberration correction, while providing high flexibility in adjusting the parameters to compensate for the aberration. high processing speed allowed us to perform a trial-and-error procedure for correction even when the amount of defocusing and spherical aberration coefficient are unknown.

In this paper, section 2 presents the phase modulation characteristics of a commercially available liquid-crystal device, and then section 3 describes a real-time interference microscope system, then applications of this method for observation dynamic domain-wall motion that an applied magnetic field induces in a thin permalloy film is presented. Section 4 describes the system for correcting the wave function of an electron lens, and experimental results is also shown.

2. PHASE MODULATION CHARACTERISTICS OF LC-PANEL

The structure and the principal operation of a twisted nematic liquid crystal (TNLC) device are shown in Fig. 1. In the TNLC device, a thin layer of nematic liquid crystals is sandwiched between two parallel glass plates, on which transparent electrodes are pasted. The surfaces of the electrodes are coated with liquid crystal aligning films. When no electric field is applied, the liquid crystal molecules turn continuously up to 90° from one electrode to the other, as shown in Fig. 1. If a modulating electric field is applied across the liquid crystal layer through the transparent electrodes, the crystal molecules turn lengthwise along the field, as illustrated in Fig. 1. Since the effective refractive index of the liquid crystal depends on the angle of the molecules relative to the direction of incident light, the phase of the light transmitted through the LC device can be controlled by the applied electric field. The problem associated with using the TNLC panel as a spatial phase modulator is that the rotation of polarization caused by the twisted nematic liquid crystals gives rise to intensity modulation. This problem can be alleviated by setting the polarization of the incident light parallel to the front director of the LC molecules and by operating the device at a low voltage, just below the optical threshold [11]. A homogeneously aligned LC panel [12] is more suitable as a spatial phase modulator because the LC molecules are arranged parallel to the glass plate and the polarization of the output light does not rotate. However, the TNLC device was used in this experiment, because that device can be easily obtained by modifying a commercially available LC television or a LC video projector.

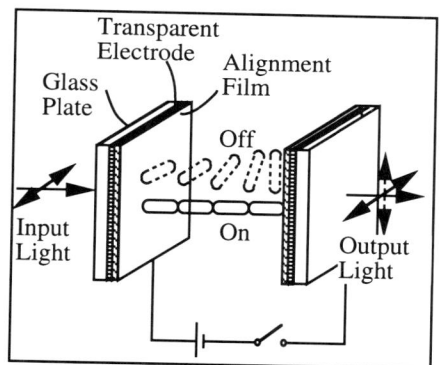

Figure 1. Structure and pricipal operation of a TNLC Panel.

Figure 2. Top view of the LC-panel used in the experiments.

The LC-panel used in our experiments of real-time electron holography was a TNLC type modified from a commercially available video projector, the Epson VPJ-2000. The plastic polarizers on both sides of the panel were removed so that phase modulation could be controlled by the video signal. Figure 2 shows a top view of the LC panel. This panel had 480 x 440 electrically addressable pixels within a 22.08 x 24.64-mm² screen. Thus the pixel dimensions were 46 x 56 μm². The LC-panel was addressed using a thin-film-transistor active matrix, and was equipped with a video terminal that allowed the electron hologram to be inputted as a video signal.

The phase modulation characteristics of the device were investigated using a standard Mach-Zehnder interferometer shown in Fig. 3. The LC panel was inserted into one arm of the interferometer, and the incident polarization of laser beam was arranged parallel to the liquid crystal molecules at the front of the panel. The voltage on each pixel was controlled by using a personal computer. To measure the dependence of the phase modulation on the video signal, a ramp pattern corresponding to a gradually varying voltage was applied to the LC-panel. The Fourier transform method was used to measure the phase distribution of the transmitted light. For this, one of the two mirrors in the interferometer was tilted to introduce the carrier fringes, and the interference fringe pattern detected with the CCD camera was fed to the computer for quantitative analysis. The measured phase change of the LC-panel is depicted in Fig. 4, which shows that the maximum phase change of the LC-panel was 0.8 π.

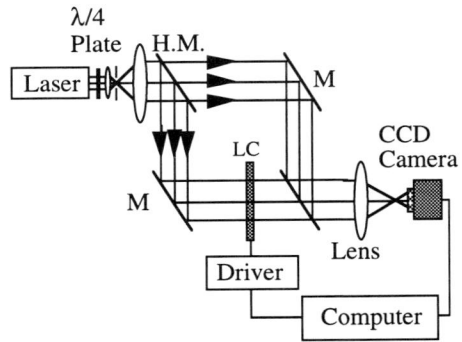
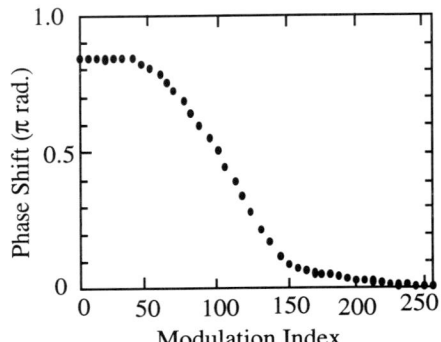

Figure 3. Mach-Zehnder interferometer used for measuring the dependence of phase shfit on an applied voltage of the LC-panel.

Figure 4. Phase-modulation property of the LC-panel.

3. REAL-TIME ELECTRON-HOLOGRAPHIC INTERFERENCE MICROSCOPY [13-16]

3.1. System

A schematic configuration of the developed real-time electron holographic interference microscope is shown in Fig. 5. This system consists of two units: an electron hologram formation unit and an interference micrograph formation unit. The electron hologram formation unit consisted of a TV detection unit and a field-emission transmission electron microscope equipped with a Möllenstedt-type electron biprism. One half of the collimated electron beam illuminated the specimen under investigation and the other half, passing through an empty area, acted as the reference beam. To prevent the magnetic field of the objective lens from influencing the structure of a magnetic specimen, the objective lens of the microscope was turned off and the intermediate lens was used as an objective lens. Applying a negative voltage biprism deflected the object and reference beams sidewise, and these deflected beams overlapped to form an interference fringe pattern in the image plane. The intensity distribution of the interference fringe pattern can be written as

$$I_h(x,y) = 1 + a^2(x,y) + 2 a(x,y) \cos[\Phi(x,y) + 2\pi f_0 x] \qquad (1)$$

where f_0 is a spatial carrier frequency introduced by the electron biprism. For an overlapping angle θ, the spatial frequency is given by $f_0 = \sin\theta/\lambda$. The term $a(x,y)$ is the amplitude modulation of the object wave, and $\Phi(x,y)$ is the phase difference between the object and the reference beams. The phase difference between the two electron beams transmitted through the magnetic field is given by

$$\Delta\Phi = -(2\pi e/h) \int B_n ds$$

, where B_n is the component of magnetic flux density normal to the surface enclosed by the electron beams, e is the electron charge, and h is Planck's constant. Therefore, the interference fringe pattern of Eq. (1) represents a contour-type interference micrograph of the projected magnetic flux. The interference fringe pattern was detected with a TV camera and transferred to the interference-micrograph formation system described in the following subsection.

In the interference micrograph formation unit, a Mach-Zehnder interferometer was used to produce properly arranged two-plane beams required for forming the interference micrograph. The LC-panel used as the rewritable medium for phase holograms was placed at the output port of the interferometer. A linearly polarized He-Ne laser (50 mW, 632.8 nm) was used as the light source. The polarization

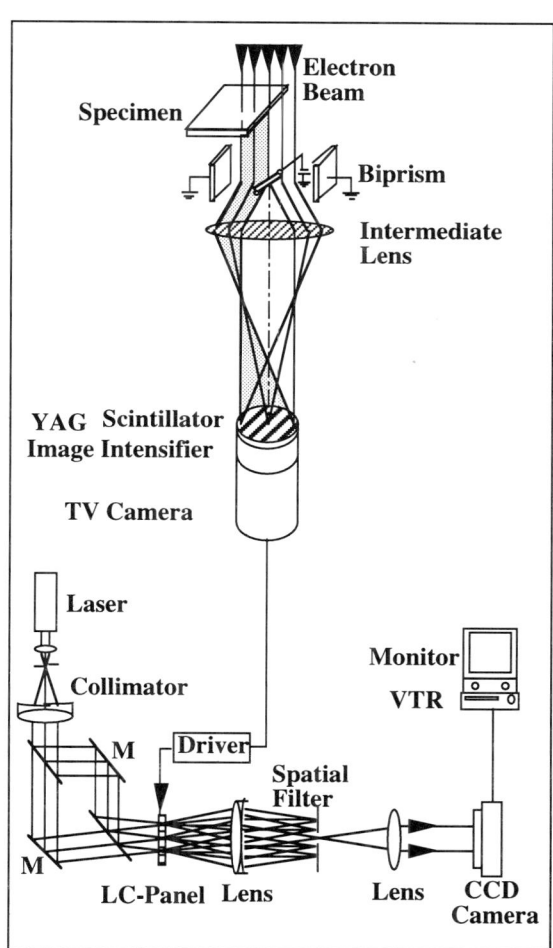

Figure 5. Schematic diagram of the real-time electron-holographic interference microscope system.

direction of the was aligned parallel to the liquid-crystal molecules at the input face of the LC-panel. The LC-panel accepted the time-sequential electron hologram as a video signal, that modulated the voltage applied to each pixel.

For simplicity, the phase modulation produced by the LC panel is assumed here to be proportional to the intensity in the interference fringe pattern. When an electron hologram signal as shown in Eq. (1) is applied to the LC-panel, the complex transmittance of the modulator can be written as

$$t(x,y) = \exp i\{1+a^2(x,y)+2a(x,y)\cos[\Phi(x,y)+2\pi f_0 x]\}$$

$$= \exp i[1+a^2(x,y)] \sum_{n=-\infty}^{\infty} i^n J_n[2a(x,y)] \exp i n[\Phi(x,y) + 2\pi f_0 x] \quad (2)$$

where J_n is the Bessel function of the first kind of order n.

In observing a magnetic field in a vacuum or a thin film of uniform thickness, the incident electron beam is mainly modulated in phase, and the amplitude modulation $a(x,y)$ can be considered as a constant. Thus the complex transmittance is given by

$$t(x,y) = \sum_{n=-\infty}^{\infty} i^n J_n(2a) \exp\{i n[\Phi(x,y) + 2\pi f_0 x]\} \quad (3)$$

where the constant phase term has been omitted.

Illuminating the LC-panel with the two beams from the Mach-Zehnder interferometer produces two sets of the reconstructed images, and the interference micrograph is obtained by making a reconstructed wave interfere with a transmitted or conjugate wave. When the first-order reconstructed object wave and the transmitted wave are used, the intensity distribution of the produced interference micrograph is given by

$$I_{0,1} = J_0^2(2a) + J_1^2(2a) + 2J_0(2a)J_1(2a)\cos[\Phi(x,y)] \quad (4)$$

On the other hand, if the first-order reconstructed object wave and its conjugate are employed, then a two-times phase-amplified interference micrograph is obtained. Its intensity distribution is given by

$$I_{0,1} = J_0^2(2a) + J_1^2(2a) + 2J_0(2a)J_1(2a)\cos[\Phi(x,y)] \quad (5)$$

A higher-order phase-amplified interference micrograph can be obtained by using the higher-order reconstructed waves; this enhances the sensitivity of the measurement.

The interference micrograph produced by the optical system was sent to a TV monitor at the side of the electron microscope, so that we could operate the holographic electron microscope by watching an interference micrograph. The interference micrograph was recorded on a video tape for further investigation.

3.2. Experimental results

To demonstrate the performance of the developed system, we used this system to observe domain-wall motion caused by an applied magnetic field. The Hitachi HF-2000 transmission electron microscope equipped with a field-emission electron gun and a Möllenstedt-type electron biprism was used as the hologram formation unit. The accelerating voltage was 200 kV, and applying a voltage of -16 V to the center wire of the electron biprism caused the reference and the object wave to overlap, thus producing an interference fringe pattern. The electron hologram was imaged with a 2000-time magnification onto the image plane by intermediate and projection lenses.

The Gatan TV system (Model 622) was used to convert this electron hologram into an electrically formatted TV image signal. In this TV system, a thin yttrium aluminum garnet (YAG) single-crystal scintillator is used as the primary screen, and the image formed at the YAG screen is fiber-optically coupled to a TV camera tube with a cadmium zinc telluride target. The usable detection area of the TV camera is 13 mm by 10 mm, and an attainable resolution is 220 line pairs per centimeter.

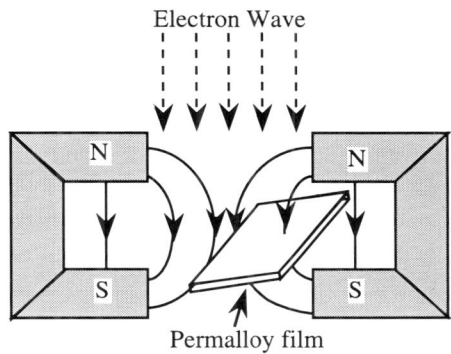

Figure 6. Experimental arrangement for observing dynamicmagnetic domain wall motion.

The thin permalloy film used in our experiments was prepared by vacuum evaporation and had a uniform thickness of about 20 nm. Figure 6 shows the experimental arrangment. The magnetic field of the weakly excited objective lens was used as the external field applied to the specimen. To apply a magnetic field parallel to the specimen, the specimen placed at the normal specimen position was tilted by 10 degrees.

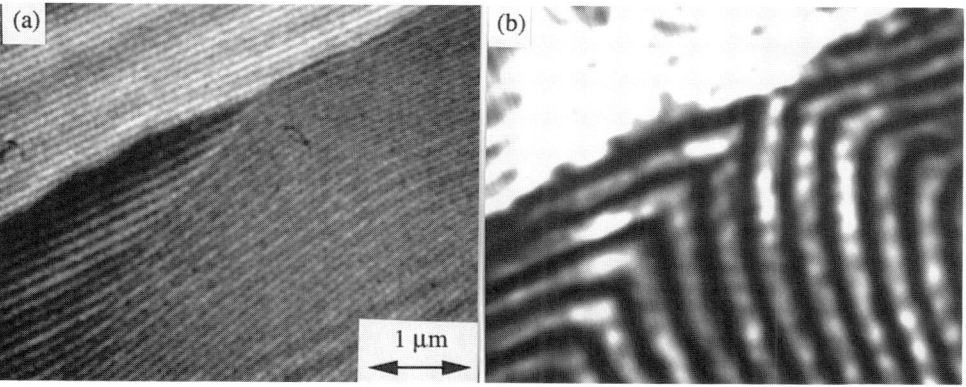

Figure 7. (a) One frame of electron hologram of a thin permalloy film, (b) interference micrograph showing the magnetic domains structure.

An electron image hologram of the specimen without an applied external field is shown in Fig. 7 (a). The magnetization distribution cannot be seen in this picture because of the carrier fringes. In the corresponding interference micrograph, however, the magnetic lines of force are directly displayed as contour-type interference fringes [Fig. 7 (b)]. Two domain walls where the magnetization direction changes abruptly can be clearly observed. These two domain walls originate from the same point at the specimen edge, and they make an angle of 90 degrees.

A permalloy film usually has magnetic-domain structures that minimize its total magnetic energy. When a external magnetic field is applied, the balance of the magnetic domains is broken and a new state is established. If the applied field is changed continuously, the domain structures will change simultaneously. To observe the motion of domain walls, we increased the current of the objective lens continuously from zero to 0.4 A and then decreased to zero. Both the time-sequential hologram and the interference micrograph were recorded on videotape, and interference micrographs showing the motion of domain-walls and the changes in the distribution of the magnetization can be observed for every 1/30 second. Four frames of the time sequential interference micrograph showing the dynamic domain-wall motion appear in Fig. 8, where (a), (b), and (c) show the changes in the domain structure when applied field increases from zero to 4000 A/m, and where (d) shows the state when the applied field is decreased to zero. As seen in this figure, the angle between the domain walls increased from 90 to 120 degrees when the field increased and it returned to 90 degrees when the field decreased to zero. The lateral shifts of the domain wall along the specimen edge were also observed.

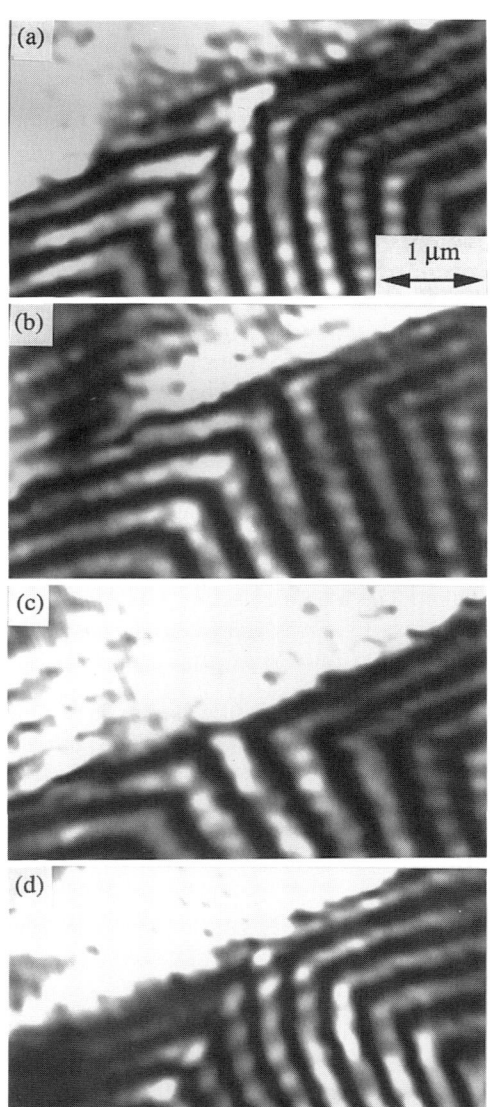

Figure 8. Interference-micrograph frames obtained in video rate showing the changes in domain structures with the changes in applied magnetic field.

A permalloy film usually has magnetic-domain structures that minimize its total magnetic energy. When an external magnetic field is applied, the balance of the magnetic domains is broken and a new state is established. If the applied field is changed continuously, the domain structures will change simultaneously. To observe the motion of domain walls, we increased the current of the objective lens continuously from zero to 0.4 A and then decreased to zero. Both the time-sequential hologram and the interference micrograph were recorded on videotape, and interference micrographs showing the motion of domain-walls and the changes in the distribution of the magnetization were observed for every 1/30 second. Four frames of the time sequential interference micrograph showing the dynamic domain-wall motion appear in Fig. 8, where (a), (b), and (c) show the changes in the domain structure when the applied field increases from zero to 4000 A/m, and where (d) shows the state when the applied field is decreased to zero. As seen in this figure, the angle between the domain walls increased from 90 to 120 degrees when the field increased and it returned to 90 degrees when the field decreased to zero. The lateral shifts of the domain wall along the specimen edge were also observed.

4. SPHERICAL-ABERRATION CORRECTION OF AN ELECTRON LENS USING A LC-PANEL [20,21]

The original objective of electron holography is to improve the spatial resolution of the TEMs by correcting the spherical aberration of the electron objective lens. When an electron hologram is recorded, the problem becomes how to compensate for the wave aberration during the hologram reconstruction. One method is to insert a phase plate with a phase distribution conjugate to the wave aberration into the Fourier spectrum plane of the hologram reconstruction system. Such a phase plate will be generated by the LC-panel, as described in the following paragraphs. The Fourier spectrum of the aberration-free image appears on the exit surface of the phase plate. Consequently, an aberration-free image can be obtained at the back focal plane of the other optical lens, when the aberration-free Fourier spectrum is exactly aligned on the front focal plane of that lens.

4.1. System

Figure 9 shows a schematic diagram of the optical system used for hologram reconstruction and aberration correction. In that system, a linearly polarized He-Ne laser (10mw, 632.8nm) is used as the light source. A half-wave plate is employed to adjust the polarization direction of the laser beam. The electron hologram placed at the front focal plane of lens L_1 is illuminated by the collimated laser beam. The Fourier spectrum of the hologram which consists of a center band and two side-bands is obtained on the back focal plane of lens L_1. A spatial filter is used to select one of the two side-bands. The LC-panel is placed just behind the filter to compensate for aberrations. The full area of the LC-panel can be used to correct aberrations, because it is not necessary to cover the center-band and the conjugate side-band. This is contrary to the digital processing system, where the area of the side-band is only a small part of the whole Fourier-spectrum. The video signal used to control the phase distribution of the LC-panel is generated by an IBM PC/AT clone and transferred through a frame memory with a resolution of 512x512x 8bits.

Unlike the situation in real-time electron holography, the LC panel used as a phase plate for aberration correction should have a phase-modulation depth larger than 2π. The LC-panel used in our experiments is a modification of a commercially available video projector, the VP-100PS (Epson corp.). The offset bias of the video signal was removed to get a large phase modulation with a small intensity variation. One of three LC devices for RGB outputs disassembled from the projector was used as a spatial phase modulator. The LC device consists of a TN mode LC display panel (off-state twist: 80°) and two plastic polarizers. These polarizers and the color filters were removed from the device to achieve phase modulation behavior. The panel has 320 horizontal and 220 vertical pixels in a display area of 25.6x19.8 mm². Thus, the pixel dimensions are 80x90 µm². The LC-panel is addressed using a thin-film transistor - active matrix and is equipped with a video terminal which allows the phase response to be written electronically with a computer through a frame memory. Electrical cross-talk between adjacent pixels in such an active matrix type LC-panel is far smaller than that in a simple multiplexed-type LC-panel.

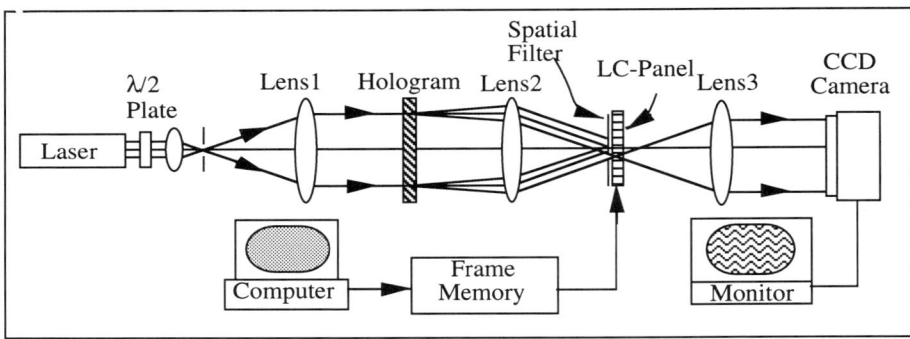

Figure 9. Schematic configuration of the optical system used for hologram reconstruction and aberration correction with LC-panel.

The dependence of the phase modulation on the applied voltage is measured using a Mach-Zehnder interferometer, as shown in Fig. 3. Quantitative mesurement was made by using the phase-shifting interferometry. The dependence of the phase modulation on the applied voltage is shown in Fig. 10. The maximum phase modulation obtained was about 3.1π. Therefore, a phase modulation of 2π can be achieved with this LC-panel. We also investigated the intensity modulation of the LC-panel and found that the intensity change was small over a low voltage range.

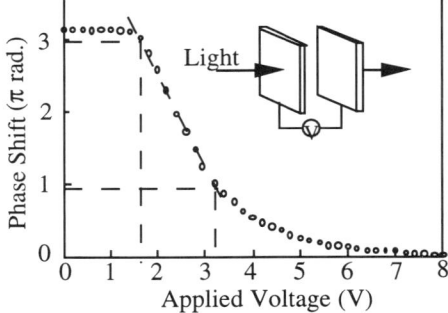

Figure 10. Phase-modulation property of the LC-panel used for aberration correction.

4.2. Experimental results

To demonstrate the performance of the developed system, the following experiments were carried out. Fine gold particles on a carbon film were used as specimens because of the suitability of crystalline particles for checking the spherical aberration correction. The electron hologram was previously taken with a Hitachi TEM HU-12A [17]. The spherical aberration coefficient of the objective lens was 1.7 mm. The accelerating voltage used to take the electron hologram was 80 kV (l=4.2 pm). The wave aberration of the objective lens under the in-focus observation condition is shown in Fig. 3. For 0.24-nm lattice fringes, we have to compensate for a wave aberration as large as 20π. Figure 11(a) shows the reconstructed image of the electron hologram. Due to the large spherical aberration, the Bragg-reflected waves were imaged far from the Gaussian image, a distance of about 10 nm.

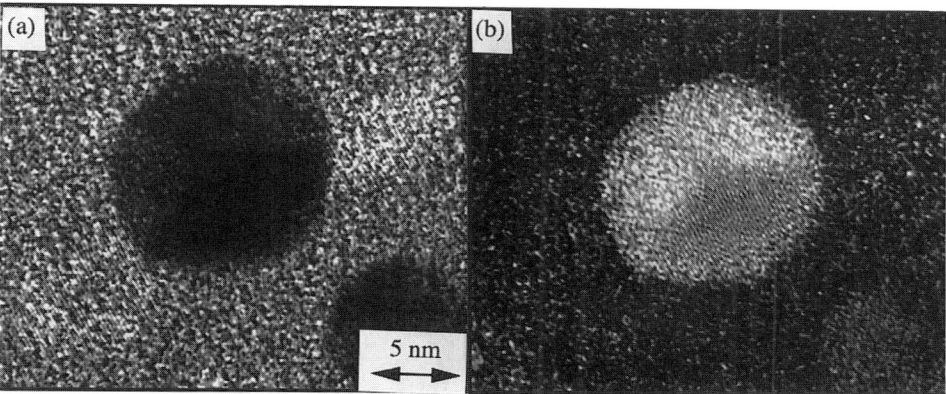

Figure 11. (a) Reconstructed image of the electron hologram, (b) Phase contrast image of the corrected image wave.

Figure 11(b) shows the phase distribution of the corrected image wave. In the image corrected for spherical aberration, the Bragg-reflected images coincide with the Gaussian image. The effect of the aberration has been successfully removed in spite of large spherical aberration. When the aberration is properly corrected, the contrast of the corrected image is very weak. Fortunately, the Zernike phase-contrast method can be ideally realized by the same LC-panel. Here an appropriate voltage is applied on the pixels located in the center of the LC-panel to shift the unscattered wave by an optical path length of $\lambda/4$. The number of pixels used to generate a $\lambda/4$ phase plate can be changed according to the experimental conditions. Both positive and negative phase contrasts can be easily produced. The quantitative phase distribution of the corrected image wave can be measured by phase-shifting interferometry, if necessary.

5. CONCLUSIONS

Application of electron holography has been severely limited by time-consuming reconstruction processes. We have shown that an video-rate electron-holographic interference microscopy can be created by using an LC-panel as rewritable medium for the electron hologram. By eliminating the time-consuming photographic process involved in conventional interference microscopy, this technique make it more convenient to operate a holographic electron microscope by watching a live interference micrograph. It also makes possible the observation of dynamic events. We have shown experimental results obtained while observing the dynamic domain-wall motion that a magnetic field induces in a thin permalloy film. This technique has wide applications in basic and practical research. We have also shown that the LC-panel can also be used for correcting the spherical-aberration of an electron lens. For this, the LC-panel was used as a computer-controlled phase plate at the optical reconstruction system. It provides high flexibility in adjusting the parameters to compensate for the wave aberration. The experimental results reveal that this method can be applied to allow high-resolution observation.

6. REFERENCES

[1] D. Gabor, Proc. R. Soc. London Ser. A 197 (1949) 454.
[2] A. Tonomura, Electron holography, (Springer-Verlag Berlin, 1993).
[3] A. Tonomura, N. Osakabe, T. Matsuda, T. Kawasaki, J. Endo, S. Yano, and H. Yamada, Phys. Rev. Lett. 56 (1986) 792.
[4] T. Matsuda, A. Fukuhara, T. Yoshida. S. Hasegawa, A. Tonomura, and Q. Ru, Phys. Rev. Lett. 66 (1991) 457
[5] T. Hirayama, J. Chen, Q. Ru, T. Tanji, and A. Tonomura, J. Electron Microsc. 43 (1994) 190.
[6] For instance, G. Lai, J. Chen, K. Ishizuka, and A. Tonomura, Appl. Opt. 31 (1992) 5940.
[7] M. Takeda and Q. Ru, Appl. Optics 24 (1985) 3068.
[8] W. D. Rau, H. Lichte, E. Völkl, and U. Weierstall, J. Compu. Assis. Micros. 51 (1991) 51.
[9] Q. Ru, J. Endo, T. Tanji, and A. Tonomura, Appl. Phys. Lett. 59 (1991) 2372.
[10]N. Hashimoto, S. Morokawa, and K. Kitamura, SPIE Proc. 1461 (1991) 291.
[11] N. Konforti, E. Marom, and S. T. Wu, Optics Lett. 13 (1988) 251.
[12] J. Amako and T. Sonehara, Appl. Opt. 30 (1991) 4622.
[13] J. Chen, T. Hirayama, G. Lai, T. Tanji, K. Ishizuka, and A. Tonomura, Optics Lett. 18 (1993) 1887.
[14] J. Chen, T. Hirayama, T. Tanji, K. Ishizuka, and A. Tonomura, Opt. Commu. 110 (1994) 33.
[15] J. Chen, T. Hirayama, G. Lai, T. Tanji, K. Ishizuka, and A. Tonomura, Opt. Rev. to be published.
[16] T. Hirayama, J. Chen, T. Tanji, and A. Tonomura, Ultramicroscopy 54 (1994) 9.
[17] A. Tonomura, T. Matsuda, and J. Endo, Jpn. J. Appl. Phys. 18 (1979) 1373.
[18] H. Lichte, Ultramicroscopy 20 (1986) 293.
[19] H. Lichte, Ultramicroscopy 38 (1991) 13.
[20]J. Chen, G. Lai, K. Ishizuka, and A. Tonomura, Appl. Opt. 33 (1994) 1187.
[21]J. Chen, T. Hirayama, K. Ishizuka, and A. Tonomura, Appl. Opt. to be published.
[22] Y. Matsueda, T. Ozawa, J. Nakamura, S. Takei, H. Kamakura, N. Okamoto, Japan Display '92 (1992) 561.

Electron Holographic Computed Tomography

G. Lai[a,b], T. Hirayama[a], A. Fukuhara[c], K. Ishizuka[a], T. Tanji[a], and A. Tonomura[a,c]

[a]Tonomura Electron Wavefront Project, JRDC, c/o Advanced Research Laboratory Hitachi, Ltd., Hatoyama, Saitama 350-03, Japan

[b]Department of Electrical and Electronic Engineering, Shizuoka University Johoku 3-5-1, Hamamatsu 432, Japan

[c]Advanced Research Laboratory, Hitachi, Ltd., Hatoyama, Saitama 350-03, Japan

Methods of reconstructing three-dimensional micro electric potential and magnetic flux distributions using electron-holographic interferometry are described. First, two-dimensional phase distributions are shown to reveal the projected electric potential and magnetic flux structures. Second, algorithms for reconstructing electric potential and magnetic flux density in three dimensions from phase distributions at different viewing directions have been developed. Applications to latex fine particles and a barium ferrite particle are presented.

1. INTRODUCTION

To detect micro electromagnetic fields, a probe should be not only sensitive to them but also available itself for microscopy. The electron in an electron microscope receives phase shift when it interacts with a specimen, and therefore satisfies these requirements. Because most of the specimens in the electron microscope appear as phase objects, they should be observed under some special imaging conditions. The transmission electron microscope (TEM) can produce a phase contrast micrograph by making use of the amount of defocus and the aberration of the electron lenses. Lorentz microscopy is also a special technique for imaging magnetic fields, in which the image is defocused by 0.1 to 10 mm in order to observe the deflection effects by the magnetic force. Neither electron microscopy nor Lorentz microscopy can give the phase distribution of the specimen itself.

Electron holography, or electron interferometry has made it possible for one to measure the electron phase distribution in the TEM. Therefore it has been used to observe micro electric and magnetic fields [1]. An interference micrograph obtained by this technique gives a full image of the electromagnetic field of a two-dimensional specimen. To interpret the image of a specimen with a three-dimensional structure, electron interferometry must be used in a revised form. In this paper, electron phase distribution measured by electron holographic interferometry is used to interpret micro electromagnetic fields and reconstruct their structures in three dimensions by computed tomography (CT). Reconstruction algorithms for electric and magnetic fields, together with phase measurement techniques are described. Some experimental results of reconstructing non-magnetic fine particles and magnetic flux are presented.

2. ELECTRON PHASE SHIFTED BY ELECTROMAGNETIC FIELDS

The phase increment of an electron beam transmitted through an electromagnetic field is derived from the Schroedinger equation with WKB approximation [2],

$$S = \frac{2\pi}{h} \int_L (m\mathbf{v} - e\mathbf{A}) \cdot dt, \tag{1}$$

where the line integral is taken along the electron path, and h, m, $\mathbf{v}$, e, and $\mathbf{A}$ are the Plank constant, the electron mass, velocity, electronic charge, and the vector potential, respectively. The first term of Eq. (1) depends on the velocity of the electron passing through the specimen and thus includes the static electric potential. The second is caused by the vector potential originating from magnetic fields of the specimen. This approximation is well justified for a weak electromagnetic specimen and high speed electron beam.

For non-magnetic specimens the phase difference $\phi(x, y)$ introduced by an electric potential corresponding to the first term of Eq. (1) can be written as [3]

$$\phi(x, y) = \frac{\pi}{\lambda E} \int_L V(x, y, z) \, dt, \tag{2}$$

where $V(x, y, z)$, λ and E are the electric potential distribution, electron wavelength and the accelerating voltage, respectively. The inner electric potential distribution is responsible for phase modulation for a specimen consisting of material. The lateral beam shift when the electron passes through the micro electromagnetic fields can, in most of the cases, be neglected. Thus, the phase distribution represents the projected material distribution or the thickness distribution of the specimen.

The second term in Eq. (1) can be simplified by considering the phase difference between two points. When a plane wave, having the same phase at points A_1 and A_2, transmits through the magnetic specimen as shown in Fig. 1, a phase difference between points P_1 and P_2 is introduced. Electron-holographic interferometry measures this phase difference on the image plane of the specimen by overlapping a reference wave. Adding two additional paths A_1A_2 and P_1P_2 where the vector potential is negligible, we have as the phase difference between P_1 and P_2

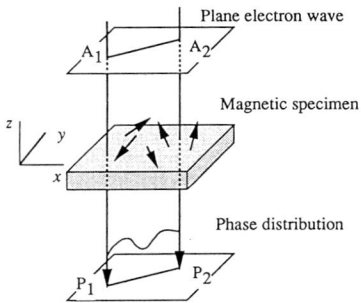

Fig. 1 Phase increment by the enclosed magnetic flux.

$$\phi_{P_2P_1} = S_{A_2P_2} - S_{A_1P_1} = \frac{2\pi e}{h} \oint_T \mathbf{A} \cdot d\mathbf{t}, \tag{3}$$

where T is the closed path of the rectangle $A_1A_2P_2P_1$. Eq. (3) directly corresponds to the magnetic flux passing through the rectangular surface by Stokes's theorem as

$$\phi_{P_2P_1} = \frac{2\pi e}{h} \int_\sigma \mathbf{B}(x, y, z) \cdot d\mathbf{S}. \tag{4}$$

If we take the surface σ in Eq. (4) perpendicular to the y axis, then the phase distribution is given by

$$\phi(x, y) = \frac{2\pi e}{h} \int_0^x d\xi \int_{-\infty}^\infty B_y(\xi, y, z)\, dz. \tag{5}$$

That is, the phase increment at $P_2(x, y)$ in the x direction is proportional to the projection of the y component of the magnetic flux density, assuming $x=0$ at P_1. In the same way, we get in the y-direction the phase increment as

$$\varphi(x, y) = -\frac{2\pi e}{h} \int_0^y d\eta \int_{-\infty}^\infty B_x(x, \eta, z)\, dz, \tag{6}$$

assuming $y=0$ at P_1. It is clear from Eqs. (5) and (6) that the x and the y components of the magnetic flux density projected along z-direction can be obtained by differentiating the two-dimensional phase distribution with respect to x and y axes, respectively. From these projections, their 3-D distribution can be reconstructed based on a CT algorithm for scalar distributions.

Furthermore, based on the characteristic of magnetic flux density that $\text{div}\mathbf{B}=0$, B_z can be calculated from B_x and B_y using the following equation:

$$B_z(x, y, z) = -\int_{-\infty}^z \left\{ \frac{\partial}{\partial x} B_x(x, y, \zeta) + \frac{\partial}{\partial y} B_y(x, y, \zeta) \right\} d\zeta. \tag{7}$$

3. COMPUTED TOMOGRAPHY FOR ELECTROMAGNETIC FIELDS

Both of the electric potential and the magnetic field give a phase shift to the electron wave, but they should be treated in different ways in reconstruction. The phase distributions by the electric specimen are the projections of the electric potential distribution. Therefore this electric potential distribution can be reconstructed from the measured phase distributions by computed tomography. Since the projections are available in the range of $120°$ in our experients, we developed an iterative algorithm. First, the projection data are backprojected for all the measured directions to obtain the first approximation of the distribution on predetermined sampling points. This approximated distribution is projected, by calculation, to get projections. The differences between the projections measured and those calculated are backprojected again to improve the distribution. This process is repeated until the total difference is smaller than a predetermined value.

We describe here a new CT algorithm that can derive the magnetic flux directly from the phase distribution. When a parallel projection of the two-dimensional function is given, its one-dimensional Fourier transform gives values along the relative line across the origin in the two-dimensional Fourier plane [4], as shown in Fig. 2.

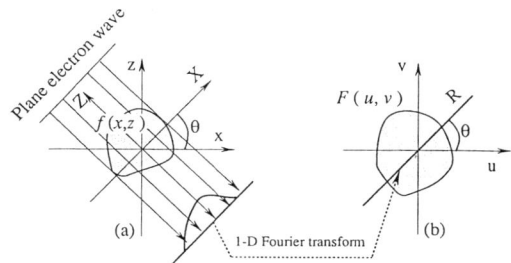

Fig. 2 Relation between the projection and the 2-D Fourier transform of a function.

Using this relationship, various algorithms have been developed for reconstructing the original function from its projections. The reconstruction by the filtered back-projection algorithm [5] is given by

$$B_y(x, y, z) = \int_0^\pi d\theta \int_{-\infty}^\infty F_X\{ P_\theta [B_y] \} |R| \exp[2\pi i R(x\cos\theta + z \sin\theta)] \, dR, \tag{8}$$

where $F_X\{\}$ and $P_\theta[\ ]$ represent the Fourier transformation of X and projection operator in the θ direction, respectively. According to Eq. (5), the projections of B_y are the X derivatives of the phase distributions. Therefore,

$$B_y(x, y, z) = \int_0^\pi d\theta \int_{-\infty}^\infty F_X\{ \frac{\partial}{\partial X}\phi(X, y) \} \left(\frac{h}{2\pi e}\right) |R| \exp[2\pi i R(x\cos\theta + z \sin\theta)] \, dR$$

$$= \int_0^\pi d\theta \int_{-\infty}^\infty F_X\{ \phi(X, y) \}\left(\frac{h}{e}i\right) R|R| \exp[2\pi i R(x\cos\theta + z \sin\theta)] \, dR \tag{9}$$

Thus, the three-dimensional distribution of the y component is calculated from the phase distributions obtained by rotating the specimen about the y axis. In the same way, the B_x distribution is calculated from the phase distributions obtained by rotating the specimen about the x axis. Based on Eq. (9), lines of the measured two-dimensional phase distributions corresponding to a section of constant y are digitally Fourier-transformed, filtered by a filter function,

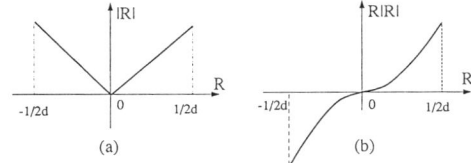

Fig. 3 Filter functions for (a) the conventional and (b) the developed magnetic algorithms.

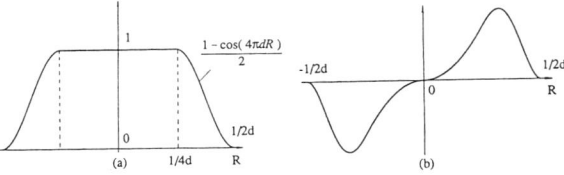

Fig. 4 Window and effective filter function.

and then back-projected for angles θ to reconstruct the section. The three-dimensional distribution $B_y(x, y, z)$ is obtained by calculations for each y section. The difference between the algorithm discussed here and the conventional filtered back-projection is the filter functions acting on the one-dimensional Fourier transform shown in Fig. 3. They are both sensitive to high frequency noise. In a practical implementation of the algorithm, we found that noise could be efficiently reduced in the reconstruction by applying a window on the filter function. Figure 4 shows the window and the effective filter acting on the Fourier transform before back-projection.

4. PHASE MEASUREMENT TECHNIQUES

Interference of the electron wave in the electron microscope is produced by using an electron biprism (Fig. 5). A collimated electron beam illuminates the specimen, with half of the beam passing through the object as the object wave and the half being the reference wave. When these two parts interfere on the image plane of the object, the phase difference between them determines the interference pattern. If the object modulates the electron wave with amplitude $a(x, y)$ and phase $\phi(x, y)$, the wave function just behind the specimen is given by $a(x, y) \exp[i \phi(x, y)]$. Since the illumination beam has an incident angle θ to the optical axis, an additional linear phase factor is introduced, and the wave functions of the object and the reference waves are given by

$$o(x, y) = a(x, y) \exp\{i [\phi (x, y) + \frac{2\pi}{\lambda} x \sin(\theta)]\} \quad (10)$$

and

$$r(x, y) = \exp[i \frac{2\pi}{\lambda}(x + x_0) \sin(\theta)], \quad (11)$$

where λ is the electron wavelength and x_0 is the width of the overlapped interference area between the object and the reference waves.

The subsequent electron biprism deflects the object and reference waves in opposite directions and thus introduces phase factors $\exp[i(2\pi/\lambda)x\sin(\alpha)]$ and $\exp[-i(2\pi/\lambda)x\sin(\alpha)]$, where α is the biprism deflection angle. The interference intensity between the waves is given by

$$I_k(x, y) = a^2(x, y) + 1 + 2a(x,y) \cos[\phi(x, y) + 2\pi fx + \psi_k], \quad (12)$$

where $f=2\sin(\alpha)/\lambda$ and $\psi_k=2\pi x_0\sin(\theta_k)/\lambda$, with θ_k being a beam tilt angle. The terms in the cosine function are the object phase distribution, the carrier-fringe phase introduced by the biprism, and the initial phase of the interference pattern.

To obtain the phase distribution from the interferograms, two categories of methods can be considered. One is the holographic method in which the object wave is reconstructed from the hologram sideband diffraction. It has the advantage that the phase distribution can be obtained from one interferogram. However it depends on tthe carrier frequency. The other is the phase-shifting method that calculates the object phase for each sampling point independently from the interferograms of different initial phases. It is not restricted by the carrier frequency and therefore has advantages in obtaining high resolution reconstruction. But more than three interferograms are required to obtain one phase distribution. Discussions in the following are made for the holographic method; however, some of them are the same as for the phase-shifting method. Details of the phase-shifting method are described in references [6-8].

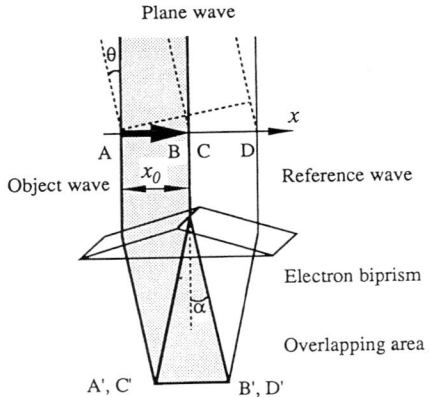

Fig. 5 Interference with an electron biprism.

Phase measurement by holographic interferometry requires recording and reconstruction of holograms. There are four variations; the hologram can be recorded on a film or captured by a TV camera, and can be reconstructed by optical or digital methods. Capturing holograms by a TV camera can provide real-time measurement, and therefore is a method appropriate for measuring a number of phase distributions. We used the holographic method with a TV camera and digital fringe processing. Because the absolute phase shift is required, the effect of the carrier frequency introduced by the electron biprism should be considered. Here, a method to determine this frequency precisely using the Fourier transform method is described and the phase distribution extracted from an electron hologram is processed to get an overall phase distribution.

4.1. Carrier frequency determination

A Fourier transform of an off-axis electron hologram consists of three terms. The first term is the autocorrelation of the wave function, which is centered at the spectrum origin. The second and third terms are the Fourier transform of the wave function and that of the conjugate, respectively, which are shifted by the carrier frequency. In phase measurement by the Fourier transform method, the second term is selected out. It is then shifted to the origin of the coordinates and inversely Fourier-transformed to obtain the recorded wave function. The phase distribution of the wave function is calculated from the real and imaginary parts for each sampling point. However, the exact value of the carrier frequency is not known beforehand. Inverse Fourier transforming of the selected second term with a carrier-frequency-locating error of (Δ_j, Δ_k) will give the original object wave multiplied by a position-dependent phase factor $\exp[-i2\pi(m\Delta_j/M+n\Delta_k/N)]$, ($m=0, ..., M-1$ and $n=0, ..., N-1$), where M and N are the sampling points in the x and y directions, respectively. A one-pixel error from the true frequency in the Fourier spectrum will cause a 2π phase variation within the sampled area. To get the pure phase distribution of the object wave, the frequency of these carrier fringes should be precisely determined.

The precise value of the carrier frequency can be determined using the fringes from the region which does not include the object wave, that is, the fringes caused by only the electron biprism. Because there is no object wave, there should be no phase variation over the measured plane. Locating a carrier frequency approximately, the phase distribution is calculated by the Fourier transform method. After it is unwrapped, the phase distribution is fitted to a plane. Phase variations coming from the tilts in x and y directions will be 2π times the frequency-locating error. For a specimen of limited area, which is the case with fine particles, the non-image area of the same hologram can be used to estimate the carrier frequency [9]. For a specimen which does not have a region for reference background, two holograms, one with and one without an object, should be available. The phase modulation caused by the carrier frequency can be completely eliminated by subtracting the phase distribution without the object from that with the object [10].

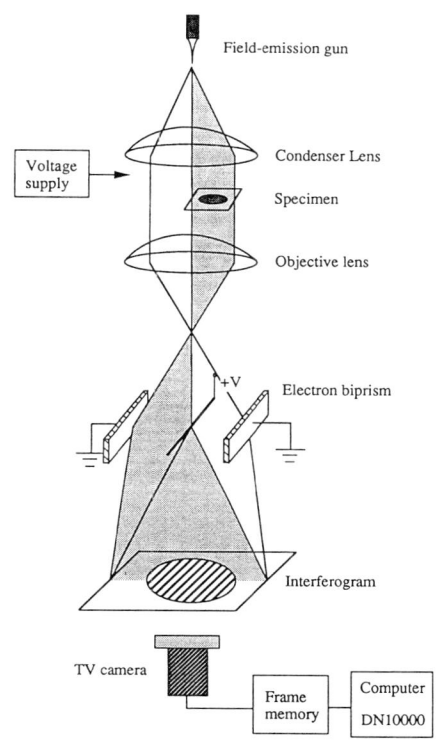

Fig. 6 Experimental system setup.

4.2. Phase unwrapping

The phase calculated by either the digital Fourier transform method or the phase-shifting method is given as the principal values between $-\pi$ and π. Therefore, the calculated phase distribution generally contains discontinuities over the measured plane, which should be unwrapped to represent the overall potential projection. For a continuous phase distribution, the assumption that the phase difference between two neighboring points is less than π can be applied to unwrap the phase discontinuities.

5. EXPERIMENTAL RESULTS

The effectiveness of the method described in the previous sections was examined for latex particles and a small ferrimagnetic particle. The experiments were performed with a field-emission electron microscope, Hitachi HF-2000, equipped with an electron biprism. Figure 6 shows schematically the system setup. The electron beam was aligned by the condenser lens for parallel illumination. The specimen

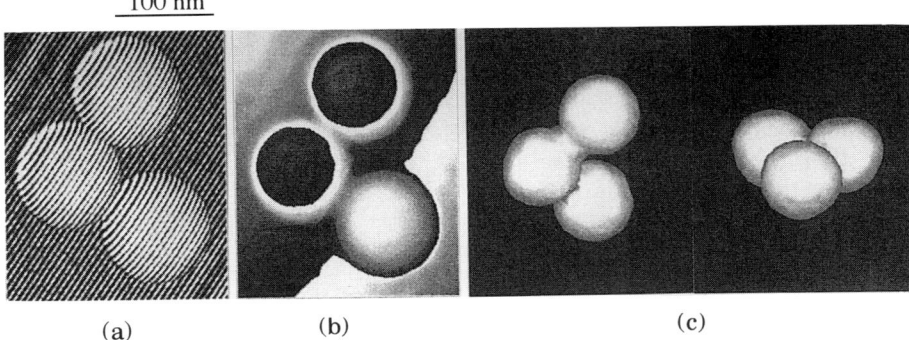

Fig. 7 Experimental results with latex particles. (a) Hologram; (b) Phase distribution; (c) two views of the reconstructed three-dimensional image.

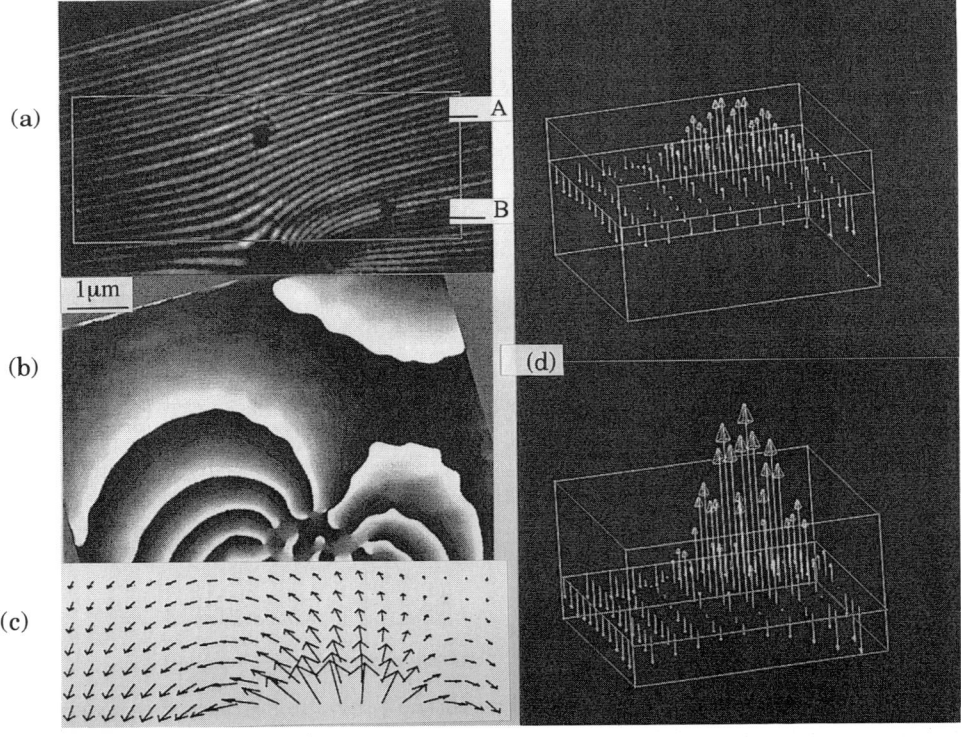

Fig. 8 Experimental results with a barium ferrite particle. (a) Hologram; (b) phase distribution; (c) projected flux density vector; (d) two sections of the three-dimentional magnetic flux density distribution in the direction.

was observed under an infocus condition using the objective and intermediate lenses. In order to avoid the magnetic effects from the microscope on a magnetic specimen, the objective lens was turned off for magnifications of several thousands. For higher magnifications a magnetically shielded objective lens may be used [11]. The object wave was overlapped with the reference wave by the electron biprism behind the objective lens. The intermediate and projection lenses form a magnified image of the specimen on the TV camera in the microscope. The TV signal of the fringe intensity is then fed to a computer for image processing.

Figure 7 shows the experimental results obtained with latex particles. The phase distribution corresponds to the projected inner electric potential distribution or thickness distribution. The three-dimensional image was reconstructed from 25 phase distributions by using a CT algorithm based on an iterative reconstruction method [9].

Results obtained with a barium ferrite particle with single magnetic domain structure are shown in Fig. 8. Figure 8 (a) shows a hologram of the barium ferrite particle, which is about 1 μm in size. Figure 8 (b) is the phase distribution calculated from the hologram by the FFT method, represented by the principal values from $-\pi$ to π. It was then unwrapped to represent an overall phase distribution appropriate for use in the reconstruction algorithm. The x and y components of the projected magnetic field shown in Fig. 8 (c) were obtained by differentiating the calculated phase distribution within the rectangle part. To study the three-dimensional distribution with the CT algorithm, the specimen was tilted about the y-axis between -60° and 60° with 5° increments. Using the described algorithm, three-dimensional distribution of the magnetic field was then reconstructed. Figure 8 (d) shows two sections of the y-component distribution near the particle (corresponding to plane A) and at a distance from it (plane B). A positive maximum magnetic flux density appears near the center and continuously changes in the negative direction. The total magnitude becomes smaller outside of the particle.

6. CONCLUSIONS

Electron phase distributions measured by electron-holographic interferometry are processed to provide projections of electric potential distribution and magnetic flux density distributions, and used to reconstruct their three-dimensional structures. Algorithms for reconstructing electric potential and magnetic flux density from phase distributions have been developed. Applications to latex fine particles and a ferrimagnetic particle show the possibility of providing three-dimensional images, and therefore the electron holography can be extended to three-dimensional applications by the computed tomography.

REFERENCES

1. A. Tonomura, Adv. in Phys. 41, 59 (1992).
2. A. V. Crewe, D. N. Eggenberger, D. N. Wall, L. N. Welter, Rev. Sci. Instrum. 39, 576 (1968).

3. A. Tonomura, T. Matsuda, J. Endo, H. Todokoro, and T. Komoda, J. Electron Microsc. 28, 1 (1979).
4. D. J. DeRosier and A. Klug, Nature, 217, 130 (1968).
5. G. T. Herman, in Computer Science and Applied Mathematics, Academic Press, 316 (1980).
6. Q. Ru, J. Endo, T. Tanji, and A. Tonomura, Appl. Phys. Lett. 59, 2372 (1991).
7. Q. Ru, J. Endo, T. Tanji, and A. Tonomura, Optik 92, 51 (1992).
8. G. Lai, Q. Ru, K. Aoyama, and A. Tonomura, J. Appl. Phys. 76, 39 (1994).
9. G. Lai, T. Hirayama, K. Ishizuka, and A. Tonomura, Appl. Opt. 33, 829 (1994).
10. G. Lai, T. Hirayama, A. Fukuhara, K. Ishizuka, T. Tanji, and A. Tonomura, J. Appl. Phys. 75, 4593 (1994).
11. T. Hirayama, Q. Ru, T. Tanji, and A. Tonomura, Appl. Phys. Lett. 63, 418 (1993).

Electron Holography
A. Tonomura, L.F. Allard, G. Pozzi, D.C. Joy and Y.A. Ono (Editors)
© 1995 Elsevier Science B.V. All rights reserved.

Practical Electron Holography: Applications of Advanced Hologram Processing Techniques to Materials Science Problems

E. Völkl[a]*, L.F. Allard[a] and B. Frost[b]

[a]Oak Ridge National Laboratory, Oak Ridge, TN 37831-6064, USA

[b]University of Tennessee, Knoxville, TN 37996-0810, USA

The use of electron holography techniques on field emission gun (FEG) electron microscopes is rapidly moving from developmental stages to practical everyday application for solving a wide variety of materials science problems. This is partly due to the increased availability of 1024 by 1024 pixel slow-scan CCD-cameras, fast and relatively inexpensive desktop computers (such as the PowerPC), and an improved understanding of the reconstruction procedures involved with electron holography that has resulted in a sophisticated software design that greatly facilitates the use of electron holography.

1. OFF–AXIS ELECTRON HOLOGRAPHY

Electron holography[1,2], especially off-axis electron holography [3], has become an easily applicable tool for the investigation of materials science problems. At the center of all holographic investigations lies the reconstruction process, using continuous wave optics as on the optical bench and/or discrete methods on the computer. A description of some of these processes can be found in [4–6]. Throughout this paper, we will deal with the purely discrete reconstruction process. The discrete reconstruction process itself can be split into two major routes. The first route is the reconstruction through Fourier space, which will be discussed in this paper. The second route is the reconstruction in real space and is described in detail in [7]. So that the hologram processing steps described herein can be more easily understood, it is useful to first review briefly off–axis holography and the reconstruction process through Fourier space.

1.1. Basic off–axis holography

In the normal imaging mode of a transmission electron microscope, the electron beam incident on the specimen is, ideally, a plane wave. In practice, this can be achieved to a reasonable approximation if a field emission electron gun is used and the illumination is spread over a large area. In this case, the object in the microscope modifies the incident

---

* This work was sponsored by the Directed R&D Program of Oak Ridge National Laboratory, managed for the DOE by Martin Marietta Energy Systems, Inc. under contract DE-AC05-84OR21400, and supported by appointments to the Oak Ridge National Laboratory Postdoctoral Research Program administered by the Oak Ridge Institut for Science and Education (E.V.) and the University of Tennessee Postdoctoral Research Program (B.F.).

plane wave $\exp(i\vec{k}\vec{r})$ to the object wave $o(\vec{r})$, which is defined as:

$$o(\vec{r}) = a(\vec{r}) \cdot e^{i\varphi(\vec{r})} \qquad (1)$$

Both $a(\vec{r})$ and $\varphi(\vec{r})$ are real, two–dimensional functions that describe the object amplitude and the object phase respectively. The vector $\vec{r}$ is a vector in the specimen plane, or the (x, y) plane. For reasons of simplicity, magnification factors and image rotations are ignored, so vectors in the image plane and the object plane coincide.

When recording the off–axis hologram, the object is, ideally, positioned exclusively on one side of the biprism. The biprism then overlaps the image wave (or object wave depending on the experimental setup) with the reference wave, i.e. the wave on the other side of the biprism. In this case, the reference wave is a plane wave. It is however not a mandatory requirement that the reference wave be a plane wave, and acceptable results may still be obtained if the reference wave passes through the specimen or is deformed by magnetic and/or electric fields. In the image plane below the biprism one finds the image intensity $I(\vec{r})$ [8]:

$$I(\vec{r}) = 1 + A^2(\vec{r}) + I^{inel}(\vec{r}) + 2\mu \cdot A(\vec{r}) \cdot \cos\left(\Delta\vec{k} \cdot \vec{r} + \Phi(\vec{r})\right) \qquad (2)$$

where the additional term $I^{inel}(\vec{r})$ takes the inelastically scattered electrons into account. The vector $\Delta\vec{k}$ is a two–dimensional vector describing direction and spatial frequency of the holographic fringes. $A(\vec{r})$ and $\Phi(\vec{r})$ are real functions describing the image amplitude and phase, respectively, which are different from the object amplitude and phase due to the aberrations of the objective lens (and other sources of disturbances). The term μ describes the contrast of the interference fringes if no object is used, i.e. $A = 1$, and $I^{inel} = 0$ and $\Phi = 0$, or constant.

1.2. Reconstruction through Fourier space

In order to extract the information about the (complex) image wave, and its amplitude and phase, a continuous Fourier transform (**FT**) of the image intensity $I(\vec{r})$ is performed:

$$\begin{aligned}\mathbf{FT}\left\{I(\vec{r})\right\} = \quad & (3)\\ & \mathbf{FT}\left\{1 + I^{inel} + A^2(\vec{r})\right\} + \\ & \delta(\Delta\vec{k} - \vec{q}) \star \mathbf{FT}\left\{A(\vec{r}) \cdot e^{i\Phi(\vec{r})}\right\} + \\ & \delta(\Delta\vec{k} + \vec{q}) \star \mathbf{FT}\left\{A(\vec{r}) \cdot e^{-i\Phi(\vec{r})}\right\}\end{aligned}$$

where $\star$ denotes convolution. The Fourier transform of the hologram of Figure 1 is displayed in Figure 2. The central area, corresponding to line 2 of equation (3) is known as the 'autocorrelation', and is equivalent to the Fourier transform of a conventional image. The 'sidebands', found in the lower left and upper right of Figure 2 (and look like ears around the autocorrelation), are characteristic of off–axis holograms and correspond to lines 3 and 4 of equation (3). Sidebands differ from the autocorrelation in general, as they show no centrosymmetry, which is required (mathematically) for the autocorrelation. In holograms from crystaline specimen, in high resolution mode, diametrically opposing

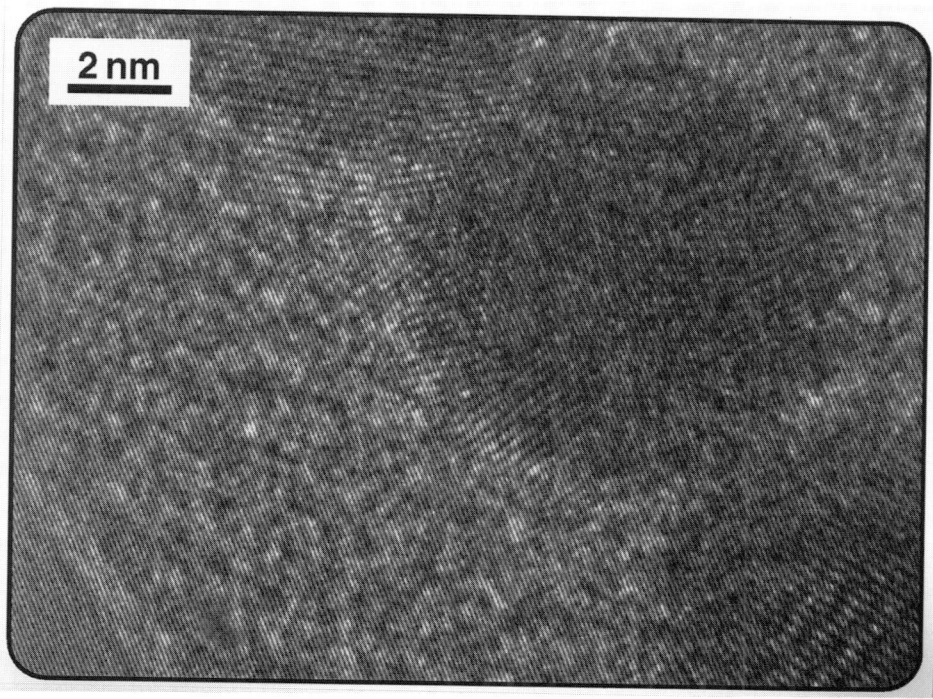

Figure 1. Hologram of Au–particles on amorphous carbon foil. Fringe spacing: 0.085 nm.

beams of the sideband do not have the same intensities (unless the crystal area is perfectly oriented), whereas opposite beams in the autocorrelation always have the same intensity.

The last two lines of equation (3) correspond to the two sidebands in Figure 2. The δ-functions describe the center of each sideband respectively. Both sidebands contain identical information; as seen from equation (3), one sideband is the complex conjugate of the other sideband. If one of the sidebands is isolated by applying an aperture, an inverse Fourier transform then yields the complex image wave. This process will be discussed in detail later. Ignoring all but the isoplanatic aberrations, the Fourier transform of the complex image wave corresponds to the Fourier transform of the complex object wave in the following way:

$$\mathbf{FT}\left\{A(\vec{r}) \cdot e^{i\Phi(\vec{r})}\right\} = \mathbf{FT}\left\{a(\vec{r}) \cdot e^{i\varphi(\vec{r})}\right\} \cdot e^{i\chi(\vec{q})} \qquad (4)$$

The term $\chi(\vec{q})$ describes the isoplanatic wave aberrations, where $\vec{q} = (u, v)$ is a 2-dimensional vector in Fourier space, i.e. the back focal plane of the objective lens. It should be noted that $\Delta \vec{k}$ lies completely in the back focal plane (u, v).

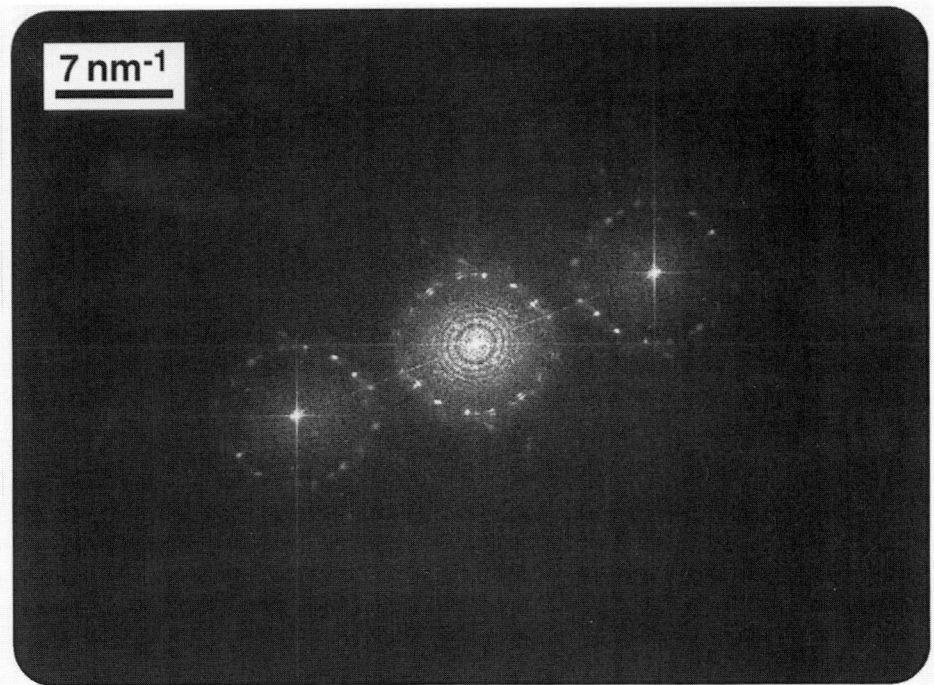

Figure 2. Logarithm of the modulus of the Fourier transform of hologram as in Fig. 1.

1.3. Basic applications

It is obvious from equation (4) that if $\chi(\vec{q})$ is known to a certain degree, this information can be used to compensate the effect of $\chi(\vec{q})$, which results in an improvement of the point resolution of the electron microscope as discussed in [9–13].

The direct accessibility of the image phase is another important feature of electron holography. In contrast to the conventional contrast transfer function, i.e. the $\sin\chi$–function, the transfer function of the object phase into the image phase is described by the $\cos\chi$–function [14]. The advantage of the cos–type transfer function is that large-area contrast (details of size >1 nm) is not decreased and, apart from a slightly decreased point resolution, the total amount of information is usually higher in the image phase than in the image intensity [15,16]. This is one of the reasons why a direct display of the image phase is preferable over the image intensity in many cases.

2. RECONSTRUCTION THROUGH FOURIER SPACE

The 'real' wave optics in an electron microscope deviates from the description given in section 1.1 for many reasons. First, the illumination may not be a plane wave. Second, Fresnel fringes from the biprism are artefacts inherent to standard off–axis electron holography. Third, image distortions that often go by unnoticed in conventional imaging mode can have a strong effect on the reconstructed image phase. Finally, the center of the sideband may not fall exactly on one of the display-points (k,l) (corresponding to $(u_k, v_l) = (\frac{k}{Nd}, \frac{l}{Nd})$) in discrete Fourier space. An uncentered sideband not only causes a distracting phase–wedge in the reconstructed image phase but, more importantly, contributes to strong artefacts in the reconstructed image amplitude [17]. Using special procedures when recording holograms and/or using special image processing techniques, some of the artefacts can be compensated and, at the same time, implemented as automated processes. This is state of the art in the software package HoloWorks[2] available as a plug–in module for DigitalMicrograph® [20,21].

Basically, HoloWorks permits the operator to choose among three different reconstruction procedures: 1) a completely automated reconstruction process that is optimized for speed and convenience; 2) an interactive reconstruction process that allows to center the sideband correctly and to chose the size of the final image; and 3) a reconstruction process that makes use of a 'reference' hologram. A reference hologram is a hologram that shows no contribution from any object. However, it contains information on image distortions, Fresnel fringes and the deviations from the ideal plane wave scenario of a two beam interference pattern. The reference hologram is generally taken immediately after the hologram is recorded to preserve the present wave-front situation of the microscope. This is a stringent condition due to instrument instabilities.

So far we described the use of continuous functions. These however are not available from the experiment. Assuming the holograms are recorded with an ideal CCD-camera, the information available is in fact described by the values of the M by N pixels p_{mn} of the CCD-camera:

$$p_{mn} = \frac{1}{d^2} \int_{(m-1)d}^{md} \int_{(n-1)d}^{nd} I(x,y)\,\mathrm{d}x\,\mathrm{d}y \tag{5}$$

for $m = 1,\ldots M$, $n = 1,\ldots N$ and d the size of one pixel. We will assume in this paper that all images are square (i.e. $M = N$) and of the power of two, so that the fast Fourier algorithm can be used in image processing. The integration process over a finite area causes a decrease of the amplitude of spatial frequencies with increasing frequency. For a 'real' CCD-camera, many more processes contribute to this effect [22–24]. The decreasing signal amplitude as a function of increasing spatial frequency is described by the modulation transfer function (MTF). The MTF however can be compensated [25,26] and will be ignored in this paper. If, instead of a CCD-camera, photographic film is used, it is advisable to take into account the additional nonlinearities involved in this process, as described in [27].

2.1. Fully automated reconstruction

The fully automated reconstruction has its main use as a preview for the reconstructed image amplitude and image phase. The process is straightforward: perform a (fast) Fourier transform of the p_{mn} which yields the Fourier transform of the p_{mn}, t_{kl}, according

---

[2] ©[1993] Martin Marietta Energy Systems, Inc.

to:

$$t_{kl} := \mathbf{FT}\{p_{mn}\} = \sum_{m,n=1}^{N} p_{mn}\, e^{-2\pi i(km+ln)/N} \qquad (6)$$

with $k, l \in [-N/2+1, N/2]$. As center of the sideband, the display point is chosen with the highest intensity value (k_0, l_0) in the upper half of the **FT** (exactly in the area $k = \{-N/2+1, 0\}$ for $l = \{-N/2+1, -N/32\}$, $k = \{-N/2+1, -N/32\}$ for $l = \{-N/32+1, N/32\}$ and $k = \{-N/2+1, -1\}$ for $l = \{N/32+1, N/2\}$. Obviously, $R = \sqrt{k_0^2 + l_0^2}$ defines the distance of the center of the sideband from the center of the Fourier transform at $(0, 0)$. The complex data of a selected area of size $N/4$ by $N/4$ around (k_0, l_0) are then copied into a new image. An aperture is applied to this new image to isolate the sideband. As diameter D for the aperture, $D = R$ is chosen if $D < N/4$, or else $D = N/4$ is used. An inverse Fourier transform is then applied to this new image which results in the complex image wave (or its complex conjugate). The modulus and phase of the complex image wave is then displayed.

This process yields in many cases acceptable reconstructed images. The final images do have a quarter the size of the original image as default, which does not decrease the information in the reconstructed images. This is because the image information in the original hologram is oversampled to allow for a sufficient sampling rate of the interference fringes. For reliable results however, a more complex reconstruction process is necessary. Such a process is designed to be user–interactive.

2.2. Interactive reconstruction with no reference hologram

This reconstruction method should be used if no usable reference hologram is available. This is for example the case if the beam-shift deflectors or the biprism are not stable. If an incorrect reference hologram is used, the image phase will not become flat in the vacuum and hologram artefacts will not be properly compensated. In this case it is important to optimize the reconstruction process described in 2.1 to minimize artefacts.

As there is a finite set of discrete pixel positions in Fourier space, the center of the sideband (at $\pm\Delta\vec{k}$) will, in almost all cases, not coincide with a pixel position (k, l) in discrete Fourier space. The general (and average) position of the center of the sideband will be $(u, v) = \left(\frac{k+\Delta k}{Nd}, \frac{l+\Delta l}{Nd}\right)$ with $\Delta k, \Delta l \neq 0$, and will therefore fall between regular display points. This problem should not be neglected, as a misfit in centering the sideband can cause severe artefacts in the reconstructed image amplitude, as an artificial signal is created that can reach up to 10% of the actual amplitude signal [17,18]. The effect in the reconstructed image phase is rather simple: it adds a linear wedge to the image phase. An example of the effect of such a phase-wedge is displayed in Figure 3. This wedge-effect can be compensated in a simple manner after the reconstruction process by interactively adding the inverse wedge. In order to automate the compensation process, however, additional processing steps are involved that can become complex and time consuming.

The artefacts that appear both in the reconstructed amplitude and phase due to a deviation from perfect centering of the sideband can be minimized by using the Extended Fourier transform (**EFT**) [17–19]:

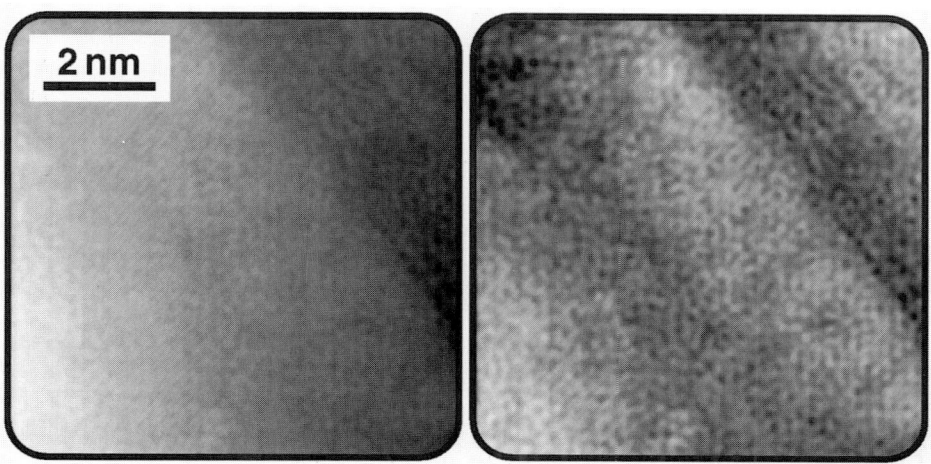

Figure 3. The fully automated reconstruction of a hologram (left) of nanotubes yields a distracting wedge in the image phase, whereas if the **EFT** algorithm is used for reconstruction the wedge is not present (right).

$$\mathbf{EFT}\{p_{mn}\} = \sum_{m,n=1}^{N} p_{mn} \, e^{-2\pi i((k+\Delta k)m+(l+\Delta l)n)/(N\tau)} = t_{k'l'} \qquad (7)$$

$$\stackrel{\tau=1}{=} \sum_{m,n=1}^{N} p_{mn} \, e^{-2\pi i(\Delta k\, m + \Delta l\, n)/N} \, e^{-2\pi i(k\, m + l\, n)/N} = \mathbf{FT}\{p_{mn} \, e^{-2\pi i(\Delta k\, m + \Delta l\, n)/N}\}$$

for reconstruction instead of the conventional Fourier transform **FT**, with $k', l' \in [-N/2+1, N/2]$ and $\tau = 1$. The additional terms Δk, Δl in the **EFT** shift all of the conventional pixel positions (k, l) that correspond to the positions $(\frac{k}{Nd}, \frac{l}{Nd})$ in Fourier space of the conventional Fourier transform to the new display points $(k+\Delta k, l+\Delta l)$ that correspond to $(\frac{k+\Delta k}{Nd}, \frac{l+\Delta l}{Nd})$. Therefore, the center pixel at $(0,0)$ in Fourier space now appears at $(\Delta k, \Delta l)$.[3] This causes intensive streaking in Fourier space that will decrease the data integrity of the sideband. A simple remediation of this effect can be achieved by first subtracting the average value of the hologram from itself and then performing the **EFT**.

---

[3] The same effect can be obtained by multiplying the hologram with the complex phase plate $\exp(-2\pi i(\Delta k\, m + \Delta l\, n)/N)$ [29]. The advantage of this method is that the conventional **FT** can be used instead of converting the fast **FT** algorithm to the **EFT** algorithm. The disadvantage of this method is that the speed of the reconstruction process is decreased and the requirements for RAM memory is increased considerably. Also, the memory of a parallel array-processor may become insufficient for a full complex 1024 by 1024 pixel **FFT**. This is true e.g. for the 80 MFlop array processor provided by Gatan.

So far we have assumed that the exact position of the center of the sideband is known. The objective to find this center is somewhat complicated as it is defined for an undisturbed interference pattern, e.g. one without object and in the absence of image distortions. In the worst case, where electric or magnetic fields contribute significantly to phase changes beyond the specimen area, the best estimation for the position of this center can be obtained only from a reference hologram. In all other cases, the area of the hologram with no object contribution usually contains the least disturbed information on the center of the sideband. A reconstruction program should therefore be sensitive to selected areas in the hologram, which need to be of a size of the power of two for the following processing steps: from the selected area, the position of $\Delta \vec{k}$ and therefore the values for k_0, l_0, Δk and Δl can be estimated automatically. A description of the algorithm itself is beyond the scope of this paper but can be found in [30].

After the position of the sideband has been calculated, the **EFT** algorithm is used to center the sideband correctly. For further processing however, it is of advantage to substitute the values $(\Delta k, \Delta l)$ with the values $k_0 + \Delta k$ and $l_0 + \Delta l$ in the **EFT**. Then, the center of the sideband appears at the center of the **EFT** which not only simplifies the reconstruction procedure but is also a necessary step if sampling rates beyond the Nyquist limit are planned for use, as suggested in [31].

The next step in the reconstruction procedure is the separation of the sideband from the autocorrelation. Although the theory is very clear on this requirement (see for example [8]), the best choice for the size of the aperture is subject to the experience and intention of the operator. This suggests that any reconstruction software should (unless its main purpose is speed) offer the possibility to interactively select the optimum size for the aperture (for different types of apertures, see [32]). An example of an interactive mode as used in the software package HoloWorks is displayed in Figure 4. Here, the size of the aperture can be interactively adapted by using the up/down arrow keys on the keyboard to increase/decrease the size of the aperture. A small side-window gives information on the current radius of the aperture and the theoretical values for the aperture for the general object situation as well as for the weak object approximation.

When the final size of the aperture has been chosen, the remaining part of the reconstruction procedure can be automated completely. In the HoloWorks software, pressing the space bar on the keyboard then displays the reconstructed (complex) image wave, image amplitude and image phase.

2.3. Reconstruction with reference hologram

This reconstruction method has become our most used function for the reconstruction of image amplitude and image phase. As has already been discussed, deviations from the ideal wave front situation in the electron microscope cause artefacts in the reconstructed image phase and image amplitude. These artefacts are preserved in the reference hologram. It is therefore obvious to use a reference hologram to compensate artefacts in the reconstructed image amplitude and phase.

As the fringe spacing is the same both in the reference hologram and the regular hologram, it is sufficient to evaluate the position of the sideband in Fourier space from the reference hologram. Two complex images are obtained from an inverse Fourier transform after the regular and reference hologram are isolated from the autocorrelation, as dis-

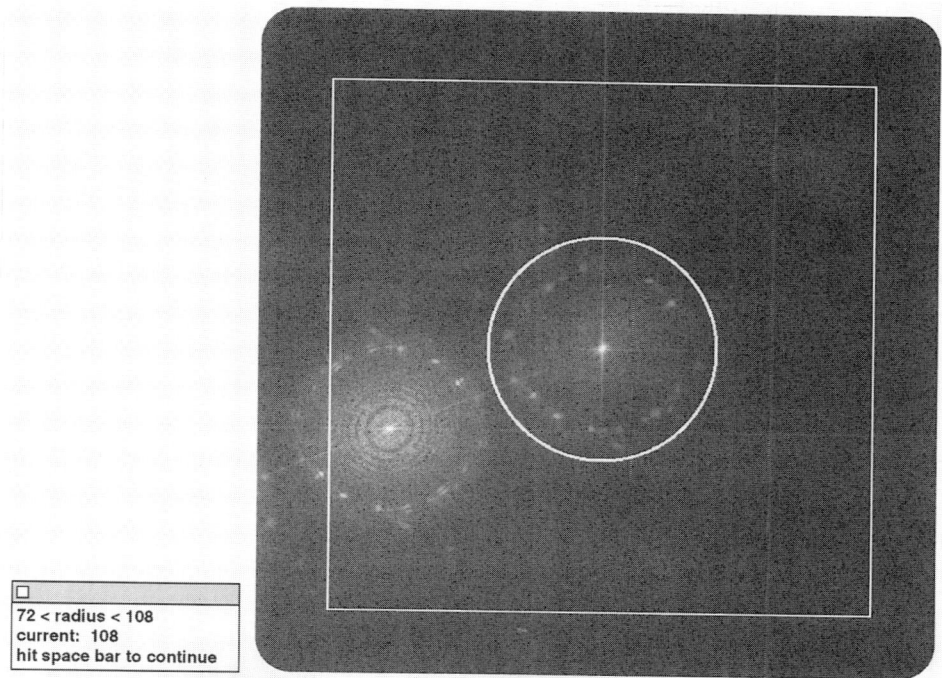

Figure 4. Left: Control-window suggests range for aperture radius and informes on current radius. Right: Center of sideband coincides with center of display. ↑/↓ keys on keyboard allow to adjust the radius of the aperture (white circle). White square outlines size of final images.

cussed in section 2.2. An accuracy of $\pm 1/2$ pixel or even less is suitable for the selection of the center of the sideband in this case, as the effects caused by a sideband not centered perfectly will be counterbalanced between the reference and the conventional hologram in the last processing step.

Dividing the reconstructed complex image by the (reconstructed) complex reference image yields the final complex image wave. In the ideal case, this improved (complex) image wave is free of distortions, flat in the vacuum (unless electric or magnetic fields are present) and free of Fresnel fringes. Unfortunately, the Fresnel fringes do not always disappear, as will be discussed in section 3.1.

For further image processing, e.g. for correction of aberrations, the division of the (complex) reconstructed regular image by the (complex) reconstructed reference image should always be the last step. Both holograms are processed the same way until this final step. A more detailed discussion of this subject however will be published later [33].

2.4. Processing time required

Working on an everyday basis with electron holography, the convenience of the software as well as the performance of the computer will influence the efficiency of the user. We have tested the performance of HoloWorks on three different computer systems. Because HoloWorks is a plug-in module for DigitalMicrograph, it will run at the present only on a Macintosh computer. A parallel processor board is available that runs at 80 MFlops for most Macintosh Quadras but not for the PowerPC. It supports the computation of the **FFT** (both inverse and forward). Although the PowerPC is a very fast computer, for some applications, e.g. the display of live–time diffractograms, the Quadra 840AV with the parallel processor board may still be the best choice. The performance of the PowerPC is very impressive though, as can be seen from Table 1.

Reconstruction of 1k by 1k images	fully automated	with ref. hologram	with EFT algorithm
Quadra 950 +*	20s	27s	2m20s†
Quadra 840 AV	50s	85s	3m30s
PowerPC 8100/80	22s	34s	1m22s

Table 1
Time needed for different reconstruction processes on different computer systems (all Apple Macintosh). *This includes the 80 MFlop array processor available through Gatan.
†Time should be 1m20s. The longer time is caused by insufficient memory on the array processor for a full complex Fourier transform of 1k by 1k images.

3. APPLICATION TO MATERIALS SCIENCE PROBLEMS

In this section we will give a description of some of the applications of off–axis electron holography to materials science problems. The materials science problems will be discussed only briefly, as they are described elsewhere [34,37]. The problems chosen will be used mainly to highlight the advantages and limitations of the present hologram processing programs.

3.1. Investigation of electric charge effects on field emitter tips

Small needles of SiC produced as potential field emitters for flat screen monitor applications were investigated. Some of these needles were coated with a thin layer of Co or Ni to improve their characteristics and were investigated to understand and predict their behaviour. Figure 5, top, shows the phase image of several needles. The needles themselves were too thick to be penetrated by electrons and therefore, within the needles, only artificial noise appears. The vacuum appears remarkably flat and the Fresnel fringes are absent, showing that the reconstruction with a reference hologram was successful. Darker areas that appear around some parts of these needles are a characteristic sign of electric charging. To better image this feature, the image phase was amplified 5 times,

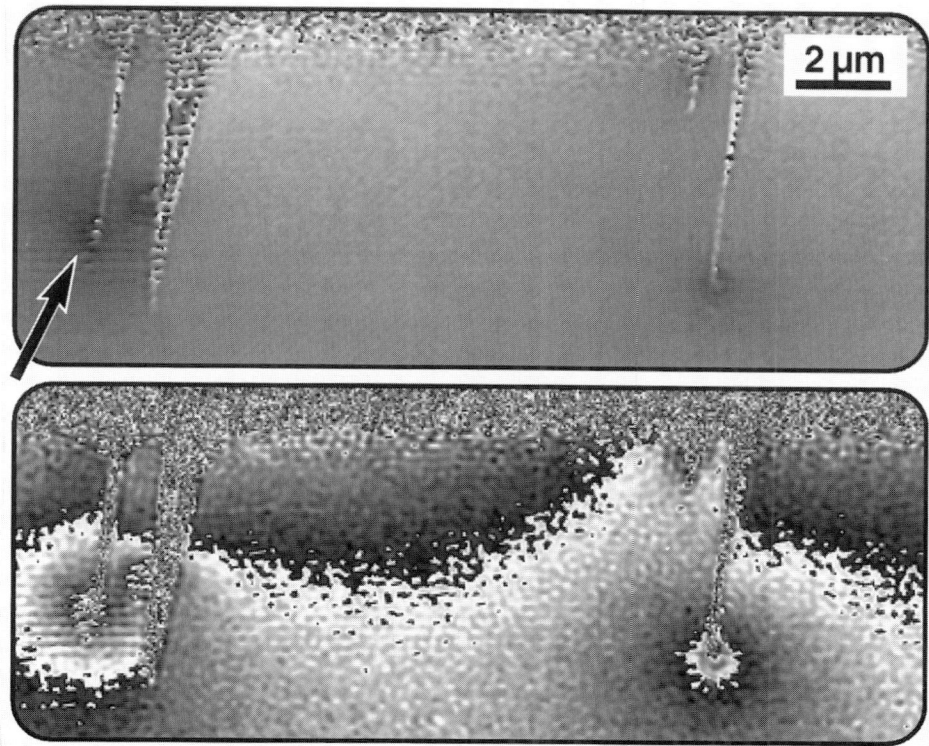

Figure 5. Information on the electric potential distribution around SiC needles as found in the image phase (top) is recognized more clearly from a phase amplified image (bottom). Phase amplification in this image is 5 times.

as displayed in Figure 5, bottom.[4] The phase jumps occuring in this image outline the equipotential lines of the electric field around the needles. Although this process yields no new information, it gives a better display for some of the information contained in the image phase.

A closer examination of the area around the center of one of the charges reveals a structure (see e.g. area with arrow) which is caused by Fresnel fringes that could not be compensated using a reference hologram. Because the biprism is positioned above (or below) an image plane, a phase shift that is present in the object plane can offset the Fresnel fringes of the biprism as well as the interference fringes. Therefore, the use of

---

[4] A phase amplification is 5 means that, the phase jumps generally appearing at $\pm j\pi$ with the integer j, will now appear at $\pm 5 j\pi$.

a reference hologram for reconstruction can actually amplify instead of compensate the occurence of Fresnel fringes. This is especially true for image areas where the gradient of the local phase is high and causes the Fresnel fringes to become a rather disturbing artefact in the evaluation of small details of the phase/amplitude images. A more detailed discussion of this problem can be found in [35,36].

3.2. Morphology of nanometer–sized crystals

The high surface-to-volume ratio of nano-scale ceramic particles allows the surface of these particles to have a strong influence on some of their materials properties. Recent investigations have shown that electron holography can provide direct insight on the morphology of nanocrystals and can even unambiguously characterize internal voids [38]. This is because the mean inner potential of any specimen causes a phase shift of the incident electron beam. For very thin crystals, this phase shift is linearly related to the specimen thickness. Therefore, a thickness profile can, to a reasonable good approximation, be extracted from the image phase of a nanocrystal. To obtain a good quantitative determination of the angles between surface facets, it is however required that variations in the mean inner potential with crystallographic orientation be considered [39,40].

A demonstration of the information obtainable from the image phase is displayed in Figure 6. The regular high resolution image, re-calculated from the hologram (and using a reference hologram), is displayed in the top/left part of this image. In the center part of that image, a nano-scale crystal shows clear facets at its borders. Due to the darker contrast, most of the other specimen areas appear to be thicker than the center crystal. The image phase as reconstructed from the hologram is displayed on the top/right of Figure 6. It is obvious that the darker contrast in the upper part of the image has been caused by diffraction contrast; it has apparently the same thickness as the center particle, as indicated by the equivalent gray value in the reconstructed phase. Due to the nearly constant gray level of the center particle, its overall thickness appears to be nearly constant. A much improved display of this information can be obtained from a three dimensional view of the image phase. The lower part of Figure 6 displays a stereo pair of the three dimensional display of the image phase, after a slight smoothing, and as calculated using [41]. Care however must be taken with the interpretation of such images. This is not a three-dimensional display of the particles themselves, but their thickness profile. For example, a perfect sphere will appear as a half sphere in a thickness profile. Despite this precaution, a phase profile, together with additional information on possible angles of interfaces can already be sufficient to uniquely determine the three-dimensional shape of nano-scale particles without further specimen tilt.

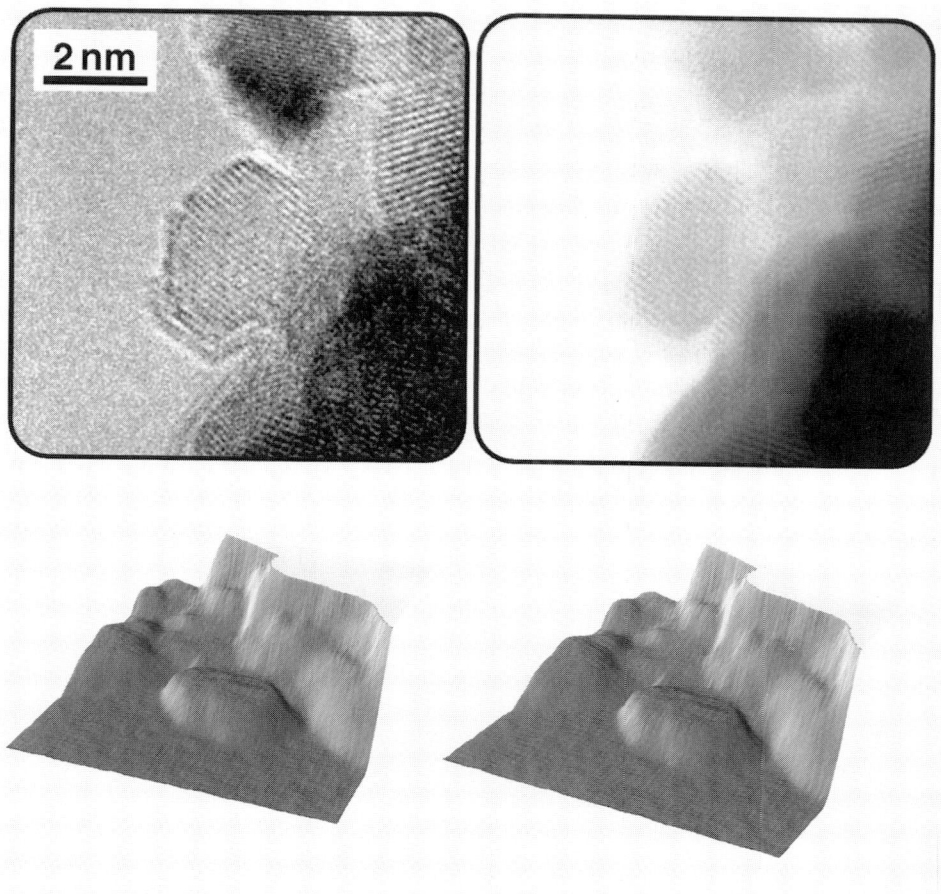

Figure 6. Investigating a small nm-sized particle of ZrO_2 (center area). Top left: The regular high resolution image. Top right: The image phase after removing phase-jumps. Bottom: Stereo-pair as prepaired from the image-phase stressing information on morphology.

REFERENCES

1. D. Gabor, Nature 161(1948),777.
2. J. Cowley, Ultramicroscopy, 41(1992)335.
3. G. Möllenstedt and H. Düker, Z. Physik, 145(1956)377.
4. E.N. Leith, J. Upatnieks, J. Opt. Soc. Am., 52(1962)1123.
5. F.J. Franke et al., Scanning Microscopy Supplement 2(1988)59.
6. J. Chen et al., Opt. Lett. 18(1993)1887.
7. M. Lehmann et al., Ultramicroscopy 54(1994)335-344.
8. E. Völkl and H. Lichte, Ultramicroscopy, 32(1990)177.
9. A. Tonomura et al., Jpn. J. Appl. Phys. 18(1979)1373.
10. A. Tonomura, Springer–Verlag Berlin Heidelberg, 1993.
11. H. Lichte, Ultramicroscopy 20(1986)293.
12. Q. Fu et al., Phys. Rev. Letters Vol. 67, No. 17(1991)2391
13. K. Ishizuka et al., Ultramicroscopy 55(1994)197
14. L. Reimer, *Transmission Electron Microscopy*, (Springer Verlag Berlin, 1989), p224
15. U. Weierstall M.S. thesis, Tübingen, 1989.
16. H. Lichte, Ultramicroscopy 38(1991)13.
17. E. Völkl et al., Ultramicroscopy, in press
18. E. Völkl and L.F. Allard, MSA Bulletin, Vol.24, No.2 (1994) 466-471
19. E. Völkl, L.F. Allard, Proceedings MSA(1994), San Francisco Press, 918
20. E. Völkl et al., (1994) Journal of Microscopy, in press
21. DigitalMicrograph, available through Gatan [28].
22. W.D. Rau et al., J. Comput. Assist. Microsc. 3(1991)51.
23. P.E. Mooney et al., Proc. XIIth Int. Cong. Elec. Micros.,San Francisco, (1990)164.
24. I. Daberkov et al., Ultramicroscopy 38(1991)215.
25. P.E. Mooney et al., Proc. 51st Ann. EMSA meet., (1993)262
26. W.J. deRuijter and J.K. Weiss, Rev. Sci. Instrum., 63(1992)4314.
27. E. Völkl et al., Ultramicroscopy 55(1994)75.
28. Gatan Inc., Owens Drive, Pleasanton, CA 94588, USA
29. M. Lehmann, private communications.
30. W.J. deRuijter et al., Scanning Microsc. Supp. 6(1992)347.
31. K. Ishizuka, Ultramicroscopy 53(1994)297
32. G. Ade, Optik 62(1982)67.
33. E. Völkl et al., to be published.
34. J. Hren et al., proceedings of MRS(1994).
35. B. Frost et al., this proceeding.
36. B. Frost et al., to be published.
37. A. Carim et al., this proceeding.
38. L.F. Allard et al., J. Mater. Sci., 29 (1994) 5612-5614
39. M. McCartney and M. Gajdardziska-Josifovska, Ultramicroscopy, 53(1994)283
40. M. Gaijdardziska-Josifovska et al., Ultramicroscopy, 50(1993)285
41. Mathematica, Wolfram Research Inc., 100 Trade Center Dr., Champaign, IL 61820, USA.

RETRIEVAL OF ATOMIC DISPLACEMENTS FROM RECONSTRUCTED ELECTRON WAVES AS AN ILL-POSED INVERSE PROBLEM

Kurt Scheerschmidt and Frank Knoll

Max Planck Institute of Microstructure Physics, Weinberg 2, D-06120 Halle, Germany

The imaging of crystal defects by using high-resolution transmission electron microscopy or the electron diffraction contrast technique is well known and routinely applied. A direct and phenomenological analysis of electron micrographs, however, is mostly not possible, thus requiring the application of image simulation and matching techniques (see, e.g. [1]). On the other hand, electron holography and other wave reconstruction techniques allow one to directly determine the scattered wave function at the exit surface of an object up to the information limit of the electron microscope (see, e.g. [2,3]). Applying such a wave reconstruction should enable an object retrieval, i.e. the determination of the object potential or the positions of the atomic scattering centres directly from the wave function reconstructed instead of using trial-and-error simulation techniques. Up to now, direct solutions have been given for very thin objects (phase grating approximation, [4]) or for the assumption that the crystal potential of the perfect structure is known and solely the atomic displacements owing to a crystal lattice defect should be determined using the dependence of the three components in the case of plane strains or stresses [5]. Based on the knowledge of the reconstructed complex electron wave and using a discretized form of the diffraction equations, an alternative method is developed [6], which, in principle, enables the direct retrieval of the atomic displacements, caused by a crystal lattice defect, relative to the atom positions of the perfect lattice. A special inverse problem of electron scattering can be deduced considering solely those atomic displacements, which are given by the zeros of a function with an incompletely known Fourier spectrum. The fundamental relations are described, with the problems of solving the ill-posed Fourier transform being discussed.

1. Wave Reconstruction and Electron Diffraction

The electron microscope imaging is mainly determined by two processes: First, by the electron diffraction owing to the interaction process of the electron beam with the almost periodic potential of the matter and, second, by the interference of the plane waves leaving the specimen and being transferred by the microscope. Images are modelled by calculating both processes, they are fitted to the experiment by varying the defect model and the free parameters. This trial-and-error image matching technique is the indirect solution to the scattering problem applied to analyse the defect nature under investigation.

The principles of image formation in the electron microscope have been well established: According to the lens aberrations and the microscope instabilities the higher spatial frequencies are transferred by alternating phase shifts and increased damping, respectively. In addition, owing to the recording of solely the image intensity modulus and phase of the electron wave are always mixed. Holography with electrons offers one of the possibilities of increasing the resolution by avoiding the microscope aberrations. It also enables the complete complex object wave to be restored. Image plane off-axis holograms are recorded in a microscope which is equipped with a Möllenstedt-type electron biprism inserted between the back focal plane and the intermediate image plane of the objective lens [2,7,8]. The object is arranged so that both the reference wave and the object wave are transferred through the microscope, and owing to a positive voltage of the biprism both waves mutually overlap in the image plane creating additional interference fringes. The intensity of the latter is modulated by the modulus of the object wave, whereas their position is varied by the phase of the object wave. Thus the recorded interference pattern is an electron hologram from which both modulus and phase of the object wave can be reconstructed by optical diffraction or numerical reconstruction. A Fourier transform of the intensity distribution of the hologram generates three distinct spectral patterns if the carrier frequency is sufficiently high. In the central region of the spectrum zero peak and autocorrelation occur, representing the conventional diffractogram of the object intensity, completely identical with that obtained from a corresponding HREM micrograph. The sidebands represent the Fourier spectrum of the complete complex image wave and its conjugate, respectively, from which the object wave o(x,y) can thus be reconstructed by separating, centring, and applying the inverse Fourier transform with the complex conjugate phase filter because of the always linear transfer to the sideband.

The interaction of electrons with a crystalline object is described assuming a periodic potential with the electron structure factors as the expansion coefficients and the Bloch-wave method of solving the high-energy transmision electron diffraction. Different formulations can be given, using Bloch wave or plane wave representations of the scattered waves, applying direct or reciprocal space expansion and using direct integration or slice techniques, which, in principle, are equivalent descriptions [9]. The object wave in terms of modified plane waves with complex amplitudes ϕ_g yields

$$o(\mathbf{R}) = \Sigma_g \phi_g e^{2\pi i((\mathbf{k}+\mathbf{g})\mathbf{R} + s_g t)} \qquad (1)$$

with reflections **g**, excitations s, wave vector **k**, and thickness t of a parallel-sided object. Amplitudes ϕ_g are constant with respect to z in the vacuum outside the object, which means that the plane waves are the stationary solutions to the wave equation. Within the crystal, however, the amplitudes of the modified plane waves ϕ_g are z-dependent according to the Ewald pendel solution as described by the Bloch waves, which are the stationary solution to the periodic potential. Using furthermore the deformable ion approximation a crystal lattice defect can be included by its elastic displacement field as a phase shift of the Fourier spectrum of the crystal potential. The evaluation of the quantum-theoretical scattering problem by the high-energy forward scattering approximation (see, e.g. [10,11]) yields a parabolic differential equation system for the complex amplitudes of the elastically scattered electron waves:

$$\partial \phi_g / \partial z = \{ik_z \nabla^2 - 2(\mathbf{k+g})\nabla\}\phi_g / 2k'_z + i\sigma \Sigma_h V_{g-h} \phi_h e^{i\alpha_{gh}} \qquad (2)$$

where $\sigma = 2\pi me/h^2 k_z k'_z$, $\nabla = (\partial/\partial x, \partial/\partial y, 0)$, $k'_z = k_z + g_z + s_g$ and $\alpha_{gh} = 2\pi[(s_h - s_g)z + (\mathbf{g-h})\mathbf{v}(x,y,z)]$ with the elastic displacement field **v** and the potential $V = V' + iV''$ including the lattice potential V' and the absorption V'' (one electron-optical potential approximation of inelastic scattering).

Boundary and initial conditions have to be applied, too: The linearized high-energy approximation directly fits $\phi_g(\mathbf{R},t)$ at the crystal exit face to $\phi_g(\mathbf{R})$ outside, demanding $|\phi_g(\mathbf{R},0)| = \delta_{g0}$ at the entrance face, whereas the continuity of the derivatives has to be omitted in the linearized case. Instead of the boundary conditions one can assume a periodic continuation for large extended crystal slabs, i.e. $\phi_g(x,y,z) = \phi_g(x+X,y,z)$ and $\phi_g(x,y,z) = \phi_g(x,y+Y,z)$, with slab extensions X,Y tending to infinity.

Fig. 1 shows the modulus (left) and phases (right) of the exit wave function in reciprocal- (Fig. 1a) and real- (Fig.1b) space representation, simulated for a spherical inclusion with the Ashby-Brown displacement field and linear displacements within the defect. The wave function is calculated using the EMS package for multi-slice simulations [12]; the data correspond to a 400 kV microscope and [011]-silicon of high symmetric incidence having a sample thickness of t=8.4nm.

Fig. 2 shows a calculated hologram (hol) assuming a perfect microscope without aberrations (a) and its Fourier transform to demonstrate the reconstruction of the wave function in reciprocal space. Fig 2. (b) demonstrates the intensity (diffraction pattern) and phase of the sideband selected. Furthermore, the corresponding modulus (mod) and phases (pha) of the complete reconstructed sideband are shown in (c), which should be equivalent to the exit wave function. The holograms are generated assuming a reference beam with a damping of 0.2 and a carrier frequency of 13.2 nm^{-1} (i.e. located approximately at 1/10(-44,43,-43) in diffraction).

Fig. 3 shows the modulus (mod) and phases (pha) of the particular reflections selected of type 000, {200}, {022} and {111}, thus demonstrating the reconstruction of the corresponding amplitudes ϕ_g from the holograms. The reconstruction of the {400} reflections is impossible here because of the overlap of the autocorrelation and the sideband. Thus, aperture and damping are chosen to exclude the {113} reflections, which also omits the dumbbells in the HREM reconstruction resulting in differences between the original HREM images and the corresponding reconstructions.

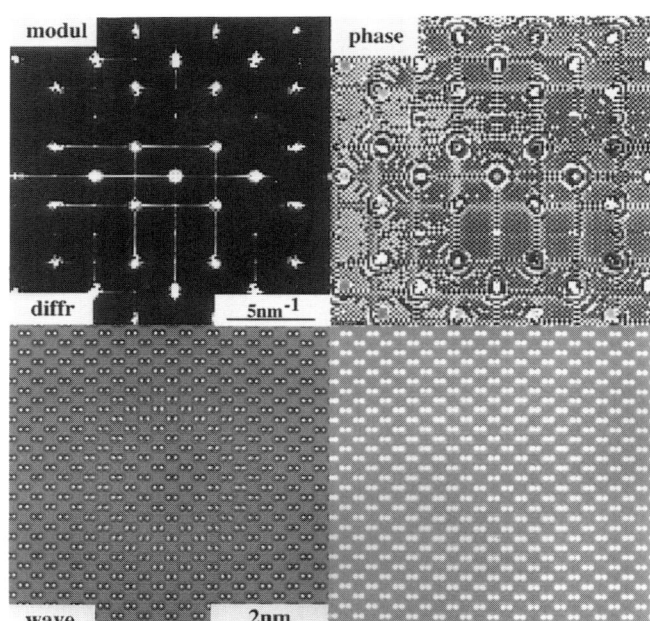

Fig. 1: Modulus (left) and phases (right) of the exit wave function simulated for a spherical inclusion:
a) reciprocal-space (diffraction), b) real-space representation (U=400kV, sample thickness t=8.4nm, sphere radius R_0= 1.6nm $(12,8\sqrt{2},\sqrt{2}/2)$-[011]-Si-supercell)

2. Forward-Backward Iteration

The differential equations (2) allow the diffusion-like interpretation and can be discretized using standard difference algorithms [14]. With the help of

$$\nabla^2\phi=[\phi(x+\Delta x,y,z)-2\phi(x,y,z)+\phi(x-\Delta x,y,z)]/\Delta x^2+[\phi(x,y+\Delta y,z)-2\phi(x,y,z)+\phi(x,y-\Delta y,z)]/\Delta y^2$$
$$\partial\phi/\partial z = [\epsilon(\phi(x,y,z+\Delta z)-\phi(x,y,z))+(1-\epsilon)(\phi(x,y,z)-\phi(x,y,z-\Delta z))]/\Delta z \quad (3)$$

an algebraic equation system results for the complex amplitudes and the elastic displacements at the (xyz)-grid points (i,j,k), (i±1,j,k), (i,j±1,k), and (i,j,k±1). Using the abbreviations

$$A^{\pm}=\pm\sigma k_{z}z_{0}I^{2}/(x_{0}^{2}K)+2\pi(k_{x}+g_{x})(\pm\varepsilon_{x}-1/2),\tag{4}$$
$$B^{\pm}=\pm\sigma k_{z}z_{0}J^{2}/(y_{0}^{2}K)+2\pi(k_{y}+g_{y})(\pm\varepsilon_{y}-1/2),$$
$$C=A^{-}-A^{+}+B^{-}-B^{+}+1/2-\varepsilon_{z}, \quad \alpha(i,j,k)=2\pi[(s_{h}-s_{g})kz_{0}/K+(\mathbf{g}-\mathbf{h})\mathbf{v}(i,j,k)]$$

and denoting the maximum number of grid nodes in x,y, and z-direction by I,J,K yields

$$(1/2-\varepsilon_{z})\phi_{g}(i,j,k-1)-(1/2+\varepsilon_{z})\phi_{g}(i,j,k+1)=A^{+}\phi_{g}(i+1,j,k)-A^{-}\phi_{g}(i-1,j,k)$$
$$+B^{+}\phi_{g}(i,j+1,k)-B^{-}\phi_{g}(i,j-1,k)+C\phi_{g}(i,j,k)-i\sigma(z_{0}/K)\Sigma_{h}V_{g-h}\phi_{h}(i,j,k)e^{i\alpha(i,j,k)} \tag{5}$$

which is equivalent to forward (k+1) and backward (k-1) integration with respect to the beam propagation, i.e. to ε_{z}= 1/2 or -1/2, respectively.

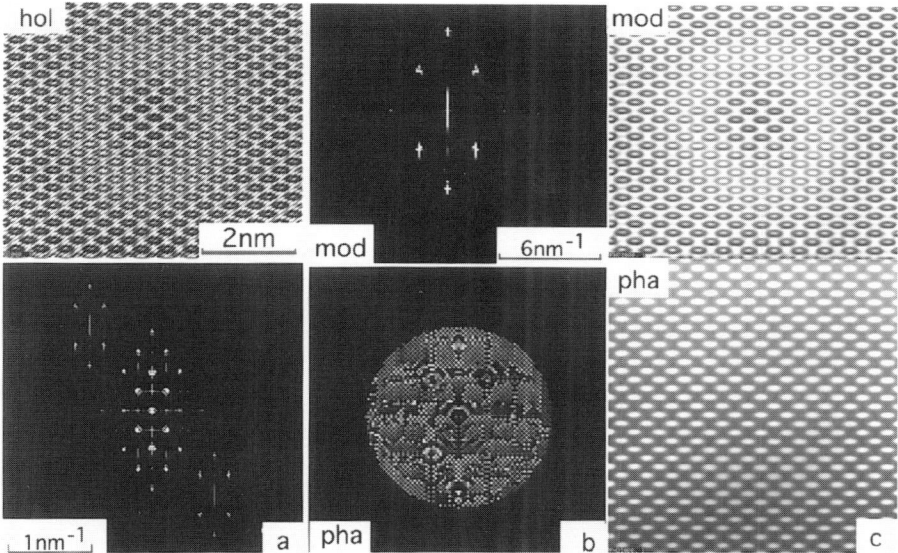

Fig.2: Calculated hologram (hol) and its Fourier transform (a; diffraction pattern with autocorrelation and sidebands) assuming a perfect microscope without aberrations for a spherical inclusion (U=400kV, α=20nm^{-1}, (12,8$\sqrt{2}$,$\sqrt{2}$/2)-[011]-Si-supercell, t=8.4nm, R$_0$= 1.6nm), the selected sideband with intensity and phase (b), and its complete reconstruction (c) with modulus (mod) and phases (pha) representing the perfect exit wave function (reconstructed HREM-image). The hologram is generated assuming a reference beam with a damping of 0.2 and a carrier frequency of 13.2 nm^{-1}.

The periodic boundary conditions and the initial conditions may simply be written $\phi_{g}(i,j,k)=\phi_{g}(i+I,j,k)$, $\phi_{g}(i,j,k)=\phi_{g}(i,j+J,k)$, and $|\phi_{g}(i,j,0)|=\delta_{0g}$, $\phi_{g}(i,j,t)=F_{g}(i,j)$, respectively, with F_{g} being known from the wave reconstruction for a certain number of reflections.

The difference equations (5) are equivalent for backward (k-1) and forward (k+1) integration, thus being insufficient for determining both the wave amplitudes $\phi(i,j,k)$ and the elastic displacement field $\mathbf{v}(i,j,k)$ at the grid points (i,j,k) considered. One of the difference equations, however, can be replaced as follows: While the optical potential in reciprocal space representation is generally non-Hermitian, the hermiticity of potential V' and of „absorption" V" yields the equation of continuity for the whole current $I=\phi_g\phi^*_g$. The continuity equation can be written as

$$\partial I/\partial z = \vartheta I - 2\Sigma_g V"_{gh}\phi_g\phi^*_h e^{i\alpha_{gh}} \tag{6}$$

yielding the abbreviation $\vartheta I = \Sigma_g [k_z(\phi_g\nabla^2\phi^*_g - \phi^*_g\nabla^2\phi_g) + 2(\mathbf{k}+\mathbf{g})\nabla(\phi_g\phi^*_g)]/k'_z$.

The equation of continuity can be discretized by analogy with the discretization of the differential equations above. The differential operator ϑ, however, yields mixed terms with respect to different nodes (i,j,k) and (i±1,j±1,k). By analogy with the Gelfand-Levitan-Algorithm (see, e.g., [13]) an additional equation results, which is a kind of completeness relation, yielding

$$\Sigma_g Q_g e^{2\pi i \mathbf{g}\mathbf{v}} = 0. \tag{7}$$

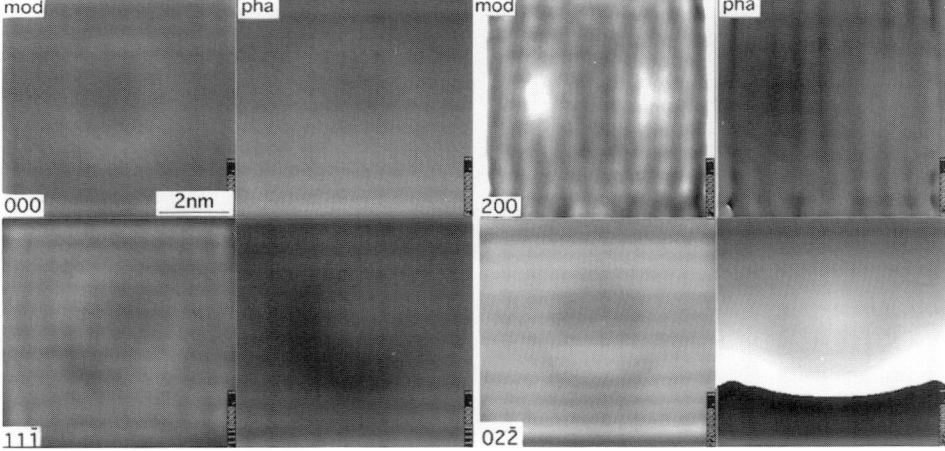

Fig. 3 Modulus (mod) and phases (pha) of the single reflections 000, {200}, {220}, and {111}, i.e. the corresponding plane wave amplitudes, reconstructed from sideband (b) of the hologram (a) of Fig. 2 by filtering and centring of the corresponding reflections.

The coefficients $Q_\mathbf{g}=\Sigma_h k_z/k'_z V"_{g,g-h}\phi_g\phi^*_{g-h}\exp[2\pi i(s_g-s_{g-h})z]$ for $\mathbf{g}\neq(000)$, and correspondingly $Q_{000}=-\vartheta I$ are given for the nodes (i,j,k) from eqs. (6) in forward scattering.

Eq.(7) can replace one of eqs.(5) in backward integration enabling the determination of displacements **v** at (i,j,k) by inverting the equation of continuity as an independent additional equation. Thus, in principle, the retrieval of the displacement is given by the remaining inverse problem (7), which is the same as to find the root of a function given by an incomplete Fourier transform.

At the exit surface a further equation is given applying the forward integration outside the crystal to determine ϕ(i,j,K+1) from ϕ(i,j,K) where the potential is assumed to be vanishing because of the vacuum propagation. The backward integration, however, using eq. (6) then enables the determination of **v**(i,j,K) at the exit surface.

3. The Remaining Inverse Problem

The inverse problem (7) is ill-posed for two reasons: Only one equation has to be solved for the vectorial root **v**(i,j,k) at node (i,j,k), which describes three unknown quantities by two conditions, and spectrum $Q_\mathbf{g}$(i,j,k) is incomplete and noisy. This results in unstable numerical solutions using standard algorithms to find the roots, owing to the existence of a large number of subsidiary roots. Different algorithms are tested, viz. the Newton-Raphson algorithm itself to solve eq. (7), and of transform eq.(7) in an iterative form as a kind of quasi-regularization, e.g. using relations for the arguments yielding

$$v_x^{n+1} = 1/2\pi \{\arg[Q\exp(2\pi i v_x^n) + \Sigma_\mathbf{g} Q_\mathbf{g} \exp(2\pi i \mathbf{g} \mathbf{v}^n)] - \arg[Q]\} \tag{8}$$

and similarly for u_y, u_z. Both algorithms demand the iteration for linear independent coefficients **g**, thus coplanar vectors **g** leave one component unconsidered.

Analytical solutions of eq.(7) can be performed if four terms at a maximum are considered, which are revealed most easily by interpreting equation (7) in the complex plane as the summation of rotating vectors $Q_\mathbf{g}$ as a function of root **v**(i,j,k) at node (i,j,k).

For non-vanishing Q_{000}, $Q_{\lambda 00}$, for instance, the system is over-determined demanding $|Q_{000}| = |Q_{\lambda 00}|$ and resulting in $v_x = 1/2\pi\lambda \{\arg(Q_{000}) - \arg(Q_{\lambda 00}) + (2k+1)\pi\}$.

Otherwise, for non-vanishing Q_{000}, $Q_{\lambda 00}$, $Q_{0\mu 0}$, and with the other coefficients being neglected, one yields

$$v_x = 1/2\pi\lambda \{\arg(Q_{000}) - \arg(Q_{\lambda 00}) + \arccos[(|Q_{\lambda 00}|^2 - |Q_{0\mu 0}|^2 - |Q_{000}|^2)/2|Q_{000} Q_{\lambda 00}|]\} \tag{9}$$

similarly also for u_y, with u_z being arbitrary, however. Thus, considering solely three terms provides an exact solution to the problem if $||Q_{0\mu 0}| - |Q_{\lambda 00}|| \leq |Q_{000}| \leq |Q_{0\mu 0}| + |Q_{\lambda 00}|$.

The solution based on four non-vanishing terms, in principle has the same structure, enabling the determination of the component v_z, too, but having to fulfil further restrictions.

The system, however, is under-determined, resulting in free parameters as, e.g. for
$\arccos\{[|Q_{000}|^2+|Q_{\lambda 00}|^2-(|Q_{0\mu 0}|-|Q_{00\eta}|)^2]/2|Q_{000}Q_{\lambda 00}|\} \le \arg(Q\lambda 00) \le$
$\arccos\{[|Q_{000}|^2+|Q_{\lambda 00}|^2-(|Q_{0\mu 0}|+|Q_{00\eta}|)^2]/2|Q_{000}Q_{\lambda 00}|\}$.

Using additional terms of (7) restricts the free parameters as to provide solutions probably unique as they have components that depend on the linearly independent terms v_x, v_y and v_z.

4. Conclusions

The retrieval of the atomic displacements from a reconstructed electron wave function at the exit surface of an object results in the algebraic equation system (5) and the particular inverse problem (7) with the difficulties of finding the roots as discussed. The procedure described has thus transformed the difficulties of solving the direct scattering problem to the mathematical problem of determining the roots of a function with an incomplete Fourier transform. Furthermore, there are certain restrictions on the existence of solutions that would enable the construction of numerical algorithms as, e.g., generic ones. From the mathematical point of view the retrieval procedure is an ill-posed inverse problem requiring additional information about the unknown reconstructed displacements in order to make the process stable and continuous, to avoid singularities, and to restrict the manifold set of solutions possible. Open questions arise, e.g., with respect to the assumptions of cyclic boundary conditions, the applicability of the completeness relation to the backward iteration and to depths, where the equations for the displacement retrieval cannot be inverted because of singular coefficients.

References

[1] K. Scheerschmidt, R. Hillebrand: Image interpretation in HREM: Direct and indirect methods, Proc. 32nd Course Int. Centre of Electron Microscopy "High-Resolution Electron Microscopy - Fundamentals and applications", Halle 1991, p. 56
[2] H. Lichte, Ultramicroscopy **20** (1986) 293
[3] W. Coene, G. Janssen, M. Op de Beeck, D. van Dyck, Phys. Rev. Letters **69** (1992) 3743
[4] D. van Dyck, W. Coene, Scanning Microsc., Suppl. **2** (1988) 131
[5] A.K.Head, Aust. J. Phys **22** (1969) 43
[6] K. Scheerschmidt, F. Knoll, ICEM-13, Paris 1994, Vol.1, p. 333; and Phys. Stat. Sol. (1994), in print
[7] H. Lichte, Advances in optical and electron microscopy **12** (1991) 25
[8] H. Lichte, E. Völkl, K. Scheerschmidt, Ultramicroscopy **47** (1992) 231
[9] D. van Dyck, Advances in optical and electron physics **65** (1985) 295
[10] G.R. Anstis, Computer Simulation of Electron Microscope Diffraction and Images, Eds.: W Krakow, M.O' Keefe, The Minerals, Metals & Materials Society, 1989, p. 229
[11] A. Howie, Z.S. Basinski, Philos. Mag. **17** (1962) 1039
[12] P. Stadelmann, Ultramicroscopy **21** (1987) 131
[13] B.N. Zakhariev, A.A. Suzko, Direct and Inverse Problems, Springer Vlg., Bln. 1990.
We are grateful to the Volkswagenstiftung for financial support.

Interference- and Lorentz-Image Simulations of Vortices in Superconductors

G. Pozzi[a], J.E. Bonevich[b,c], K. Harada[b], H. Kasai[b], T. Matsuda[b], T. Yoshida[b], and A. Tonomura[b]

[a]Department of Physics, University of Bologna, via Irnerio 46, I-40126 Bologna, Italy

[b]Advanced Research Laboratory, Hitachi, Ltd., Hatoyama, Saitama 350-03, Japan

[c]Lawrence Berkeley Laboratory, Materials Science Division, MS72-150, 1 Cyclotron Road, Berkeley, CA 94720 USA

Recently superconducting vortices have been observed by means of Lorentz microscopy and by electron holography methods. The interpretation of the main features of the out-of-focus patterns and amplified contour map has been made so far using the London model for describing the topography of the magnetic field around the core. However, especially for the case of Niobium, a better description of the core is advisable. Such a description is given by the Clem model. In this work the predictions of the Clem model for the interpretation of experimental results are presented and compared with those obtained by the London one. Some prospects for discriminating between the two models and for obtaining information about the vortex inner core structure are discussed.

1. INTRODUCTION

Observations of quantized flux lines, or vortices, in thin superconducting specimens have been successfully carried out in transmission electron microscopy by means of standard methods of Lorentz microscopy [1] and also by electron holography [2], see also these proceedings. Although a former theoretical analysis (reviewing also the work done until 1972) carried out for the case of vortices in thin specimens lying perpendicular to the electron beam predicted the feasibility of such experiments [3], no successful attempts have been reported until recently.

One reason for the recent breakthrough in observing vortices is due to the fact that a new geometry has been introduced, namely the specimen is observed tilted with respect to both the vertically incident electron beam and the horizontal ancillary magnetic field used to introduce and stabilize the vortices.

From the waveoptical point of view this case is complicated because the electrons experience both the internal and external field and the resulting phase shift is two-dimensional. Nonetheless it has been shown that when the vortex is modeled by a flux tube of negligible radius, a simple analytical solution exists for the field and for the phase shift suffered by the electron beam [4]. This elementary solution can then be used as a building block to investigate more realistic models; e.g. taking into account the vortex core structure by approximating it with a suitable arrangement of flux tubes.

It has been shown that the most relevant features of the experimental results obtained by the out-of-focus method (Lorentz microscopy) [5] and by electron holography [2] can be interpreted by modeling the magnetic field around the vortex core by means of the London model, corresponding to a flux line having an infinitely small normal core [6]. In addition to being described by an analytical expression, this model has the advantage that a single parameter, the London penetration depth, completely characterizes the superconducting vortex.

However, Clem has proposed another more realistic model for the vortex core that removes the unphysical limitation of the infinitely small normal core and can also be described by an analytical expression depending on two parameters: the London penetration depth λ_L and the core radius ξ_v, which is of the same order of magnitude as the coherence length ξ [6].

The aim of this work is to present the results of simulations of interference [7] and Lorentz [8] images of vortices in superconductors using the Clem model and to compare them with the predictions of the London one. Also some prospects for discriminating between the two models, thus obtaining some insight on the vortex inner core structure, are discussed.

2. THE FLUX-TUBE MODEL

In the standard high energy or phase-object approximation, only the phase distribution of the incoming electron wavefunction is modified when electrons pass through a magnetic specimen [9]. That is, a vortex is a perfect phase object and the magnetic phase shift is given by;

$$\phi = (-2\pi e/h) \int_L \mathbf{A} d\mathbf{s} \tag{1}$$

where the integral is taken along a straight line L corresponding to the classical electron trajectory, $\mathbf{A}$ is the magnetic vector potential and e and h the absolute value of the electron charge and Planck constant, respectively.

It is worthwhile to point out that this two-dimensional phase distribution, completely describing the interaction between the vortex and the electron beam, is defined on a plane, the observation plane, perpendicular to the electron beam and whose position along the z axis is not rigorously determined, but is taken coincident with the specimen plane.

The difference of the phase shift between two classical trajectories L_1 and L_2, passing through points 1 and 2 in the object plane, can be expressed by Stoke's theorem as;

$$\delta\phi = (-2\pi e/h) \int_{S_{12}} B_n dS \tag{2}$$

Since B_n is the component of the magnetic flux density normal to the surface S_{12} enclosed by L_1 and L_2, the integral stands for the magnetic flux passing through S_{12}. This second form of the phase shift has been used in reference [4] to calculate analytically the phase shift for a flux tube of negligible radius carrying the quantum flux $\Phi = h/2e$ lying perpendicular to the two surfaces of an ideal superconducting specimen of thickness t, inclined at an angle α with respect to the optic axis z, Figure 1 (a).

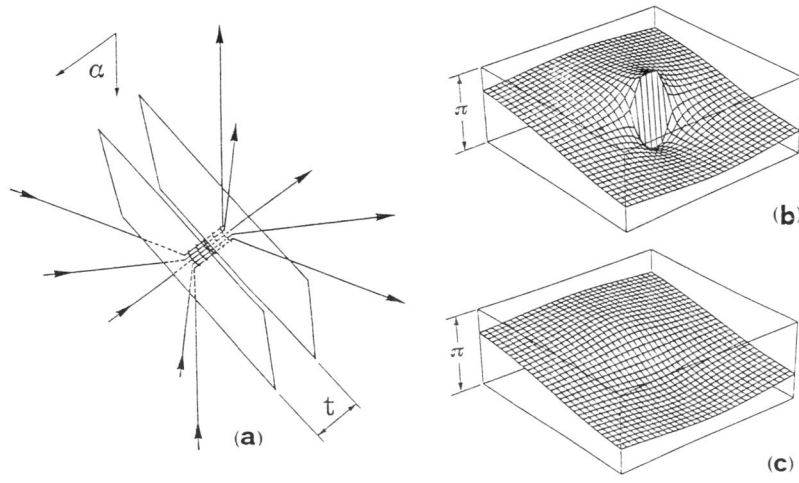

Figure 1. (a) The experimental set-up: a specimen of thickness t containing a single superconducting vortex is inclined by an angle α to the optical axis. (b) Corresponding phase shift of the flux tube, in a specimen of thickness $t = 2\lambda_L$, $\alpha = 45°$. (c) Phase shift for the Clem model with $\xi_v = 1.33\lambda_L$.

In this case the magnetic field in the upper and lower half-spaces is equivalent to that produced by two magnetic poles of strength $\pm\Phi$ respectively, whereas in the perfectly diamagnetic specimen the field is everywhere zero except in the flux tube core. The corresponding phase shift for a flux tube in a specimen of thickness $2\lambda_L$, tilted at $\alpha = 45°$ is reported in Figure 1 (b). It should be noted that the maximum phase difference across the flux tube is π, corresponding to the Aharonov-Bohm effect due to the quantized flux $\Phi = h/2e$.

By overlapping a suitable distribution of flux tubes we can approximately model any flux core structure, in the same spirit as in electrostatics where a given field can be approximated by a suitable arrangement of point charges. Figure 1 (c) reports the phase shift for the Clem model with $\xi_v = 1.33\lambda_L$. In this case, for a specimen of finite thickness, the broadening due to to the finite core severely dampens the maximum phase difference and smoothes the slope at the core, with the results that the phase difference no longer reaches the value of π corresponding to the flux quantization, but drops to values slightly above $\pi/2$. At large distances from the core, the phase difference becomes $\pi/2$ irrespective of the core structure. Geometrical reasoning [2] shows that this phase difference (which in the more general case of tilt by an angle α is given by α itself) is still present even if the distance between the two pole vanishes, which corresponds to a specimen whose thickness is much lower than λ_L. This is an unexpected effect arising from the unusual set-up and confirms that the main contribution to the phase shift is due to the external fringing field extending in the two half-spaces above an below the perfectly diamagnetic superconducting film.

3. LONDON VS CLEM MODELS FOR THE VORTEX CORE

Two models for the vortex magnetic core are compared in this work, the first is the London one, in which the field is described by the following equation [6],

$$B = \frac{\Phi}{2\pi\lambda_L^2} K_0(r/\lambda_L) \qquad (3)$$

where K_0 is the zero-order Bessel function, and r the radial distance from the vortex axis. The second is the Clem model, whose magnetic field distribution is given by [6]

$$B = \frac{\Phi}{2\pi\lambda_L\xi_v} \frac{K_0(R/\lambda_L)}{K_1(\xi_v/\lambda_L)} \qquad (4)$$

where $R^2 = r^2 + \xi_v^2$, and K_1 is the first order Bessel function.

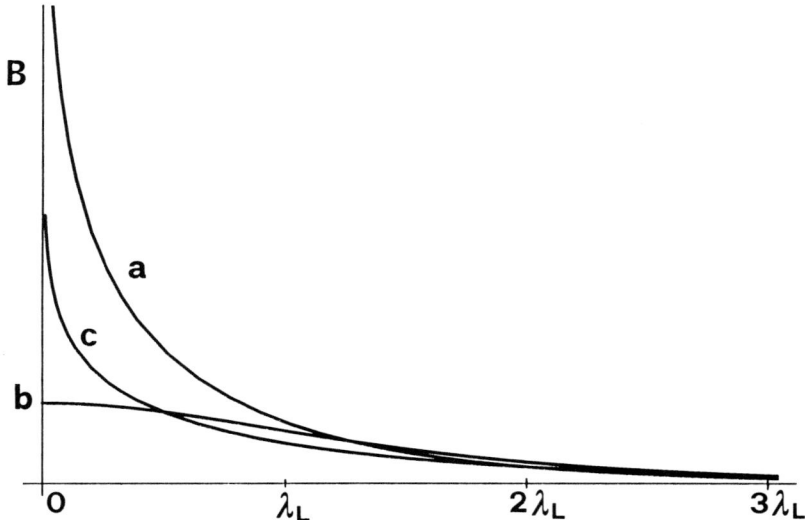

Figure 2. Relative trends of the magnetic field B (arbitrary units) as a function of the radial distance for the London and Clem models. (a) London model; (b) Clem model with $\xi_v = 1.33\lambda_L$; (c) London model with $\lambda'_L = \sqrt{\lambda_L^2 + \xi_v^2}$

It should be noted that the London topography is a function of a single parameter, the London penetration depth λ_L, whereas the Clem one depends on the additional parameter ξ_v which is of the same order of magnitude as the coherence length ξ. The relative trend of the z-component of the magnetic field in the radial direction from the vortex axis is reported in Figure 2 for the London model with penetration depth $\lambda_L = 30$ nm (a) and $\lambda_L = 50$ nm (b), and for the Clem model with $\lambda_L = 30$ nm and $\xi_v = 40$ nm (c). Also the case with $\lambda_L = 30$ nm and $\xi_v = 0.3$ nm (d) has been calculated, and the curve is identical to that corresponding to the London model (a) within the given coordinate ranges, thus confirming the fact that the Clem model converges to the London one for small values of ξ_v, except at the axis, where the London model diverges.

At this point the object phase shift corresponding to a more realistic vortex core model can be calculated numerically by convolving the phase shift due to the flux tube with the two-dimensional projections in the object plane of the London or Clem models.

4. INTERPRETATION OF LORENTZ IMAGES

The principle of the out-of-focus observation is very simple: the specimen, illuminated by a coherent electron beam emitted by a field emission gun, is observed out-of-focus, so that phase contrast is generated in the image.

The image wavefunction is given by the Kirchhoff-Fresnel integral

$$\psi(X,Y,Z_0) = \frac{exp(i\beta)}{\lambda Z_0} \int\int \exp\left\{\frac{i\pi}{\lambda Z_0}[(x-X)^2 + (y-Y)^2 + \phi(x,y,\alpha,t,\lambda_L,\xi_v)]\right\} dx\,dy \quad (5)$$

where x and y are the coordinates in the object plane, X and Y the coordinates in the out-of-focus plane at a defocus distance Z_0 from the object plane, λ the deBroglie wavelength of the incident electrons and β a phase factor of no value in the present case as here only the intensity in the image plane is relevant which is proportional to $|\psi|^2$.

The object phase shift $\phi(x,y,\alpha,t,\lambda_L,\xi_v)$ in eq. 5, has been calculated by means of FFT methods, on a square region centered at the vortex having side 1.2 μm, with 256 x 256 sampling points. FFT methods have been also employed to calculate the image wavefunction in the defocus plane, and a linear phase term has been subtracted from the phase shift in order to avoid Fresnel diffraction effects from the edge parallel to the vortex axis, where the phase has a $\pi/2$ jump owing to the truncation. Test calculations done with doubled side and sampling points show no relevant differences. Figure 3 shows a focal series calculated for the case of a specimen having London length $\lambda_L = 30$ nm (of the same order as that of Nb), $\xi_v = 3$ nm, thickness 60 nm, i.e. 2 λ_L, at the following values of the defocus: (a) 1 mm; (b) 2.5 mm; (c) 5 mm; (d) 10 mm; (e) 15 mm and (f) 20 mm. The accelerating voltage has been taken equal to 300 kV, corresponding to a deBroglie wavelength of 1.968 pm.

It can be ascertained that the calculated image has the appearance of a tiny globule, with two halves of bright and dark intensity aligned along the vortex axis. Also low contrast fringes surrounding the globules are present. Both the contrast and dimension of the globules increase with defocus distance: however if the patterns are scaled proportionally to $\sqrt{\lambda_L}$ it can be ascertained that, apart from the contrast, they look very similar. As in the geometric optical approximation the dimensions of the image, being proportional to the Lorentz deflection at the specimen, should increase linearly with the defocus distance, this means that the image contrast and appearance is mainly a waveoptical effect due to Fresnel diffraction.

No differences can be ascertained between the predictions of the London and Clem model with a narrow core, see f.i. Figure 6 of reference [5].

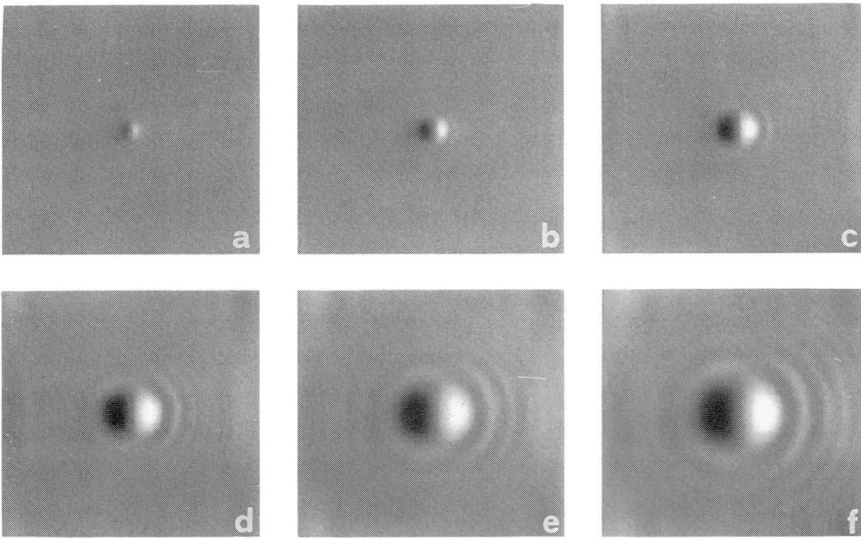

Figure 3. Vortex at different defocus distances, described by the Clem model with narrow core, $\lambda_L = 30$ nm, $\xi_v = 3$ nm, in a specimen of thickness $t = 60$ nm, tilted at $\alpha = 45°$. (a) 1 mm; (b) 2.5 mm; (c) 5 mm; (d) 10 mm; (e) 15 mm and (f) 20 mm.

Figure 4 reports the results calculated for the realistic case of the Clem model with $\lambda_L = 30$ nm and $\xi_v = 40$ nm for the same defocuses as Figure 3.

Comparing Figures 3 and 4 it can be seen that the introduction of a finite coherence length of the same order as the penetration depth has the main effect of lowering the contrast of the fringes surrounding the central globule, leaving the contrast of the globule itself only slightly affected. This effect is shown more clearly in the intensity line scan across the globules reported in Figure 5, for the two defocuses of 10 mm and 20 mm. The line scans at the left refer to the case of large core, whereas those at the right to the case of narrow core, which cannot be distinguished from the corresponding results obtained by the London model.

The conclusion is that it is extremely difficult to discriminate between the London and Clem models on the basis of their out-of-focus patterns; brighter prospects to achieve such discrimination are linked to the electron holography analysis.

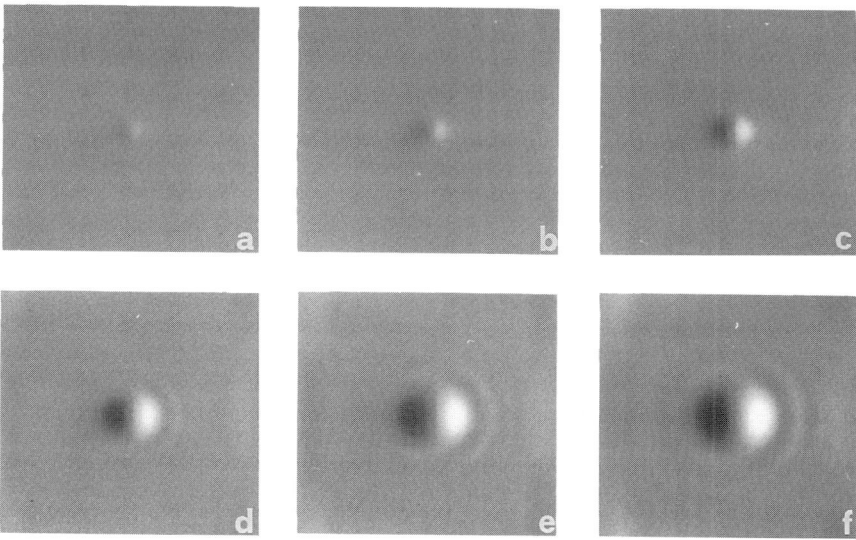

Figure 4. Vortex at different defocus distances, described by the Clem model with large core, $\lambda_L = 30$ nm, $\xi_v = 40$ nm, in a specimen of thickness $t = 60$ nm, tilted at $\alpha = 45°$. (a) 1 mm; (b) 2.5 mm; (c) 5 mm; (d) 10 mm; (e) 15 mm and (f) 20 mm.

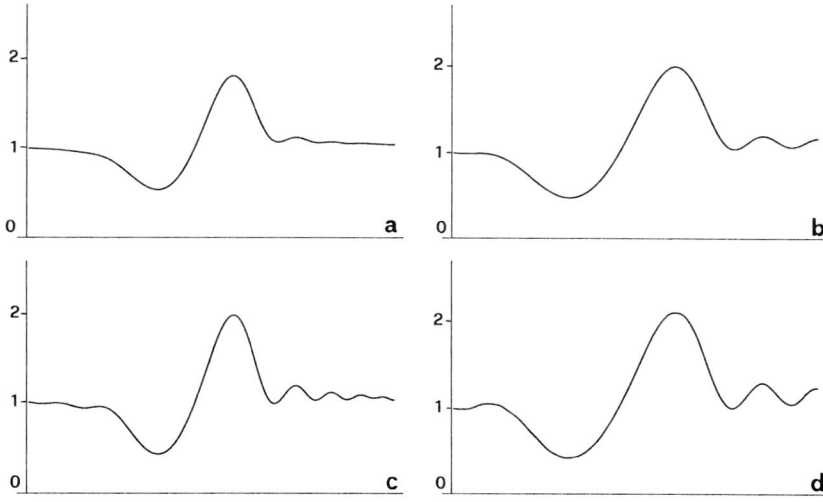

Figure 5. Intensity line scans of (a) Fig. 4 (d) ; (b) Fig. 3 (d); (c) Fig. 4 (f); (d) Fig. 3 (f).

5. INTERPRETATION OF INTERFERENCE IMAGES

The superiority of electron holography with respect to the standard phase-contrast methods grouped under the heading of Lorentz microscopy [10] lies in the fact that at the end of the reconstruction process the true object phase difference is available in numerical or optical form. In this latter case, the results are presented in the form of n-times amplified contour density maps or interference images which, in addition to presenting vividly the information on the phase difference values and topography, have the advantage that they can be directly compared with the theoretical simulations, calculated according the following equation

$$I(x,y) = 1 + \cos\left[n\phi(x,y,\alpha,t,\lambda_L,\xi_v)\right]. \tag{6}$$

The results of the London and Clem models for the case of 16x amplified contour maps are reported in Figure 6.

Figure 6 (a) reports the interference image for the London model, calculated for a penetration depth of 30 nm, in a specimen 60 nm thick tilted at 45° with respect to the electron beam and the applied magnetic field. The comparison with the map calculated from the Clem model, Figure 6 (b), same data as before plus $\xi_v = \xi = 40$ nm, shows that the latter is more broadened and can be distinguished from the first one. However, if the London map is calculated with an increased penetration depth $\lambda'^2 = \lambda_L^2 + \xi^2 = 50$ nm, Figure 2 (b), then no relevant differences are detectable, indicating that discrimination between the models based on the analysis contour maps alone is very difficult without a separate knowledge of the penetration depth.

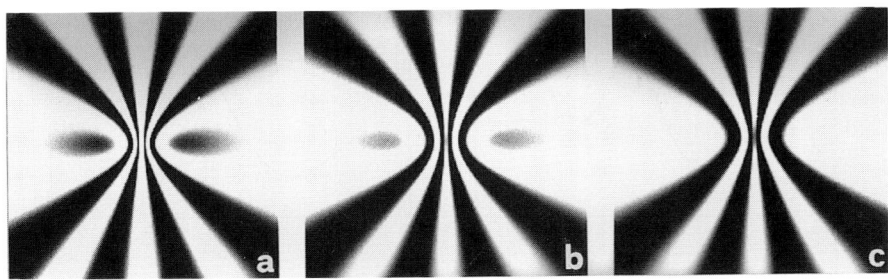

Figure 6. 16 times amplified contour maps for (a) the London model with $\lambda_L = 30$ nm; (b) the Clem model with $\lambda_L = 30$ nm and $\xi_v = 40$ nm; (c) the London model with $\lambda_L = 50$ nm

However, numerical processing of the holograms allows the extraction of the trend of the phase across the vortex with high accuracy, so that it is worthwhile to compare the three cases from this point of view. Figure 7 (a) reports the three phases across the vortex and again differences between the curves are hardly detectable as they are larger, but nonetheless of the same order of magnitude as the experimental error, but if the first derivative of these curves is made, Figure 7 (b), then the Clem model shows a marked difference.

In fact the London model predicts a cusp at the core, due to the negligible coherence length, whereas the Clem model presents a smooth maximum. The differences are even more marked if the second derivative is taken, Figure 7 (c).

It should be remarked that the curves in Figure 7 (b) are very similar to the trend of the field across the core as predicted by the two models, Figure 2, and this similarity is not entirely casual as the relation between field and phase is given by an integral/derivative.

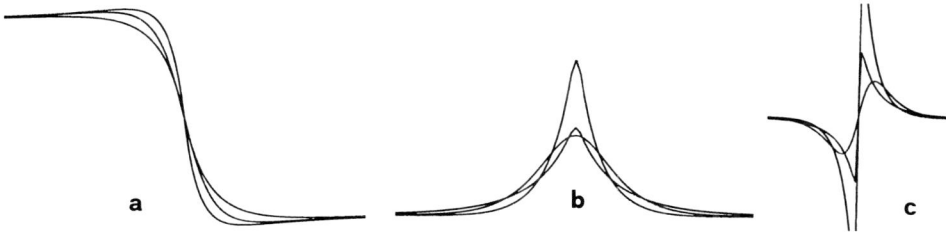

Figure 7.   Trend of the phase (a), and of its first (b) and second (c) derivatives, across the vortex core for the three cases of Fig. 6.

The above results show that, even if the penetration depth is not accurately known from separate measurements, if the experimental contour maps allow reliable extraction of the first and second derivatives from the phase across the core, it should be possible to assess whether the data are better described by the Clem or the London model. Of course, also the effect of the hologram resolution should be considered carefully, because it introduces some broadening of the phase and its derivatives which may mask the sought effect.

6. CONCLUSIONS

The results of out-of-focus image simulations of superconducting vortices by means of the Clem model have been presented and compared with those of London, showing that it is very difficult to discriminate between the two models.

Also the predictions of the two models for the case of holographic amplified contour maps have been compared. The obtained results show that for the same value of the penetration depth the contour map calculated from the Clem model is more broadened and can be distinguished from the London one. However, if the London map is calculated with an increased penetration depth, then no relevant differences are detectable, indicating that discrimination between the models based on the analysis of the contour maps alone is very difficult without a separate knowledge of the penetration depth.

Nonetheless, if the first derivative of the phase across the vortices is taken, then the

Clem model shows a marked difference. In fact the London model predicts a cusp at the core, due to the negligible coherence length, whereas the Clem model presents a round maximum. The differences are even more marked if the second derivative is taken.

In conclusion, even if the penetration depth is not accurately known from separate measurements, if the experimental contour maps allow reliable extraction of the first and second derivatives from the phase across the core, it should be possible to assess whether the data are better described by the Clem or the London model.

Acknowledgements The authors gratefully acknowledge Dr. A. Fukuhara, Dr. U. Kawabe and Dr. T. Onogi of Hitachi and Dr. Q. Ru of the Tonomura Electron Wavefront Project, ERATO-JRDC, for their stimulating discussions and Dr. J. Endo, T. Furutsu, Dr. M. Igarashi, Dr. H. Kajiyama, Dr. S. Kondo, S. Kubota, S. Matsunami, N. Moriya and Dr. N. Osakabe of Hitachi, for their technical assistance. This work has been done within a collaboration scheme of one of the authors (G.P.) with Dr. A. Tonomura and his group at the Hitachi Advanced Research Laboratory, Japan. Financial support from MURST coordinated by Consorzio INFM and CNR-GNSM is acknowledged. Useful discussions with Prof. G. F. Missiroli, Dr. G. Matteucci, Department of Physics, and Dr. A. Migliori, LAMEL-CNR, as well as the skilful technical assistance of S. Patuelli are gratefully acknowledged.

7. REFERENCES

[1] K. Harada, T. Matsuda, J. Bonevich, M. Igarashi, S. Kondo, G. Pozzi, U. Kawabe and A. Tonomura, Nature 360, 51 (1992).
[2] J.E. Bonevich, K. Harada, T. Matsuda, H. Kasai, T. Yoshida, G. Pozzi and A. Tonomura, Phys. Rev. Lett. 70, 2952 (1993).
[3] C. Capiluppi, G. Pozzi and U. Valdrè, Phil. Mag. 26, 865 (1972).
[4] A. Migliori, G. Pozzi and A. Tonomura, Ultramicroscopy 49, 87 (1993).
[5] J.E. Bonevich, K. Harada, H. Kasai, T. Matsuda, T. Yoshida, G. Pozzi and A. Tonomura, Phys. Rev. B 49 (1994) 6800.
[6] R.P. Huebener, Magnetic Flux Structures in Superconductors (Springer, Berlin, 1979).
[7] G. Pozzi, J. E. Bonevich and A. Tonomura, Proc. XIIIth Internat. Congr. on Electron Microsc. (1994), Vol 1., p. 309.
[8] G. Pozzi, J. E. Bonevich and A. Tonomura, Proc. XIIIth Internat. Congr. on Electron Microsc. (1994), Vol 1., p. 339.
[9] A. Fukuhara, K. Shingawa, A. Tonomura and H. Fujiwara, Phys. Rev. B 27, 1839 (1983).
[10] J.N. Chapman, J. Phys. D 17, 623 (1984).

Magnetic field observation of vortices in superconductors by electron holography

J.E. Bonevich[a][b], K. Harada[a], H. Kasai[a], T. Matsuda[a], T. Yoshida[a], G. Pozzi[a][c] and A. Tonomura[a]

[a]Hitachi Advanced Research Laboratory, Hatoyama, Saitama 350-03, Japan

[b]Lawrence Berkeley Laboratory, Materials Science Division, MS72-150, 1 Cyclotron Road, Berkeley, CA 94720 USA

[c]Dept. of Physics, University of Bologna, via Irnerio 46, I-40126 Bologna, Italy

The magnetic flux-line lattice penetrating superconducting Nb thin foils have been investigated by means of electron holography. A field-emission TEM with a specially constructed cold stage was used to cool the Nb down to 4.5 K and apply magnetic fields up to 200 G. The phase distribution maps of electrons transmitted through the specimen were quantitatively measured. Reconstructed interference micrographs revealed tiny regions of rapid phase change that coincided spatially with the spot-like contrast observed by Lorentz microscopy and were found to be quantized vortices containing a flux of $h/2e$. The experimental results agreed well with those predicted by theoretical simulations. Vortex diameters were also observed to broaden with increasing temperature, in agreement with theoretical prediction. Experiments exploring the vortex inner core structure at high spatial resolution are presented.

1. INTRODUCTION

The behavior of the magnetic flux-line lattice (FLL) can illuminate many important fundamental aspects of superconductivity. Although until recently the direct FLL observation has proved experimentally cumbersome. For example, the traditionally used Bitter technique [1,2] images the static FLL magnetic fields that emanate from the surface of a superconductor as a replica and, consequently, is not suitable for dynamic studies. And while neutron diffraction can detect the FLL [3], the information is averaged over the whole specimen. Magneto-optical techniques can sense dynamic FLL behavior, however, individual vortices can not be resolved [4]. Scanning electron microscopy [5], scanning tunneling microscopy [6] and Hall probes [7] have been used to examine the FLL, although only the magnetic fields outside the specimen are detected. Previous attempts [8-12] to observe the FLL by transmission electron microscopy (TEM) were stymied by the lack of suitable coherent electron sources with high brightness, as well as the experimental difficulties of stabilizing detectable magnetic fields within the superconductor. For example, any magnetic fields aligned in-plane with thin foil specimens are highly unstable due to the large effects of demagnetization.

Our recent success in imaging the FLL directly via TEM techniques has provided a unique perspective of the behavior of vortices [13,14]. Information about vortices, contained within the electron phase, is revealed in defocused electron micrographs whereby a vortex appears as a spot of bright/dark contrast [15]. The dynamic behavior of vortices can be observed by this Lorentz microscopy method [16] and their response to changes in temperature and magnetic field as well as their interaction with specimen defects such as grain boundaries and dislocations can be discerned. One disadvantage of Lorentz microscopy, however, is that quantitative extraction of data is problematic. For instance, while Lorentz microscopy reveals the location and polarity of a vortex, the degree of flux quantization of the vortex can not be measured. Furthermore, the large lens defocus distorts the image and affects the apparent vortex size.

In contrast, electron holography [17] is highly quantitative because both the amplitude and phase information of the entire object are recorded in the hologram allowing quantitative flux measurement with higher spatial resolution than Lorentz microscopy. Electron holography was previously used to observe the surface magnetic fields emanating from superconducting films [18], however as the specimen was viewed in profile, the correlation of vortices with specimen defects was difficult. Specimen defects play an important role in the mechanism of flux pinning, as well as the development of practical applications, and thus the ability to directly observe the FLL existing within specimens has many advantages. We present here investigations of the FLL penetrating superconducting niobium thin foils by exploiting the high spatial resolution and magnetic field sensitivity of electron holography [19, 20].

2. EXPERIMENTAL METHOD

The TEM specimens were prepared by first annealing Nb (T_c = 9.2 K) that had been cold-rolled to 7 µm thickness at ~2000°C in a vacuum of 10^{-6} Pa. This oxygen-desorbing anneal resulted in a resistance ratio $R_{300\ K} / R_{10\ K}$ of ~20 and a foil with grain size 200-300 µm and a [110] surface texture. Then 2 × 2 mm sections were chemically polished (HNO_3 - 40%HF, 0°C) to final thicknesses of ~100 nm. These electron transparent thin foils were then sandwiched between two Cu grids and subsequently mounted in the TEM.

The experimental breakthrough in directly visualizing the FLL was achieved by observing the specimen tilted by 45° with respect to both the illuminating electron beam and the ancillary magnetic field used to introduce and stabilize the vortices within the superconductor, see Fig. 1. The specimen was observed via a specially developed 350 kV cold tip field-emission TEM [21] with a liquid He cold stage that could maintain stable temperatures from 4.5 K to 26 K while allowing the application of magnetic fields up to 200 G. The FE-TEM was also equipped with a rotatable electron biprism used to form holograms, by interfering electron wavefronts passing through the specimen (object wave) with those in the vacuum (reference wave), that were then recorded on photographic film. Our experimental procedure was to first observe a defocused FLL image by Lorentz microscopy to find suitable regions for holography, the image was then refocused and the holograms taken. The holograms were subsequently reconstructed, optically and numerically, to retrieve the FLL information stored in the electron phase in the form of phase-amplified interference micrographs.

In the reconstruction process, a planar comparison wave is interfered with the hologram thereby producing a contour map of the phase. The proper interference conditions may be chosen, by tilting this comparison wave, to create a phase map of the FLL. Implicit in this procedure, however, is the assumption that a plane wave reference was used to form the hologram. To satisfy in the electron microscope this condition of an unperturbed reference wave coming from the vacuum region, an interference distance as large as possible between the object and the reference wave should be chosen [22]. Furthermore, the biprism interference fringe spacing with respect to the specimen should be very fine, as this parameter affects the ultimate hologram resolution, about three times the fringe spacing. Given that the interference distance and reciprocal fringe spacing are both proportional to the biprism voltage, by increasing the latter the ideal conditions can be approached. Unfortunately, this is possible only to a limited extent due to the lateral coherence of the electron beam, which diminishes the fringe contrast and hence the hologram quality as the interference distance is increased.

A compromise between the opposing requirements of hologram contrast and resolution, the limitations set by the photographic recording medium as well as the FE-TEM mechanical and electrical stability, can be achieved by carrying out the observations with the objective lens switched-off, and imaging with the intermediate lens, so that the hologram has an overall electron optical magnification of 1800×, with a carrier fringe spacing referred to the specimen of about

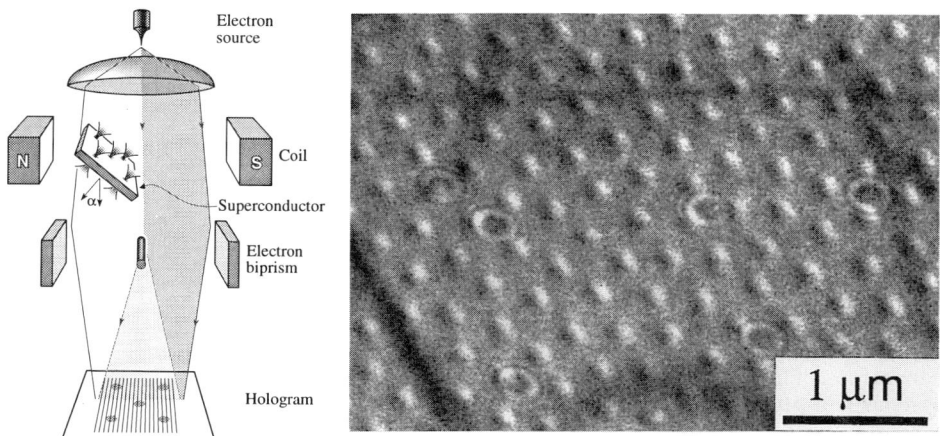

Figure 1. A tilted specimen is illuminated with coherent electron wavefronts; those passing through specimen are interfered with those in vacuum via biprism to form a hologram.

Figure 2. Lorentz image of the FLL at 100 G and 4.5 K (Δz = 20 mm). Phase shifts produced by the vortices are manifested as bright/dark spots with the core aligned along the intersection.

30 nm. These optical conditions comprise relatively low resolution electron holograms as their information limit is about 100 nm. The resolution can be increased by using the objective lens to focus on the specimen and subsequently forming a hologram in the usual fashion. Since the objective lens magnifies the specimen with respect to the biprism (by a factor of 3-4), the hologram resolution can be improved to better than 30 nm. However, there are drawbacks to the high-resolution approach: By effectively magnifying the specimen with respect to the biprism the interference width of the corresponding hologram will be reduced to ~1.5 µm. This means that the "reference" wave interfered with the object wave can no longer be assumed to be an unperturbed plane wave. In fact, the magnetic field applied to the specimen strongly perturbs the reference wave complicating the reconstruction process. Furthermore, the narrow interference width means that only those vortices near the specimen edge may be examined increasing the possibility of vortex motion and core broadening due to thin film effects, etc.

Holograms were reconstructed both optically and numerically to extract the phase information about the FLL. The optical reconstructions, performed by Mach-Zehnder interferometry, were achieved by first making a bleached hologram to improve the contrast. This hologram was then phase-amplified to increase its sensitivity to the phase distribution. Interference micrographs reconstructed from these holograms had wide fields of view revealing much of the FLL. However, for precise phase information, the original holograms were digitized and numerically reconstructed with attention focused on individual vortices. While the optical method allows observation of many vortices, the numerical method is amenable to the application of digital image processing techniques and is better suited for the extraction of phase data as well as the comparison of the experimental data with theoretical simulations. Vortex simulations are discussed in a companion paper in these proceedings.

3. VORTEX OBSERVATION

The detection of the FLL was first made via Lorentz microscopy. A typical example is shown in Fig. 2 where the Nb thin foil was cooled to 4.5 K in a field of 100 G. This Lorentz image was defocused by 20 mm to reveal the presence of the vortices.[†] Each vortex is composed of adjacent spots of bright and dark contrast with the vortex core being oriented in between them. Bend contours are present due to the slight deformation of the self-supporting thin foil. The vortices have arranged themselves along the applied field direction.

Holograms were then taken of the same specimen region under the in-focus condition. The interference micrograph shown in Fig. 3 is 16× phase amplified so that the phase difference between each dark contour line, given by $\Delta\phi = (2\pi/n)$ n being the phase amplification, is $\pi/8$. The tilt of the planar comparison wave in the reconstruction was chosen in such a way that the contour fringes were running in the same direction as the applied magnetic field with the result that fringe contours become narrowly spaced at the regions corresponding to vortex locations. Comparison with Lorentz images revealed that the spot-like

---

†The minimum defocus necessary to image the vortices can be estimated as $\sim 5(\lambda_L^2/\lambda_E)$, where λ_L is the London penetration depth and λ_E the incident electron wavelength, both in nm [15].

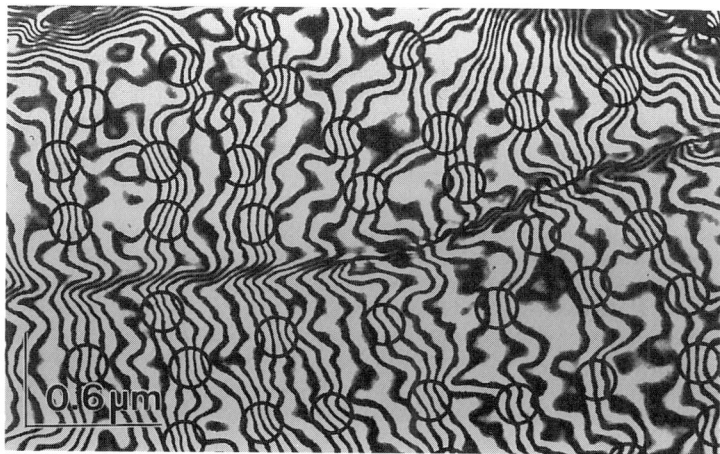

Figure 3. 16× phase amplified contour map of Nb specimen at 4.5 K and 100 G. Contours of projected magnetic lines of force become locally dense at circled regions, i.e. vortex positions.

contrast spatially coincides with the regions of finely spaced contours, the contours are widely spaced in between vortices.

A smaller region containing a few vortices was processed via a computer with dedicated software. Before digitizing the hologram, it was aligned so that the vortex axis was parallel to the vertical axis. The tilt of the wave in the reconstruction was first chosen in such a way to have the overall phase as flat as possible over the whole region. Then the vortices in the phase map appear similar to the Lorentz image, namely two adjacent bright/dark spots, see Fig. 4(a). It should be remarked that while both the Lorentz and holographic phase images look similar, the vortex size and contrast in the former are the result of waveoptical effects [15] whereas holography indicates the true vortex size. Then a further tilt was added to the reconstruction, so that the phases on both sides of the vortices were as flat as possible: in this condition the contour map, Fig. 4(b), best resembles the expected contour map and the phase difference due to a vortex can be evaluated, in this case measured phase differences were 0.55π. The vortices appear in the tilted interference micrograph as smooth phase steps where the step height indicates the amount of flux enclosed and the step width the diameter of the vortex.

3.2. Theoretical vortex model

The amount of flux detected by holography results from the Aharonov-Bohm effect [23] which predicts that a magnetic flux Φ causes a phase difference $\Delta\phi$ between two coherent electron wavefronts passing on either side of the flux as;

$$\Delta\phi = 2\pi \frac{\Phi}{h/e} \tag{1}$$

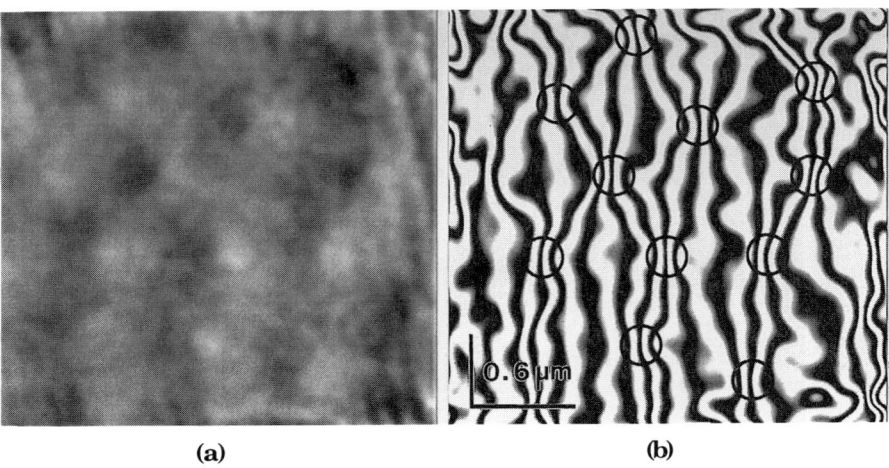

Figure 4. FLL at 4.5 K and 20 G. Flat overall phase vortices, (a), resemble Lorentz-like bright/dark contrast. Compare with the linearly tilted 12× phase amplified contours (b).

Therefore, a singly quantized flux (2.07×10^{-15} Wb, or $h/2e$) should produce an electron phase difference of exactly π. At first glance, the measured phase differences of $\sim 0.5\pi$ would appear in conflict with the theoretical expectations, but it is important to recall that the specimen is tilted to the electron beam. That is, the AB effect predicts a phase difference in proportion to the flux that is enclosed by the electron paths, i.e. only the flux that *penetrates* the plane of the electron paths will contribute to a phase difference. The discrepancy between experiment and theory arising from specimen tilt [24] can be accounted for by a simple geometrical model, see also these proceedings.

Consider the case of a single vortex within a very thin superconductor. While the total enclosed flux penetrating the superconductor is Φ_0 not all of this flux is also enclosed by the electron paths. Consequently, the phase difference measured is reduced when the specimen is tilted to the electron beam as;

$$\Delta\phi = 2(2\alpha)\frac{\Phi_0}{h/e} \quad , \tag{2}$$

where α is the angle between the superconductor normal and the electron beam as in Fig. 1. Thus when $\alpha = \pi/4$ (45°), the phase difference is expected to be $\pi/2$. Equation (2) assumes a very thin superconductor, and so the contribution from the vortex core can be neglected. However, a real specimen has a thickness on the order of the penetration depth ($t \approx 2\lambda_L$) and in this case we must consider the vortex core structure. In this situation, the presence of a finite *length* of core will increase the amount of flux enclosed by the electron paths and the resulting phase differences will be slightly greater than $\pi/2$.

3.3. Temperature dependence

It is well known [25] that the London penetration depth, λ_L, of vortices varies with temperature, T, as;

$$\lambda_L(T) = \lambda_L(0)\left[1-(T/T_c)^4\right]^{-1/2} \quad , \tag{3}$$

where $\lambda_L(0)$ for Nb is 31 nm. One expects $\lambda_L(T)$ to vary from 32 nm at 4.5 K to 47 nm at 8 K, an increase of 47%. We assume here that the coherence length, ξ, has a similar temperature dependence with $\xi(0)$ of 39 nm. According to the Clem model [26], the diameter of a vortex is $\sim 2(\lambda_L + \xi)$.[†] Thus, the resolution of electron holography, as described above, is ideally suited to observe the size of vortices as a function of temperature [20]. Holograms were taken at 4.5, 7 and 8 K under the condition of constant applied magnetic field (20 G) where the equilibrium spacing between vortices is about 1.1 µm, and they can be considered well isolated.

The linearly tilted interference micrographs (flat phase on either side of vortex) allow the phase differences due to the vortices to be properly evaluated. In this way, not only the broadening of single vortices with temperature, but their actual size at each temperature can be measured, see Fig. 5. As before, we have taken the width of the phase step to indicate the vortex diameter. Data compiled from several vortices showed average vortex diameters of 150 nm at 4.5 K, 185 nm at 7 K and 230 nm at 8 K, representing an increase of ~50%. Our values are comparable with those expected from the Clem model: 145, 172 and 212 nm, respectively. Also, while the vortices did broaden with temperature, the total flux enclosed by them was constant.

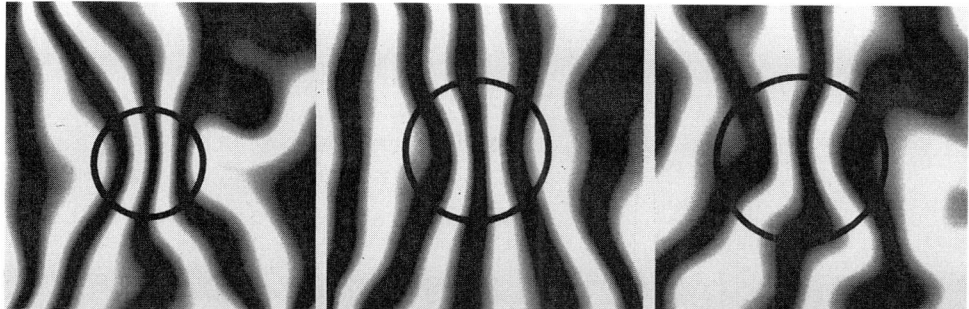

Figure 5. Interference micrographs of vortex diameters (12× phase amplified) broadening with temperature at 4.5, 7 and 8 K (at 20 G) measuring 150, 185 and 230 nm (error: ± 4 nm), respectively.

[†] An alternative way to define the vortex diameter, in terms of the magnetic field distribution B(r), is the radius intercepted by a tangential line plotted through the inflection point of B(r). That is, the r-intercept of a line of slope ∂B(r)/∂r evaluated at r* where r* is determined by the condition ∂²B(r*)/∂r² = 0.

3.4. Higher resolution holography

The holograms of the FLL in the previous sections were taken with the objective lens turned off and have a standard resolution of about 100 nm, of the same order as the vortex diameters. To explore the vortex inner-core structure, we have therefore endeavored to examine the FLL under higher resolution conditions by employing the objective as the focusing lens [27]. Despite the information gained from this analysis, it is important to remember that the interference width reduction will cause the reference wave in the hologram to be more strongly attenuated by the long-range magnetic fields surrounding the specimen.

Holograms formed under these higher resolution conditions were taken and compared with those at low resolution. The reconstruction process was as follows: a region of the hologram containing vortices was digitized with the vortex axis oriented vertically. The tilt of the planar comparison wave was then adjusted so that the phase across the entire field of view was as flat as possible (because of the attenuation of the reference wave, substantial phase modulation remained). Then this phase map was fitted to a 3rd-order polynomial: the difference between them resulting in a corrected phase map. The corrected phase map could then be evaluated for the presence of vortices.

Vortices reconstructed from a higher resolution hologram show several features, see Fig. 6. First, as expected, the equiphase contour lines flow smoothly through the vortex and spread out above and below the core. Also the vortex diameter agrees with the low resolution ones, indicating that although the narrow interference width restricts observable vortices to those near the edge where the specimen is only 30-50 nm ($\sim \lambda_L$) thick, there are no appreciable thin film effects. Moreover, the measured phase difference from the hologram is $\sim 0.5\pi$ indicating no change in the degree of flux quantization in the vortices.

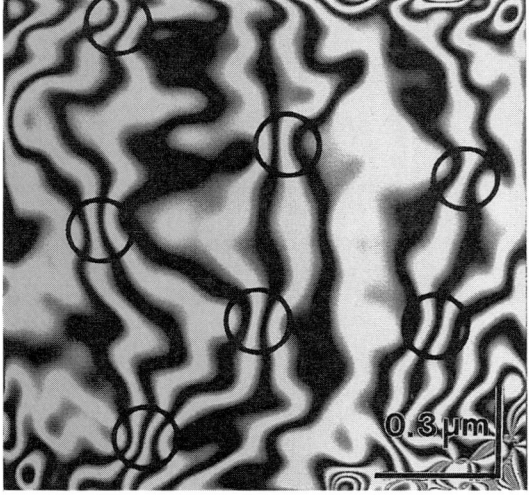

Figure 6. Vortices at high resolution (4.5 K and 10 G). 8× amplified phase was numerically fitted and corrected to compensate the perturbed reference wave.

3.5. Vortex inner-core structure

An important consideration in the study of vortices is the exact nature of the vortex inner-core, an issue that electron holography is well suited to examine. The models advanced to describe the magnetic field distribution within a vortex, e.g. the London and Clem models, differ in that the distribution in the London model is described by a modified Bessel function depending only upon the penetration depth, whereas the Clem model invokes a ratio of Bessel functions dependent upon both the penetration depth and the coherence length of the vortex [25]. High resolution electron holography offers the possibility of discriminating between these two models through the use of digital reconstruction techniques [28]. For instance, while the phase shifts caused by the vortices are identical for both models, subtle differences between them can be discerned by taking the derivatives of the phase across the vortex core. The London model is expected to have a sharp discontinuity in the 2nd derivative at the core; in contrast the Clem model reveals a smooth, continuous phase change.

Reconstructed vortex phase distributions were examined: the long-range magnetic field effects in the "reference" wave were numerically fitted and the perturbed background subtracted to reconstruct extended regions of the holograms [27]. Attention was focused upon individual vortices and numerical derivatives of the phase evaluated across the vortex core. Preliminary results suggest that the Clem model may provide a better description of the magnetic field distribution of the vortex; remaining issues are the role of shot noise in the holograms and the effect of apertures on the derivatives, as well as thin film effects on the broadening of vortices. The exact details of the phase analysis and the comparison with theoretical simulations will be published elsewhere.

4. CONCLUSIONS

In summary, the direct observation of the flux-line lattice (FLL) in superconductors with high sensitivity and spatial resolution by electron holography has opened exciting possibilities in physics and materials science. In this study, fundamental issues such as the phase distributions of single vortices in the FLL were investigated by both high and low resolution holography and the results were found to agree with theoretical simulations. Furthermore, the broadening of single vortex diameters with temperature was measured and agreed with the expected behavior. The application of high resolution holography, coupled with digital reconstruction techniques, extends the possibility of discriminating between the theoretical models of magnetic field distribution of the vortex core. Basic materials science questions, e.g. the dynamic interactions of vortices with specimen defects and the nature of pinning sites as well as the study of high temperature superconductivity, can all be resolved by electron holography.

The authors gratefully acknowledge Dr. A. Fukuhara, Dr. U. Kawabe and Dr. T. Onogi of Hitachi and Dr. Q. Ru of the Tonomura Electron Wavefront Project (ERATO) for their stimulating discussions and Dr. J. Endo, T. Furutsu, Dr. M. Igarashi, Dr. H. Kajiyama, Dr. S. Kondo, S. Kubota, S. Matsunami, N. Moriya and Dr. N. Osakabe of Hitachi for their technical assistance.

REFERENCES

1. V. Essman and H. Träuble, Phys. Lett. A 24 (1967) 526.
2. G.J. Dolan, F. Holtzberg, C. Field and T.R. Dinger, Phys. Rev. Lett. 62 (1989) 2184.
3. R. Cubitt, E.M. Forgan, G. Yang, S.L. Lee, D. McK. Paul, H.A. Mook, M. Yethiraj, P.H. Kes, T.W. Li, A.A. Menovsky, Z. Tanawski and K. Mortensen, Nature 365 (1993) 407.
4. C.A. Durán, P.L. Gammel, R. Wolfe, V.J. Fratello, D.J. Bishop, J.P. Rice and D.M. Ginsberg, Nature 357 (1992) 474.
5. J. Mannhart J. Bosch, R. Gross and R.P. Huebener, Phys. Rev. B 35 (1987) 5267.
6. H.F. Hess, R.B. Robinson, R.C. Dynes, J.M. Valles and J.V. Waszczak, Phys. Rev. Lett. 62, 214 (1989).
7. A.M. Chang, H.D. Hallen, L. Harriott, H.F. Hess, H.L. Kao, J. Kwo, R.E. Miller, R.Wolfe and J. van der Ziel, Appl. Phys. Lett. 61 (1992) 1974.
8. H. Yoshioka, J. Phys. Soc. Jpn. 21 (1966) 948.
9. M.J. Goringe and J.P. Jakubovics, Philos. Mag. 15 (1967) 393.
10. D.J. Wohlleben, Appl. Phys. 38 (1967) 3341.
11. C. Colliex, B. Jouffrey and M. Kleman, Acta Cryst. A 24 (1968) 692.
12. C. Capiluppi, G. Pozzi and U. Valdrè, Philos. Mag. 26 (1972) 865.
13. K. Harada, T. Matsuda, J. Bonevich, M. Igarashi, S. Kondo, G. Pozzi, U. Kawabe and A. Tonomura, Nature 360 (1992) 51.
14. K. Harada, H. Kasai, T. Matsuda, M. Yamasaki, J.E. Bonevich and A. Tonomura, Jpn. J. Appl. Phys. 33 (1994) 2534.
15. J.E. Bonevich, K. Harada, H. Kasai, T. Matsuda, T. Yoshida, G. Pozzi and A. Tonomura, Phys. Rev. B 49 (1994) 6800.
16. J. Chapman, J. Phys. D 17 (1956) 330.
17. A. Tonomura, *Electron Holography* (Springer-Verlag, Berlin, 1993).
18. T. Matsuda, S. Hasegawa, M. Igarashi, T. Kobayashi, M. Naito, H. Kajiyama, J. Endo, N. Osakabe, A. Tonomura and R. Aoki, Phys. Rev. Lett. 62 (1989) 2519.
19. J.E. Bonevich, K. Harada, H. Kasai, T. Matsuda, T. Yoshida, G. Pozzi and A. Tonomura, Phys. Rev. Lett. 70 (1993) 2952.
20. J.E. Bonevich, K. Harada, H. Kasai, T. Matsuda, T. Yoshida and A. Tonomura, Phys. Rev. B 50 (1994) 567.
21. T. Kawasaki, T. Matsuda, J. Endo and A. Tonomura, Jpn. J. Appl. Phys. 29 (1990) L508.
22. G. Matteucci, G.F. Missiroli, E. Nichelatti, A. Migliori, M. Vanzi and G. Pozzi, J. Appl. Phys. 69 (1991) 1835.
23. Y. Aharonov and D. Bohm, Phys. Rev. 115 (1959) 485.
24. A. Migliori, G. Pozzi and A. Tonomura, Ultramicroscopy 49 (1993) 87.
25. R.P. Huebener, *Magnetic Flux Structures in Superconductors* (Springer, Berlin, 1979).
26. J.R. Clem, in *Inhomogeneous Superconductors-1979* (Berkeley Springs, WV), ed. D.U. Gubser *et al.*, AIP Conf. Proc. No. 58 (AIP, New York, 1979), p. 245.
27. J.E. Bonevich, K. Harada, H. Kasai, T. Matsuda, T. Yoshida, G. Pozzi and A. Tonomura, in *Determining Nano-scale Physical Properties of Materials by Microscopy and Spectroscopy*, eds. M. Sarikaya *et al.*, MRS Conf. Proc. No. 332 (Materials Research Society, Pittsburgh, 1994), p. 219.
28. G. Pozzi, J. Bonevich and A. Tonomura, Proc. XIIIth International Congress on Electron Microscopy, Vol. 1 (1994) p. 309.

Electron Holography
A. Tonomura, L.F. Allard, G. Pozzi, D.C. Joy and Y.A. Ono (Editors)
© 1995 Elsevier Science B.V. All rights reserved.

Holographic studies on magnetic phenomena in small regions

T. Hirayama[*], J. Chen, Q. Ru, K. Ishizuka, T. Tanji and A. Tonomura

Tonomura Electron Wavefront Project, ERATO,
Reseach Development Corporation of Japan (JRDC),
P.O. Box 5, Hatoyama, Saitama, 350-03, JAPAN

Electron holography is applied to the observation of magnetic phenomena in small regions. Magnetic domain states of small particles of barium ferrite and iron are studied. The single domain state which is one of the proper characteristics of small particles has been successfully observed. Furthermore, a system for dynamic observation has been developed by using a liquid crystal panel to combine an electron microscope and an optical reconstruction system. With this system the motion of magnetic domains in a thin permalloy film has been clearly observed.

1. INTRODUCTION

Electron holography was invented by Dennis Gabor [1] in 1948 to correct spherical aberration of electron lenses and improve the resolution of electron micrographs. An application of electron holography to the observation of electromagnetic fields was proposed by Cohen [2] in 1967. Experimental studies were started in the early 1970's [3,4] and various magnetic substances were observed successfully in the 1980's [5-11]. Observing the electromagnetic field of materials or devices is very important for industrial purposes as well as scientific research. This is because properties of materials or functions of devices are usually related to their electromagnetic states. In our research magnetic domain states of small particles and thin films have been studied. This paper presents the principle of magnetic flux-line observation, static observation of domain states of small particles [12,13], and then, dynamic observation of domain-walls in a thin film [14,15].

2. PRINCIPLE AND METHOD OF MAGNETIC FLUX-LINE OBSERVATION

2-1 Principle
Figure 1 shows the principle of magnetic flux-line observation [13]. The phase difference between points P1 and P2 in the specimen plane is given by the following equation[16]:

$$\Delta\phi(P_1, P_2) = -\frac{e}{\hbar} \int B dS , \quad (1)$$

---

[*] Presently with JAPAN FINE CERAMICS CENTER
2-4-1 Mutsuno, Atsuta-ku, Nagoya 456
JAPAN

where e is the electron charge, $\hbar$ is Planck's constant divided by 2π, B is the magnetic flux density and S is the area enclosed by the two electron trajectories passing through points P1 and P2. This equation implies that the phase difference of the electron waves is zero provided that the two points are along a magnetic flux line as shown by P1 and P2' in Figure 1. Therefore, magnetic flux lines are directly observed as contour fringes which are equal-phase lines in interference micrographs [5,11]. Equation (1) also implies that, in the interference micrographs, there is constant magnetic flux of $h/e = 4.1 \times 10^{-15}$ Wb between two adjacent contour lines when the phase shift of the electron wave is not amplified.

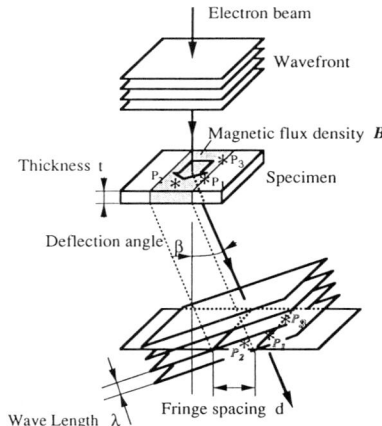

Figure 1. Principle of the observation of magnetic flux lines.

2-2. Hologram formation

Holograms were formed by the interference of an object wave and a reference wave. They were recorded on electron microscope film, Kodak 4489, using a transmission electron microscope, Hitachi HF-2000, equipped with a field-emission electron gun and an electron biprism. The accelerating voltage was 200 kV.

To observe magnetic materials, the magnetic field of the objective lens should be controlled to prevent specimens from changing their magnetic states. In the present experiment, for observing particles about 1 μm in size, the objective lens was turned off before specimens were installed into the electron microscope, and the image was focused using the first intermediate lens. Under this condition, the maximum direct-magnification of the electron microscope is about ×2,000, which is large enough to observe the interference area of more than 3 μm required to make holograms of such particles. To observe smaller particles, we employed a specially designed low magnetic-field pole piece [17] as shown in Figure 2. The specimen is set above the lens gap and this objective lens is used with an electric current of about 1 A. Under this condition, the magnetic field at the specimen position is 0.55 mT and the focal length is 8.6 mm. Using this objective lens, maximum magnification of the electron microscope becomes about ×500,000[12,13].

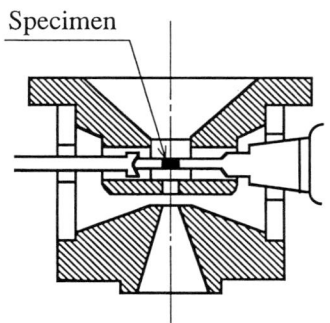

Figure 2. Cross sectional view of the low magnetic-field lens. The specimen is set above the lens gap. The focal length is 8.6 mm.

2-3. Formation of interference micrographs

The Fourier-transform method [18,19] was used to extract the phase information from the holograms. The holograms recorded on the films were re-imaged with a CCD camera and digitized with 512×512 pixels and 8 bits. The digitized data were Fourier-transformed to obtain two dimensional Fourier spectra of the holograms, in which a center band and two side-bands were seen. One of the side-bands was extracted by a side-band pass filter and it was inversely Fourier transformed to reconstruct the wave function of the object electron wave as the complex value $\Psi(x,y)$. The phase distribution $\phi(x,y)$ of the electron wave around the specimen, which is proportional to the projected magnetic-field distribution, was obtained by calculating $\phi(x,y) = \tan^{-1}(\operatorname{Im}\Psi/\operatorname{Re}\Psi)$ of the complex value $\Psi(x,y)$. Then, the interference micrograph was obtained by calculating $\cos\phi(x,y)$, which produced the equal-phase lines.

For studying particles smaller than 0.2 μm in diameter, the phase shift of the electron wave was amplified so that the magnetic domain structure could be easily observed. Otherwise, the phase shift by the magnetic field of such small particles is too small to produce even one contour line around the particles. Furthermore, the following correction processes were used to decrease systematic errors caused by Fresnel fringes of the electron biprism and/or the distortion of electron lenses[13]. After recording the electron hologram of the sample, a so-called "reference hologram" was also recorded under the same conditions but with the sample removed. The phase distribution from the hologram with the specimen and that from the "reference hologram" were calculated by the Fourier-transform method mentioned above. We call the former "the object phase" and the latter "the reference phase." All the systematic errors are also included in the reference phase. Both the object phase and the reference phase were processed by a 5×5 pixel smoothing operation. This operation removes the high-frequency noise without affecting the shape of the magnetic flux lines, because magnetic flux lines in space satisfy a harmonic function, i. e. the flux lines should not turn abruptly. The reference phase was then subtracted from the object phase in order to remove the background distortion and to obtain the true phase distribution. Finally, the true phase distribution was converted to an interference micrograph.

The polarity of the magnetic domains was found in the following procedure: 1) in the hologram or the phase distribution map the inclination of the wavefront of the electron beam is understood; 2) as shown in Figure 1, the direction of the beam deflection by the magnetic field of the specimen can be found by taking into account the inclination of the wavefront; 3) the direction of magnetic flux lines, i. e. the polarity of the domain is found by Fleming's rule.

3. STUDIES ON DOMAIN STATES OF SMALL PARTICLES

In general, magnetic substances have magnetic-domain structures to minimize their magnetic energy which is the total of magnetic domain-wall energy and magnetostatic energy. The magnetic domain-wall energy is proportional to the area (the square of diameter) of the domain wall. On the other hand the magnetostatic energy is proportional to the volume (the cube of diameter) of the particle. Therefore, as the particle size decreases, the relative contribution of the domain-wall energy to the total magnetic energy of the particle increases. Finally, the single-domain state becomes the most stable below a certain critical size. The idea of the single-domain state in a small particle was first proposed by Frenkel and Dorfman [20] in 1930 and the critical size of the single-domain particles was, then, theoretically studied by Kittel [21] in 1946, Néel [22] in 1947 and Brown [23] in 1957. Proving the existence of single domain particles is very important for industry as well as scientific research of magnetism because the maximum coercive forces are expected theoretically for all permanent magnets if these magnets are composed only of the single-domain particles of the materials. Although those theoretical studies were done many years ago, the existence of single-domain particles was not confirmed experimentally until Goto [24-26] et al. successfully observed domain structures of barium ferrite particles in 1980 by the application of the colloid-SEM method. In the present work, we first studied magnetic domain states of barium ferrite particles about 1 μm in size. We then examined the domain state of barium ferrite and iron particles smaller than 1 μm in size, which cannot be observed by the colloid-SEM method.

3-1. Barium ferrite particles

Barium ferrite ($BaO \cdot 6Fe_2O_3$) is known to be a magnetoplumbite-type oxide having a hexagonal structure ($a_0=0.5876$nm, $c_0=2.317$nm) and it is widely used for permanent magnets. The present sample was prepared by Goto et al.[24-26]. The sample was confirmed by X-ray powder diffraction to be $BaO \cdot 6Fe_2O_3$ having the hexagonal structure [25,26]. These barium ferrite particles were dispersed onto a thin carbon film and carbon was then deposited on the particles by vacuum evaporation in order to prevent charging up. Although most particles were sticking to each other to form large clusters of a few tens of micrometers because of their magnetic force, isolated particles were occasionally found. In the present experiment, some completely isolated particles were selected to study the domain structure of individual particles.

Figure 3 shows one of the typical shapes of single-domain particles observed by a scanning electron microscope [24-26]. At first we tried to observe such a particle from the [001] direction. Under this observation condition, however, no flux line was observed around the particle in the interference micrograph. The reason is as follows: The axis of easy magnetization for barium ferrite is well-known to be [001] and it was parallel to the incident electron beam under our observation condition. The magnetic field around the particle can be divided into two components: the component parallel to the electron beam and that perpendicular to the beam. The electron beam is deflected only by the latter component, whose directions above and below the particle are opposite. Therefore, the phase shift of the electron wave due to the magnetic field is completely canceled out.

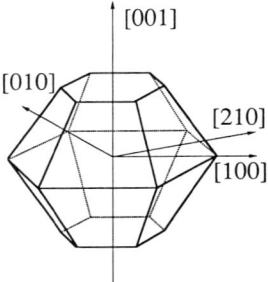

Figure 3. Typical shape of the barium ferrite particle. The particle has a flat surface of (001) plane since it is stably attached to the thin carbon film.

To observe magnetic flux lines, the specimen was tilted about 14 degrees around the tilting axis parallel to the [210] axis of the hexagonal structure of barium ferrite. The tilt direction of the specimen is shown in Figure 4a. Under this condition the electron beam is along [01$\bar{1}$]. Figures 4a and 4b show an electron micrograph and an interference micrograph, respectively. This particle is about 1 μm in size and has a well-defined crystal habit. The particle is so thick that its image contrast is entirely black. In Figure 4b magnetic flux lines emerging from one N-pole and converging into one S-pole are very clearly seen. The shape of the magnetic flux lines indicates that this particle is in the single-domain state.

To observe the particle from another angle, the specimen was tilted back to the condition where the incident electron beam is parallel to the c-axis, and then it was tilted about 24 degrees around the [010] axis. The tilt direction is shown in Figure 5a. Under this condition, the incident beam is along [$\bar{2}1\bar{1}$]. It should be noted that the direction of the magnetization in Figure 5b appears almost perpendicular to that in Figure 4b. These results suggest that the direction of the magnetization of specimens should be carefully interpreted from the three-dimensional viewpoint when studying domain structures of particles. Three-dimensional

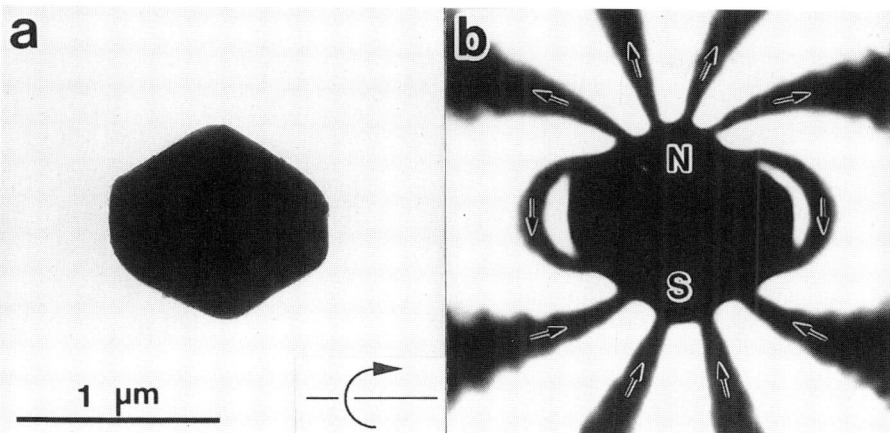

Figure 4. Isolated barium ferrite particle observed from [011]: (a) Electron micrograph. (b) Interference micrograph. Magnetic flux lines emerging from the N-pole and converging into the S-pole are clearly seen.

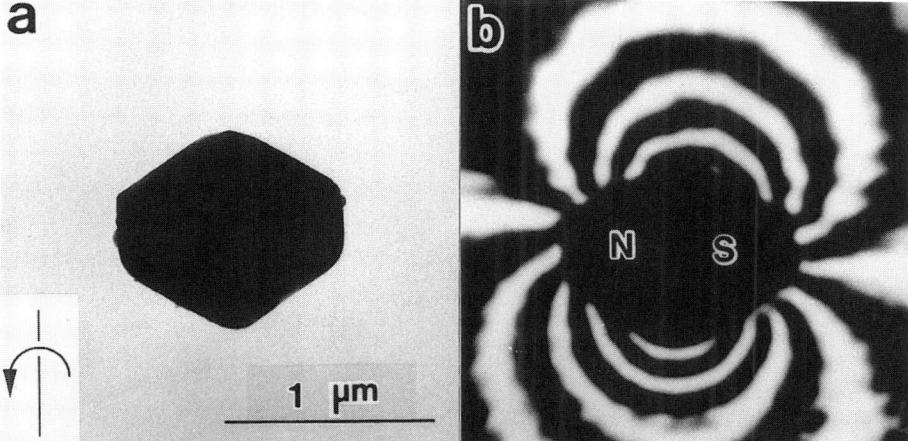

Figure 5. Isolated barium ferrite particle observed from [$\bar{2}1\bar{1}$]: (a) Electron micrograph. (b) Interference micrograph. Magnetic flux lines are clearly seen. However, the direction of magnetization appears different from that in Figure 4b.

reconstruction of the magnetic-field using electron holography [27] developed by the present authors is one of the most reliable methods.

Figure 6 shows another particle having large magnetic flux lines on the right and left sides of the particle and two small lines at the bottom left of the particle. This particle is interpreted to have the two-domain structure. Among the particles studied as described above that were about 1-2 μm in size, about half of them were in the single-domain state and the other half were in the two-domain state or could not be interpreted.

A single-domain particle about 0.15 μm is shown in Figure 7. It is an advantage of electron holography with a low magnetic field lens that magnetic flux lines around such a small particle are observed very clearly. Figure 8 shows a two-domain particle about 0.2 μm. According to the Kittel's theory, the critical diameter for the barium ferrite particle having a spherical shape is roughly estimated to be 0.3-0.9 μm using values of 2.8-9 erg/cm^2 for magnetic domain-wall energy. Experimental values of the critical diameter by Goto et al [25,26] and Kitakami et al.[28] for the barium ferrite ranged 0.6-1.8 μm. The authors present a two-domain particle about 0.2 μm in Figure 8. Therefore, it should be concluded that single-domain and two-domain particles coexist in a rather wider range than the calculated result. Actual particles have irregular shapes and/or crystal defects such as stacking faults or grain boundaries and they probably cause the expansion of the range. The relationship between domain structures and such factors is worth studying.

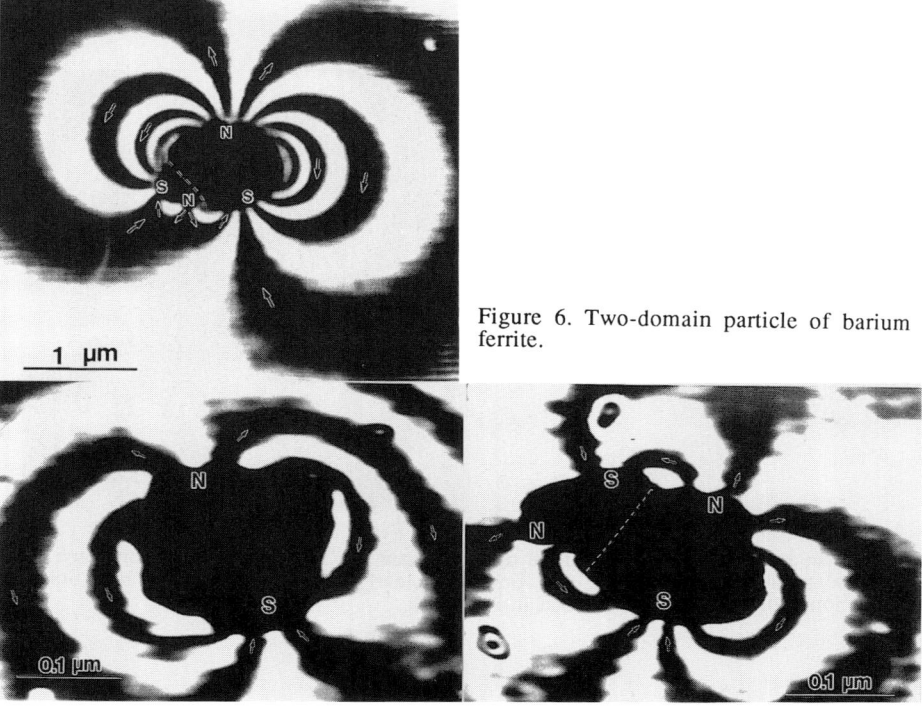

Figure 6. Two-domain particle of barium ferrite.

Figure 7. Ten times amplified interference micrograph of a small single-domain particle.

Figure 8. Ten times amplified interference micrograph of a small two-domain particle.

3-2. Iron particles

Figure 9 shows an example of the general view of small iron particles produced by a gas evaporation method[29]. The particles are sticking together to form chains, and no isolated particles are seen.

In the interference micrograph obtained directly from a hologram of the specimen, magnetic flux lines were so distorted that it was difficult to interpret the domain state. Therefore, the systematic errors caused by Fresnel fringes and/or the distortion of electron lenses were decreased using the "reference hologram". The processes to decrease the systematic errors were indispensable for detecting such a small phase distribution because the errors are comparable with the phase distribution. Figure 10a shows the corrected interference micrograph of the tip of a string of particles. The phase shift of the electron wave is amplified ten times. In this figure magnetic flux lines around the particle are clearly observed. On the other hand, magnetic flux lines inside the particle cannot be seen because of the concentric fringes which are the equal thickness lines of the spherical particle. Although flux lines inside the particle are not visible, they are deducible as illustrated in Figure 10b, considering one of Maxwell's equations, "div B = 0," which implies that magnetic flux lines should be continuous. The direction of the flux lines was found in the same way as barium ferrite.

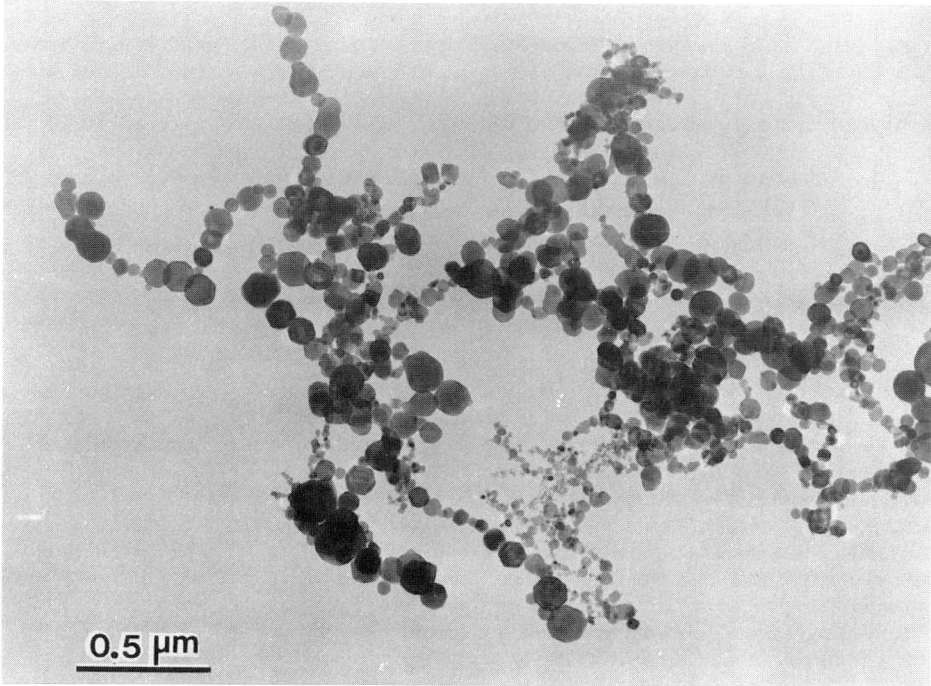

Figure 9. General view of iron particles produced by a gas-evaporation method. Particles are sticking together to form chains.

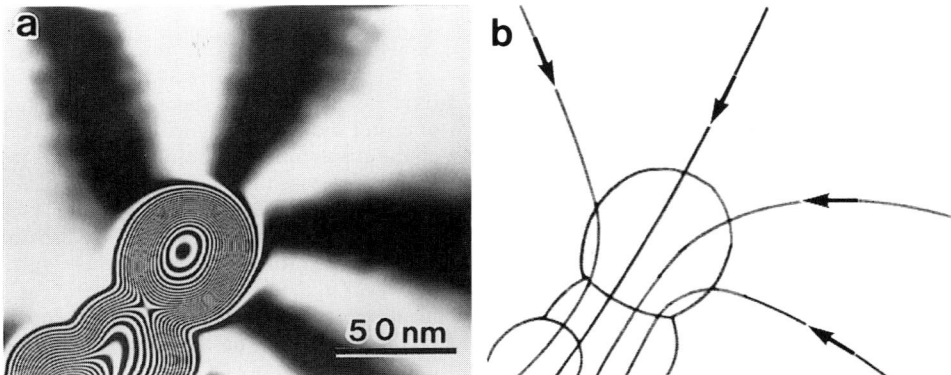

Figure 10. Tip of a string of iron particles: (a) Corrected interference micrograph. (b) Schematic of the particles and magnetic flux lines. Although flux lines inside the particle are not visible, they are deducible considering one of Maxwell's equations ,"div $B = 0$", which implies that magnetic flux lines should be continuous.

In the Fourier-transform method, the resolution of an interference micrograph is limited by the area of the sideband extracted from the Fourier spectrum of the hologram, i.e. the diffraction pattern of the hologram[18]. In this experiment, the radius of the sideband was about one half of the distance between the center of the side band and that of the center band. This suggests that, since the fringe spacing of the hologram is 2.9 nm, the resolution of the interference micrograph is 5.8 nm. Although it was easy to make finer fringes, holograms with such fine fringes were not effective in achieving a better resolution because decreasing systematic errors was actually much more important to obtain trustworthy information.

4. DYNAMIC OBSERVATION BY ON-LINE REAL-TIME SYSTEM

4-1 Approaches to real-time electron holography

Since electron holography is a two-step imaging method, magnetic flux lines could not be observed in real time. Although a few new methods of electron holography such as a phase shifting method [30,31] and computer-assisted systems [32,33] succeeded in shortening the processing time, they did not achieve real-time observation. However, observation of dynamic behavior is very important in materials research especially for studying magnetic substances and superconductors because the fundamental characteristics of such materials are strongly related to dynamic behavior of domain-walls or flux quanta (fluxons). Dynamic observations of magnetic domain-walls by Lorentz microscopy were reported [34,35] but those observations by electron holography were extremely difficult. In order to study dynamic behavior of fluxons, electron holograms of the time-varying field recorded on a videotape were reconstructed digitally frame by frame [36]. Fluxon dynamics were successfully observed using this system. However, such a frame-by-frame method involves a time-consuming process and it still remains a problem that the operator of the electron microscope cannot see the flux lines simultaneously. Therefore, on-line real-time observation is needed for further study.

4-2 Description of the system

Recently, liquid-crystal spatial light modulators (LC-SLM) have been used in optical image processing systems [37] or in optical holography [38]. We have developed an on-line real-time electron holography system by using the LC-SLM to combine an electron holographic microscope and a Mach-Zehnder interferometer.

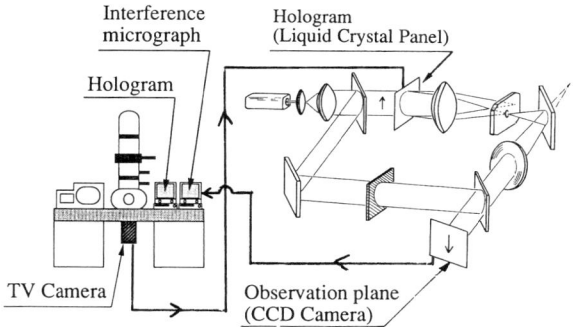

Figure 11. Schematic diagram of the on-line real-time electron holography system. The hologram formed in the electron holographic microscope is transferred, as a video signal, to the liquid-crystal spatial light modulator located in the Mach-Zehnder interferometer. The magnetic flux lines are observed on the monitor beside the microscope.

The scheme of the new system developed for real-time electron holography is shown in Figure 11. The system consists of two units: an electron holographic microscope and a Mach-Zehnder interferometer with the LC-SLM [14]. The hologram formed in the holographic microscope is detected and then converted into a time sequential video signal by a TV camera, GATAN Model 622. The video signal is transferred via a 75 Ω coaxial cable to the LC-SLM in the Mach-Zehnder interferometer. The LC-SLM works as an electrically controllable hologram. The Mach-Zehnder interferometer is used to produce an interference micrograph on the observation plane. Finally, the interference micrograph is detected by a CCD camera and is sent to a monitor beside the electron microscope. Using this system, the operator of the electron microscope can see the interference micrograph in real time.

4-3. Experimental procedure

Using this real-time system, dynamic behavior of magnetic domains in a thin permalloy film was observed [15]. The specimen was about 20 nm thick. The experiment was performed according to the following procedure. First, the objective lens current was turned off and the specimen was installed into the normal specimen position. The specimen was tilted 5-15 degrees and the image was focused by controlling the first intermediate lens. Under this lens condition, final magnification of the images on the monitor was about forty thousand times. Then, in order to apply a magnetic field to the specimen, the objective lens current was increased slowly up to about 0.1 A, which was about one fiftieth of the standard lens current. Finally, the current was decreased slowly to zero. In this range of objective lens current, focusing of the image was scarcely changed. Both the hologram and the interference micrograph of the domain motion were recorded on videotape.

The magnetic field parallel to the specimen film H_1 is given by the following equation:

$$H_1 = H_0 \sin \theta \qquad (2)$$

where H_0 is the vertical magnetic field of the objective lens and θ is the tilt angle of the specimen. H_0 was measured by a Gauss meter against the objective lens current. Maximum H_1 in our experiment was 4000 A/m.

4-4. Dynamic observation of domain-walls in a thin permalloy film

Figures 12a-12e show a series of interference micrographs of magnetic domain walls moved by the magnetic field applied to the specimen. Their domain states are illustrated in Figures 12a'-12e', respectively. Figure 12a shows the state before applying the magnetic field. In this figure, three domains I, II and III are separated by two domain walls A and B as illustrated in Figure 12a'. The domain walls A and B originate at the same point P_1 on the specimen edge

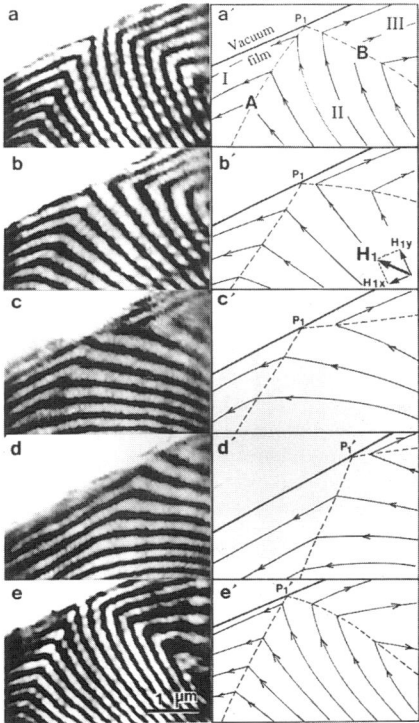

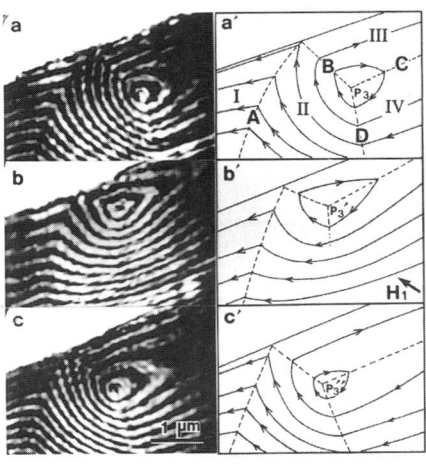

Figure 13. Complicated domain states. The point where three domain walls meet is moved by applying a magnetic field. (a) Before applying magnetic field H_1. (b) When H_1 is increased to 1600 A/m. (c) After decreasing H_1 to zero.

Figure 12. Interference micrographs of thin permalloy film: (a) State before applying magnetic field H_1. (b) By increasing H_1 up to 3000 A/m, the domain wall B moves upward. (c, d) When H_1 reaches 4000 A/m, the two domain walls are shifted abruptly to the right. This takes place in less than one tenth of a second. (e) By decreasing H_1 to zero, the domain wall B moves downward.

and extend into the inner part. The angle between the two domain walls is 95 degrees. Magnetic flux lines reaching the specimen edge turn to become parallel to the edge in order to prevent increasing magnetostatic energy. With increasing magnetic field H_1 up to 3000 A/m, domain wall B moved slowly upward while domain wall A moved only a little. The angle between the two domain walls increased up to 115 degrees as shown in Figures 12b and 12b'. Consequently, the area of domain II became larger than that in Figure 12a. This directly shows the magnetizing process of the permalloy film. This motion is interpreted as follows: H_1 can be resolved into two components, H_{1x} parallel to the specimen edge and H_{1y} perpendicular to the edge. Domain wall B was moved by H_{1y} which had almost the same direction of flux lines in domain II. H_{1x} prevented domain wall A from moving upward because H_{1x} and flux lines in domain I were in almost the same direction. During the motion of domain wall B, the meeting point of the two domain walls stayed at P_1. Therefore, P_1 is believed to be a pinning point. However, during observation of low magnification in this experiment, no cause was observed for this pinning. The more H_1 increased, the more domain wall B moved upward as shown in Figures 12c and 12c', and finally the angle between the two domain walls became 125 degrees. It should be noticed that the direction of magnetic flux lines in domain II is changed to be parallel to H_1. It is an advantage of electron holography to visualize the magnetic state inside the domain. When the magnetic field H_1 reached about 4000 A/m, the two domain walls were suddenly shifted, keeping the same angle, to the right at the same time as shown in Figures 12d and 12d'. H_{1x}, which is the driving force in enlarging domain I, caused the shifting. It took place abruptly in less than one tenth of a second. In Figure 12d the two domain walls meet at point P_1' which is about 1 μm away from P_1. Then, by decreasing the magnetic field H_1 to zero, domain wall B moved downward and the domain-wall angle returned to 95 degrees as shown in Figures 12e and 12e'. This is the observation of the demagnetizing process. However, the meeting point of the two domain walls remained at point P_1'. This shows the origin of the residual magnetization. In Figure 12e the position of P_1' looks different from that in Figure 12d because the specimen itself was shifted.

Such pairs of domain walls were observed very often in the experiment. Some of them were reversible while others did not return to the exact starting place even when H_1 decreased to zero.

A more complicated domain structure was found as shown in Figure 13. In Figure 13a four domains I, II, III and IV divided by four domain walls A, B, C and D are seen as illustrated in Figure 13a'. The three domain walls B, C and D meet at point P_3 and magnetic flux lines rotate around this point. The meeting point moved upward to point P_3' as shown in Figures 13b and 13b', stepping several times, by increasing H_1 up to 1600 A/m . In this figure the domain wall D is vague. By decreasing H_1 to zero, the meeting point moved downward to point P_3'', which is a little above point P_3, as shown in Figures 13c and 13c'. This is also the observation of the residual magnetization. In this figure domain wall D appears clearly again.

In order to discuss the time resolution of the present system, holograms on the videotape and their corresponding interference micrographs on the other videotape were checked frame by frame. It has been proved that the time resolution of the image-reconstruction system is one thirtieth second or better than that. Nevertheless, by checking the holograms very carefully, it has also been proved that the TV system sometimes shows afterimages several times longer than one thirtieth second. Therefore, the time resolution of the overall system depends on that of the TV system.

As described above, the new system developed for real-time electron holography has established the way to dynamic observation of magnetic phenomena which are strongly related to important properties of magnetic substances.

5. CONCLUSIONS

Magnetic domain states of small particles of barium ferrite and iron were studied. The single domain state which is one of the proper characteristics of small particles has been successfully observed. By using the low magnetic-field lens and digital reconstruction, the observation of magnetic phenomena of particles smaller than 0.1μm has been realized. Furthermore, a system for dynamic observation has been developed by using a liquid crystal panel to combine an electron microscope and an optical reconstruction system. Dynamic behavior of magnetic domain walls during the magnetizing or demagnetizing processes in a thin permalloy film was clearly observed. Magnetic phenomena indicating residual magnetization were also observed. The authors believe that the on-line real-time electron holography system is very useful for studying basic properties of magnetic substances and developing new magnetic devices.

ACKNOWLEDGMENT

We would like to thank Professor Kimiyoshi Goto and Mr. Tomoaki Sakurai of Tohoku University for providing the barium ferrite samples and for valuable suggestions.

REFERENCES

1. D. Gabor, Nature 161 (1948) 777.
2. M. Cohen, J. Appl. Phys. 38 (1967) 4966.
3. A. Tonomura, Jpn. J. Appl. Phys. 11 (1972) 493
4. G. Pozzi and G. F. Missiroli, Journal de Microscopie, 18, 103 (1973).
5. A. Tonomura, T. Matsuda, J. Endo, T. Arii and K. Mihama, Phys. Rev. Lett. 441(1980) 430.
6. A. Tonomura, T. Matsuda, H. Tanabe, N. Osakabe, J. Endo, A. Fukuhara, K, Shinagawa and H. Fujiwara, Phys. Rev. B 25 (1982) 6799.
7. N. Osakabe, K. Yoshida, Y. Horiuchi, T. Matsuda, H. Tanabe, T. Okuwaki, J. Endo, H. Fujiwara and A. Tonomura, Appl. Phys. Lett. 42 (1983) 746.
8. T. Matsuda, A. Tonomura, R. Suzuki, J. Endo, N. Osakabe, H. Umezaki, H. Tanabe, Y. Sugita and H. Fujiwara, J. Appl. Phys. 53 (1982) 5444.
9. A. Tonomura, Advances in Physics 41 (1992) 59.
10. A. Tonomura, Rev. Mod. Phys. 59 (1987) 639.
11. A. Tonomura, Electron Holography, Springer-Verlag, Heidelberg, 1993.
12. T. Hirayama, Q. Ru, T. Tanji, A. Tonomura, Appl. Phys. Lett. 63 (1993) 418.
13. T. Hirayama, J. Chen, Q. Ru, K. Ishizuka, T. Tanji and A. Tonomura, J Electron Microsc 43 (1994) 190.
14. J. Chen, T. Hirayama, G. Lai, T. Tanji, K. Ishizuka and A. Tonomura, Opt. Lett. 18 (1993) 1887.
15. T. Hirayama, J. Chen, T. Tanji and A. Tonomura, Ultramicroscopy 54 (1994) 9.
16. Y. Aharanov and D. Bohm, Phys. Rev. 115 (1959) 485.
17. K. Shirota, A. Yonezawa, K. Shibatomi and T. Yanaka, J. Electron Microsc. 25 (1976) 303.
18. M. Takeda and Q. Ru, Appl. Opt. 24 (1985) 3068.
19. Q. Ru, T. Matsuda, A. Fukuhara and A. Tonomura, J. Opt. Soc. Am. A 8: (1991) 1739.
20. J. Frenkel and J. Dorfman, Nature 126 (1930) 274.
21. C. Kittel, Phys. Rev. 70 (1946) 965.
22. L. Néel, CR Acad. Sci. 224 (1949) 1488.

23. W. Brown, Phys. Rev. 105 (1957) 1479.
24. K. Goto and T. Sakurai, Appl. Phys. Lett. 30 (1977) 355.
25. K. Goto, M. Ito and T. Sakurai, Jpn. J. Appl. Phys. 19 (1980) 1339.
26. K. Goto, J. Jpn. Soc. Powd. Powd. Metall. 36 (1989) 761.
27. G. Lai, T. Hirayama, A. Fukuhara, K. Ishizuka, T. Tanji and a. Tonomura, J. Appl. Phys. 75 (1994) 4593.
28. O. Kitakami, K. Goto and T. Sakurai, Jpn. J. Appl. Phys. 27 (1988) 2274.
29. R. Uyeda, Crystallography of metal smoke particles, Terra Scientific Publishing Company, Tokyo, 1987.
30. Q. Ru, J. Endo, T. Tanji and A. Tonomura, Appl. Phys. Lett. 59 (1991) 2372.
31. Q. Ru, T. Hirayama, J. Endo and A. Tonomura, Jpn. J. Appl. Phys. 31 (1992) 1919.
32. Q. Ru, T. Matsuda, A. Fukuhara and A. Tonomura, J. Opt. Soc. Am. A 8 (1991) 1739.
33. W. D. Rau, H. Lichte, E. Volkl and U. Weierstall, J. Compu. Assis. Micros. 3 (1991) 51.
34. D. Watanabe, T. Sekiguchi, T. Tanaka, T. Wakiyama and M. Takahashi, Jpn. J. Appl. Phys. 21 (1982) L179.
35. J. N. Chapman, J. Phys. D 17 (1984) 623.
36. T. Matsuda, A. Fukuhara, T. Yoshida, S. Hasegawa, A. Tonomura and Q. Ru, Phys. Rev. Lett. 66 (1991) 457.
37. N. Konforti, E. Maron and S. T. Wu, Opt. Lett. 13 (1988) 251.
38. N. Hashimoto, S. Morokawa and K. Kitamura, Proc. SPIE 1461 (1991) 291.

Electron Holography
A. Tonomura, L.F. Allard, G. Pozzi, D.C. Joy and Y.A. Ono (Editors)
© 1995 Elsevier Science B.V. All rights reserved.

Electron holography of magnetic and electric microfields

G. Matteucci [a], M. Muccini [b] and D. Cavalcoli [a]

[a]Department of Physics, University of Bologna, via Irnerio 46, 40126 Bologna, Italy.

[b]Institute of Molecular Spettroscopy-CNR, Via Gobetti 101, 40129 Bologna, Italy.

Transmission electron holography is discussed for the study of the leakage field around small magnetic particles and thin magnetic tips, the latter used in magnetic force microscopy. Since the reference wave is affected by this stray fields, problems related to the recording of reliable information are discussed. The possibility of investigating the potential distribution of linear charged dislocations lying perpendicular or parallel to the electron beam direction is reviewed.

1. INTRODUCTION

The success of electron holography has been demonstrated in a variety of investigations which range from the mapping of electric [1-7] and magnetic fields [8-11] at the micrometer scale and below, up to the achievement of atomic resolution [12,13].

As is well known, an electron hologram is an interference pattern obtained by the superposition of an object and a reference wave. The latter is usually assumed to be a plane wave. However, in many applications we should consider that the reference wave is affected by the fields which extend around the specimen under investigation. The violation of the forementioned assumption has far-reaching consequences, some of which have been investigated in previous works [1-7, 9]. Other important limitations arise in the optical processing step where, in order to display the phase information stored in the hologram, a plane parallel interferometric wave is coherently superposed on the reconstructed object wave (contour mapping). The fringe system displayed on the reconstructed image maps either the lines of force of the magnetic field or the projected potential distribution of the electric one. During this procedure it is not always possible to find an objective criterion for the setting of the parallelism between the object and the interferometric wave [6,9]. Therefore, an uncritical application of electron holography to the mapping of these fields can give rise to serious reliability problems as the useful information is to be extracted by holograms.

A completely different situation happens when the field under investigation is strictly confined within the specimen. The reference wave is not affected, therefore only phase variations of the object wave can be displayed.

The aim of this contribution is to discuss the general problems underlying the recording of electron holograms when the fields under investigation extend in the space outside the specimen so as to affect the reference wave.

Examples on magnetic specimens will be reported to show that the experimental results can be duly interpreted whenever a model for the field source is known.

Starting with the simulation of the field generated by a micro-magnetic dipole, it will be shown how different the phase distribution is in the final maps, whether or not the effect of the external

field on the reference wave is considered.
It will be also described how this dipole model can be used to account for the experimental data obtained in the study of the magnetic field around small particles which are used in magnetic recording technology.
A further application concerns the study of the leakage field of ferromagnetic microprobes of the type used in magnetic force microscopy [9].
 As an example of the capability of transmission electron holography in the investigation of electric microfields [2,6,7] we will report a theoretical study on the possibility to reveal the electric potential distribution arising from charged dislocations placed perpendicular or parallel to the electron beam direction.

2. EXPERIMENTAL METHODS : RECORDING AND PROCESSING OF ELECTRON HOLOGRAMS

The electron optical set-ups used for the study of electric and magnetic fields can be based both on amplitude [14-16] and wave front division devices [17]. The former requires less stringent brightness conditions of the illuminating source but the holographic fringe spacing is fixed by the lattice parameter of the single crystal used. The latter needs a field emission source but allows a free variation of the holographic fringe spacing, hence the final resolution can be chosen (within certain limits) by the microscope operator.
 The most diffuse and versatile type of electron interferometer is the Moellenstedt-Dueker electron biprism which is realized by placing a thin charged wire between two earthed plates and locating it under the back focal plane of the objective lens.
 Our off-axis image electron holograms are recorded by means of a Philips EM 400 T electron microscope equipped with a field emission source and an electron biprism inserted at the selected area level.
 For the investigation of magnetic specimen, the microscope operated in the diffraction mode with the objective lens switched off (it being an immersion lens) so as not to perturb the object field. The diffraction lens was used to focus the object under investigation onto the final screen and provided, together with the remaining lenses, an ultimate magnification of about 2000X .
 On the contrary, when electric fields are investigated, the microscope can operate with the magnetic objective lens switched on , hence higher magnifications are obtained for the final image.
 The retrieval of the phase information related to the distribution of magnetic and electric microfields can be performed both with optical [8,9,18] or digital means [19,20]. The former require lengthy and sometimes tedious procedures, in particular when phase difference amplification maps are needed. The latter methods allow the extraction of the final maps in a much shorter time and with improved accuracy.
 With the increasing demand for information about the configuration of fields on micrometer scale with a trend towards the nanometer one, a technique which is worthwhile to mention is that based on the time reversal [21]. When electric and magnetic field sources are present in the same specimen, this method allows the extraction of separate maps for the electric or the magnetic field.

3. STUDY OF MICRO-MAGNETIC FIELD DISTRIBUTIONS

We consider a microscopic magnetic dipole of length $L = 0.5$ µm whose axis is along the x one and lying in the x,y plane.
With one of the poles centred in the origin of the coordinate axis, the associated magnetic

scalar potential is given by:

$$V_m(x,y,z) = \frac{Q_m}{4\pi}\left\{\frac{1}{[x^2+y^2+z^2]^{\frac{1}{2}}} - \frac{1}{[(x+L)^2+y^2+z^2]^{\frac{1}{2}}}\right\} \quad (1)$$

where Qm and L are the magnetic charge and the dipole length, thus we can calculate the three-dimensional components of the external field. The phase difference between the object and reference waves, revealable in the recording plane, can be written as:

$$\Delta\varphi(x,y) = \int_{-\infty}^{+\infty} dz \int_{x_1}^{x_2} B_y(x,y,z)\,dx \quad (2)$$

A computer simulation of the field arising from such a dipole can be obtained by exploiting the holographic recording principle and introducing suitable boundary conditions.

The result of this procedure is reported in Figure 1, where an IBM PC/AT equipped with a video board able to display 512x512 pixels at 256 grey levels was used to simulate the lines of force around the North, N and South, S poles in the (x,y) plane.

Figure 1. Computer simulation of the field around a magnetic dipole.

The loci of points with constant phase are displayed as a set of curves with a phase difference of 2π between two successive black and white ones which enclose a magnetic flux of h/e. The number of lines in the map can be varied by changing the magnetic dipole charge.

However, in experiments we must consider that the reference beam, travelling at a distance of a few microns, is modulated by the leakage dipole field. Therefore, the reconstructed hologram will show the loci of constant phase difference between the perturbed reference wave and the object wave.

In the present case, the dipole length being much shorter than the interference field width, the configuration of the dipole field is not significantly affected. Therefore, this procedure can be used, as shown in Figure 2, to display the external field of small magnetic particles whose profiles are represented by the darkened central area drawn around the North and South poles location labelled as white spots.

Figure 2. Possible configuration of the field around a single magnetic-domain particle.

We will show now how the simple dipole model can be exploited also to map the field leaking out of a thin bulk magnetic tip.

We start by approximating the tip apex with a macroscopic magnetic dipole of length 20 μm. According to the existing experimental data such a dipole length seems realistic for soft magnetic sensor tips [22-23]. Following the same procedure described above we can calculate the phase shift.

The result is that the holographic method allows the recording of bi-dimensional maps arising as a projection of the investigated total field. An ideal electron hologram will display the phase difference between the object wave, which travels along the z axis and is influenced by the whole leakage field, and a reference wave which is unperturbed (that is one which moves at an infinite distance from the apex). Following the same procedure described above we can simulate the in plane projection of the three-dimensional structure of the leakage field around the tip apex.

The computer simulation of the contour map of the stray field obtained by an ideal hologram is reported in Figure 3a.

The dark lines are once more equiphase lines with a phase difference of 2π between two successive ones. The number of lines in the map can be varied by changing the magnetic dipole charge. Since the interference region width displayed in Figure 3 a is much smaller than the dipole length, as expected the lines fan out more or less radially from the tip apex T. However, in experiments, we must consider that the reference beam, travelling at a distance of a few microns, is modulated by the leakage field of the tip, hence the reconstructed hologram will show the loci of constant phase difference between the affected reference wave and the object wave. Accordingly, Figure 3b shows the simulated phase difference distribution.

This is also the kind of map we expect by processing experimental holograms: the lines, starting from the tip apex T, assume a rounded shape and then join the tip itself.

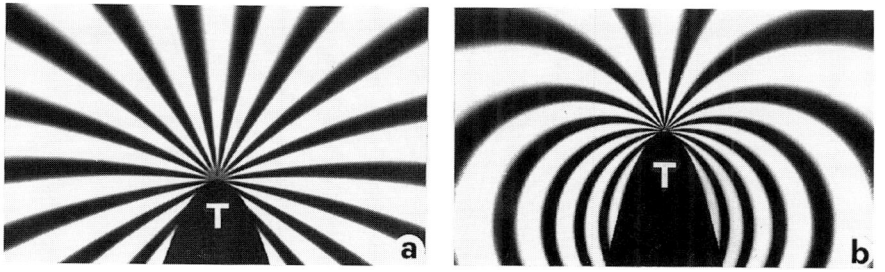

Figure 3. (a) Simulation of the leakage field of a magnetic macro-dipole; (b) phase difference trend arising when a perturbed reference wave is used (Courtesy of Physical Review B,1994).

It is worthwhile to point out once more the striking difference between the expected maps for the leakage field, Figure 3b, and the assumed dipole trend of the field.

In order to record reliable phase difference maps of the leakage field, we used the double exposure technique which also allows the shortening of the whole holographic process.

Figure 4a reports a double exposure hologram in which the reference wave was modulated by the stray field. The interference field width is 4 µm. The region displayed above the apex of the tip T is crossed by black and white fringes which represent the loci of equal phase difference between the two exposures. The lines on the left side show a different curvature with respect to those on the right side. This is due to the presence of the magnetic tip which, in order to perform the second exposure, was moved along its axis, towards the bottom of the picture, and slightly shifted to the left side. The phase-difference lines do not directly display the magnetic stray field around the apex.

In order to obtain the actual field trend we must first show that, starting from the field model, we can simulate the experimental phase distribution of Figure 4a. The resulting pattern is shown in Figure 4b. As can be seen the trend of the equiphase lines is in good agreement with the experimental results. Therefore, the leakage field of the tip can be inferred, with fairly accurately, as that produced by a magnetic macro-dipole. It can also be shown that measurements of the leakage flux can be performed on the simulated map of the dipole field associated to the tip apex.

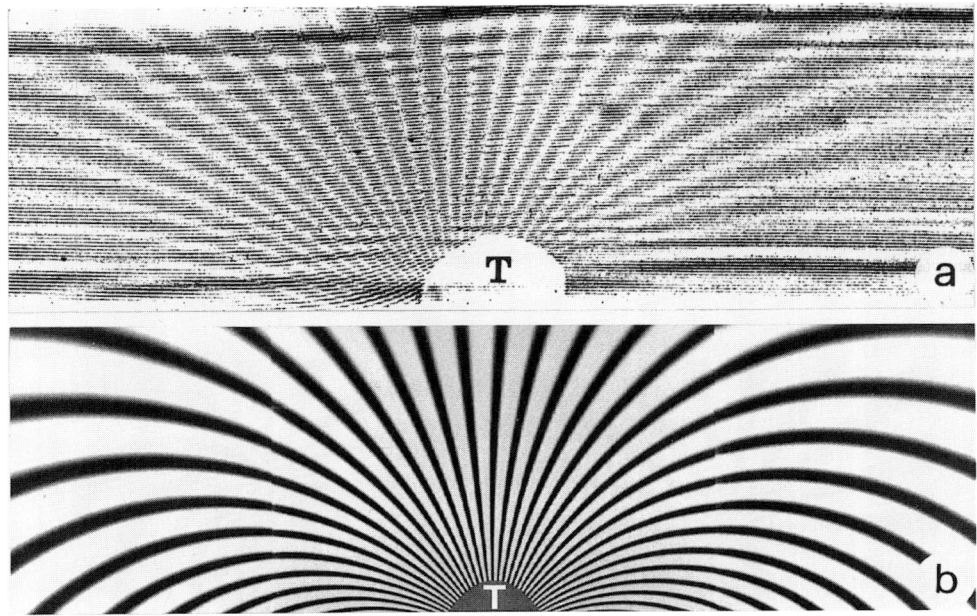

Figure 4. (a) Experimental double exposure hologram of the phase difference map around a bulk magnetic tip T; (b) computer simulation of (a). (Courtesy of Physical Review B, 1994).

Also the magnetic flux density can be evaluated once the distance between the tip and the specimen under investigation has been fixed [24].

4. STUDY OF THE MICRO-ELECTRIC FIELD OF LINEAR CHARGED DISLOCATIONS

In this section a review of the possibility to investigate charged dislocations in an n-type silicon sample is presented [25, 26].

Straight dislocations can be produced in silicon by plastic deformation. Since these defects break the translational symmetry of the crystal, deep levels in the energy gap must be expected. Dislocations, therefore, behave as electrically active defects and for many years have been widely investigated by using a number of techniques in order to understand their electrical properties. Among the many unsolved problems, the determination of a dislocation charge is still a challenging subject [27].

An approximate method for the evaluation of the electronic charge and potential at the dislocation can be deduced by the Read model [28, 29].

If a neutral dislocation, assumed to have n_d rechargeable acceptor centers per unit length becomes negatively charged, the conduction electrons will be scattered away in the proximity of the dislocation line. As a consequence, a cylinder of ionized positive atoms will be formed around the dislocation. The relation between line charge $q=n_d e$ and the potential V at a distance r from the line, in the frame of this model, is given by:

$$V(r) = \frac{q}{\varepsilon \varepsilon_0 \pi R^2} \left\{ \frac{R^2 - r^2}{4} + \frac{R^2}{2} \ln\left(\frac{R}{r}\right) \right\} \quad (3)$$

where R is the Read cylinder radius, $R = (q/\pi e N_D)^{1/2}$, ε and ε_0 are the silicon and vacuum dielectric constants and N_D is the silicon doping.
A local measure of V can thus give an evaluation of the line charge q at the dislocation, i.e. of the linear density of the dislocation related centers. For the application of this model, some restrictions concerning the overlapping of the Read cylinder of different dislocations, effects of the local doping, etc. must be fulfilled [27,29].
Computer simulations can be performed of the phase shift undergone by the electron wave which crosses the potential distribution generated by a charged dislocation. The simulations have been made by means of a PC equipped with a video board able to display 512x512 pixels at 256 grey levels.
In the following two different geometries for dislocations within the sample will be analyzed : dislocation line perpendicular or parallel to the electron beam.
Let us suppose a dislocation lies parallel to the x,y plane and perpendicular to the electron beam direction z as sketched in Figure 5a. The dislocation core is represented by the solid line S inside the charged Read cylinder C.

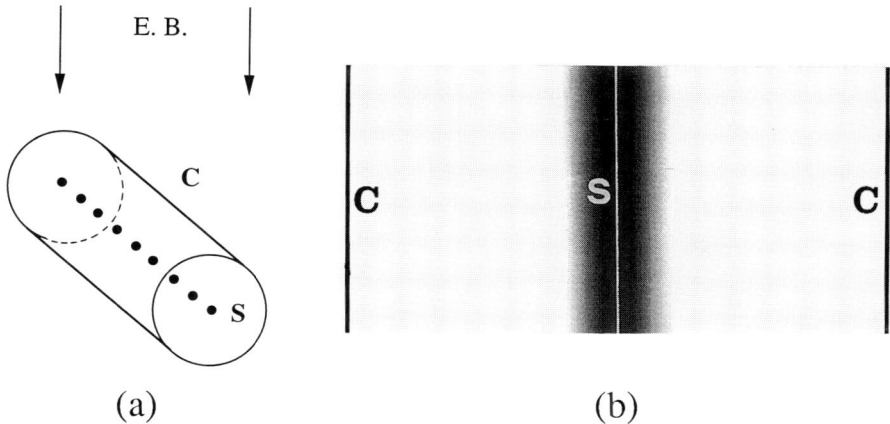

Figure 5. (a) Sketch for a dislocation perpendicular to the electron beam direction E.B.
(b) Calculated contour map (phase amplification=120) of the projected potential of a dislocation as shown in (a)

The part of the electron wave crossing the Read cylinder (object wave) will be phase modulated and the phase shift f(r) recorded in the hologram is given by:

$$f(r) = \frac{\pi}{\lambda E} \int_{-R}^{R} V(r)\, dr \qquad (4)$$

where λ is the electron wavelength, E the accelerating potential, $r=(y^2+z^2)^{1/2}$, and V(r) is the potential according to equation 3.

The reference wave can be taken as a part of the beam travelling through the vacuum outside the specimen rim. The absolute phase of the object is therefore recorded.

Since the phase shift due to a charged dislocation is rather weak, phase difference amplification techniques are needed [30].

In Figure 5b, are shown the intensity variations of a 120 times phase difference amplified contour map, although other maps have shown that a factor 100 would be enough to display the effect.

The charged cylinder is confined within the two vertical solid lines C-C. The dislocation negative charge is along the white central line S.
The two dark bands around S display equiprojected potentials. This map has been simulated by taking 100 kV electrons and a Read radius of 100 nm calculated for an n-type silicon sample with a doping of 5.0×10^{21} m^{-3} and a line charge density Nd of 1.0×10^{8} m^{-1}.

In a more convenient geometry the dislocation line is oriented parallel to the electron beam direction z as shown in Figure 6a.

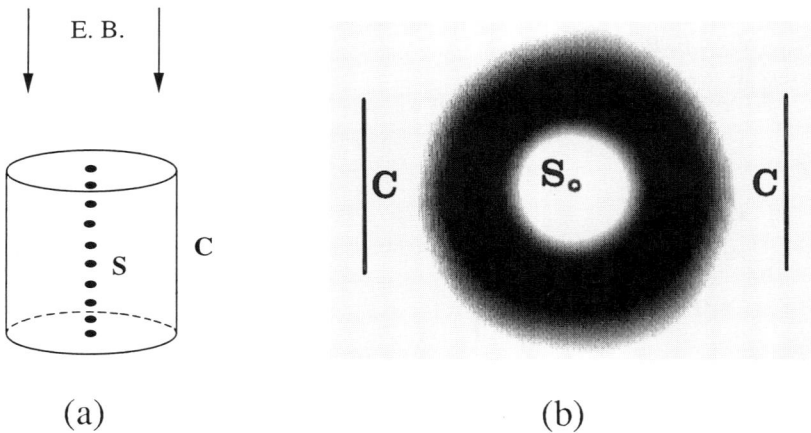

(a) (b)

Figure 6. (a) Sketch for the dislocation parallel to the electron beam direction E.B. (b) Calculated contour map (20 times phase amplification) of the potential distribution of the dislocation as in (a).

In this case the phase shift is given by:

$$f(r) = \frac{\pi}{\lambda E} \int_0^L V(r)\,dz = \frac{\pi}{\lambda E} V(r) L \tag{5}$$

where $r=(x^2+y^2)^{1/2}$, L is the line charge length that equals the specimen thickness.
In Figure 6 b, is shown a 20 times amplified map of the potential of a linear charged dislocation, with L=200 nm and R=100 nm.

The white central spot corresponds to the in plane projection of the vertical charged dislocation core S. The border of the charged cylinder is represented by the two vertical solid lines C-C. Comparing of Figure 5 b and Figure 6 b, it results that the case of the dislocation parallel to the electron beam should be more sensitive and efficient in revealing phase contrast variations linked with the potential distribution of the dislocation. In these calculations we assumed an accelerating voltage of 100 kV which is probably too low for this kind of investigation. With higher energetic beams a smaller but still revealable phase shift is expected.

Finally, since the Read model is based on the equilibrium condition for the line charge, calculations show that the concentration of electron-hole pairs induced by the electron beam is low enough to avoid perturbation effects on the dislocation potential distribution.

5. CONCLUSIONS

We have demonstrated that electron holography can be applied successfully to the study of stray magnetic fields which affect strongly the reference wave.

A practical example and simulations show how with a simple magnetic dipole model for the field source, the fields around small magnetic particles and the apex of thin bulk tips can be suitably interpreted. It has been underlined that without a theoretical model for the field source it is not possible to decide whether or not final interferograms represent maps of the field configurations.

A possible application of electron holography is suggested also for the investigation of the electric micro-field arising from a linear charged dislocation.

In this case the field is strictly confined within the sample and in the hologram the absolute phase is directly recorded. For charged dislocations the retrieval of the phase information requires the use of the most sophisticated phase amplification techniques possibly coupled with a high voltage holographic electron microscope.

The development of a new generation of dedicated electron holographic instruments opens the possibility to perform such measurements on a microscopic scale thus improving the understanding of the electronic properties of these charged defects.

Acknowledgements. We are grateful to Professors. G.F. Missiroli and G.Pozzi for the many useful discussions. Prof. U. Hartmann (an expert of MFM) is acknowledged for providing the magnetic tips and for the stimulating discussions. The skilful technical assistance of S.Patuelli is gratefully acknowledged. This work has been supported by funds from MURST coordinated by Consorzio INFM and CNR-GNSM.

6. REFERENCES

1. S. Frabboni, G. Matteucci, G. Pozzi, and M. Vanzi, Phys. Rev. Lett. 55 (1985) 219.

2. S. Frabboni, G. Matteucci and G. Pozzi, Ultramicroscopy 23 (1987) 29.
3. G. Matteucci, A. Migliori, G. Pozzi and M. Vanzi, EUREM 1988, Inst. Phys. Conf. Ser. No.93 (IOP, York, 1988) Vol.1, p.195.
4. G. Matteucci, G.F. Missiroli, J.W. Chen and G. Pozzi, Appl. Phys. Lett. 52 (1988) 176.
5. J.W. Chen, G. Matteucci, A. Migliori, G.F. Missiroli, E. Nichelatti, G. Pozzi and M. Vanzi, Phys. Rev. A 40 (1989) 3136.
6. G. Matteucci, G.F. Missiroli, E. Nichelatti, A. Migliori, M. Vanzi, and G. Pozzi, J. Appl. Phys. 69 (1991) 1835.
7. G. Matteucci, G.F. Missiroli, M. Muccini and G. Pozzi, Ultramicroscopy 45 (1992) 77.
8. A. Tonomura, Rev. Mod. Phys. 59 (1987) 639.
9. G. Matteucci and M. Muccini, Ultramicroscopy, 53 (1994) 19.
10. J.E. Bonevich, K. Harada, T. Matsuda, H. Kasai, T. Yoshida, G. Pozzi and A. Tonomura, Phys. Rev. Lett. 70 (1993) 2952.
11. T. Hirayama, J. Chen, T. Tanji and A. Tonomura, Ultramicroscopy 54 (1994) 9.
12. H. Lichte, in Advances in Optical and Electron Microscopy, edited by T. Mulvey (Springer, Berlin,1991) Vol. 12, p. 25.
13. T. Kawasaki and A. Tonomura, Phys. Rev. Lett. 69 (1992) 293.
14. G. Matteucci, G.F. Missiroli and G. Pozzi, Ultramicroscopy 6 (1981) 109.
15. G. Matteucci, G.F. Missiroli and G. Pozzi, Ultramicroscopy 8 (1982) 403.
16. Q. Ru, N. Osakabe, J. Endo and A. Tonomura, Ultramicroscopy 53 (1994) 1.
17. G. Moellenstedt and H. Dueker, Z. Phys. 145 (1956) 377.
18. K. J. Hanszen, Adv. Electron. Electron Phys. 59 (1982) 1.
19. H. Lichte, E. Voelkle and K. Scheerschmidt, Ultramicroscopy 47 (1992) 231.
20. G. Ade and R. Lauer, Optik 88 (1991) 103; Optik 91 (1992) 5.
21. A. Tonomura, T. Matsuda, J. Endo, T. Arii and K. Mihama, Phys. Rev. B 34 (1986) 339.
22. D. Rugar, H.J. Mamin, P. Guethner, S.E. Lambert, J.E. Stern, I. McFadyen and T. Yogi, J. Appl. Phys. 68, (1990) 1169.
23. C. Schönenberger and S.F. Alvarado, Z. Phys. B 80 (1990) 373.
24. G. Matteucci, M. Muccini and U. Hartmann, Phys.Rev.B 50 (1994) 6823.
25. G. Matteucci and M. Muccini, Electron Microscopy 1994, Proceedings of the 13th International Congress on Electron Microscopy, Vol. 1, B. Jouffrey and C. Colliex (eds), (Le Edition de physique, France, 1994).
26. D. Cavalcoli, G. Matteucci and M. Muccini, Ultramicroscopy (in press).
27. Exaustive literature in: H. Alexander and H. Teichler, Dislocations, in: "Material Science and Technology", Vol. 4: "Electronic Structure and Properties of Semiconductors", eds. R.W. Cahan, P. Haasen and E.J. Kramer, Volume editor W. Schroeter, p.249 (VCH Publ. Weinheim, Germany, 1991).
28. W.T. Read, Jr., Phil. Mag. 45 (1954) 775.
29. T.S. Fell, P.R. Wilshaw and M.de Coteau, Phys. stat. sol. (a) 138 (1993) 695.
30. S. Hasegawa, T. Kawasaki, J. Endo, A. Tonomura, Y. Honda, M. Futamoto, K. Yoshida, F. Kugiya and M. Koizumi, J. Appl. Phys. 65 (1989) 2000.

Electron Holography
A. Tonomura, L.F. Allard, G. Pozzi, D.C. Joy and Y.A. Ono (Editors)
© 1995 Elsevier Science B.V. All rights reserved.

Holography of electrostatic fields

B.G.Frost, L.F.Allard, E.Völkl, and D.C.Joy

High Temperature Materials Laboratory, Oak Ridge National Laboratory,
Bethel Valley Road 1, Oak Ridge, TN 37831-6064, USA

Key words
Alignment of the microscope, phase shift by a charged sphere, influence of the field on the reference wave, measurement of the interference width, simulation of equipotential lines, evaluation of number of charges, Fresnel fringes, fringe branching.

1. EXPERIMENTAL SETUP

A variety of samples over a wide size range can electrically charge in the beam of the electron microscope. In order to study this effect by image plane off-axis electron holography we use a Hitachi HF-2000 field emission microscope equipped with a Möllenstedt [1] type biprism positioned between the first and second intermediate lenses.
Two completely different alignments of the microscope allow us to match the magnification of the sample appropriately with the required interference width of the hologram [2]. The low magnification condition is achieved by switching off the objective lens and imaging the sample with the intermediate lens, as shown by the ray diagram in Figure 1, left side. Since the biprism is above the image lens we have to apply a negative voltage on the fiber to get an interference pattern. Taking a hologram with these lens data and a fiber voltage of $-30V$ results in 100 fringes in an interference width of $5.36\,\mu m$. In a first approximation, the width Δ of this pattern, the number n of its fringes and the spacing s_o of its fringes relative to the sample depend on the fiber voltage U_f and are given by

$$\Delta = c_1\, U_f \qquad s_o = \frac{c_2}{U_f} \qquad n = \frac{c_1}{c_2} U_f^2 \qquad (1)$$

The constants c_1 and c_2 result to $0.18\mu m/V$ and $1.59\mu m\,V$ respectively. Varying the excitation of the projector lenses varies the magnification of the hologram. It does not change either the interference width or the fringe spacing relative to the sample. Therefore a different width or a different fringe spacing is obtained either by using a different fiber voltage or by increasing the magnification using the objective lens.

A ray diagram for an alignment with the objective lens excited is shown in Figure 1, right side. With this geometry, the biprism is below the image lens, so we have to apply a positive voltage on the biprism fiber to get an interference pattern. In order to obtain a medium magnification the objective lens is excited in such a way that its front focal plane is close to the specimen. Even small changes of this distance drastically change the magnification of the sample. This enables us to continuously change the magnification in the image plane with a nearly constant interference width with respect to this plane, and a constant number of fringes. Since the distance from the sample to the focal plane is a constant for a desired magnification, we have to focus the image with the first intermediate lens. The equations (1) are still valid but with different constants for every magnification. For example, Table 1 shows three sets of lens conditions that result in a large change in interference width with essentially no change in number of fringes.

Replacing the biprism fiber usually results in a different diameter ϕ_f of the fiber. According to equation (2) which follows from geometrical considerations on Figure 1, a change in the diameter of the fiber results in a change in the interference width. In order to obtain an identical width as before we have to change the fiber voltage according to equation (1).

$$c\,\delta\Delta = \delta\phi_f \qquad (2)$$

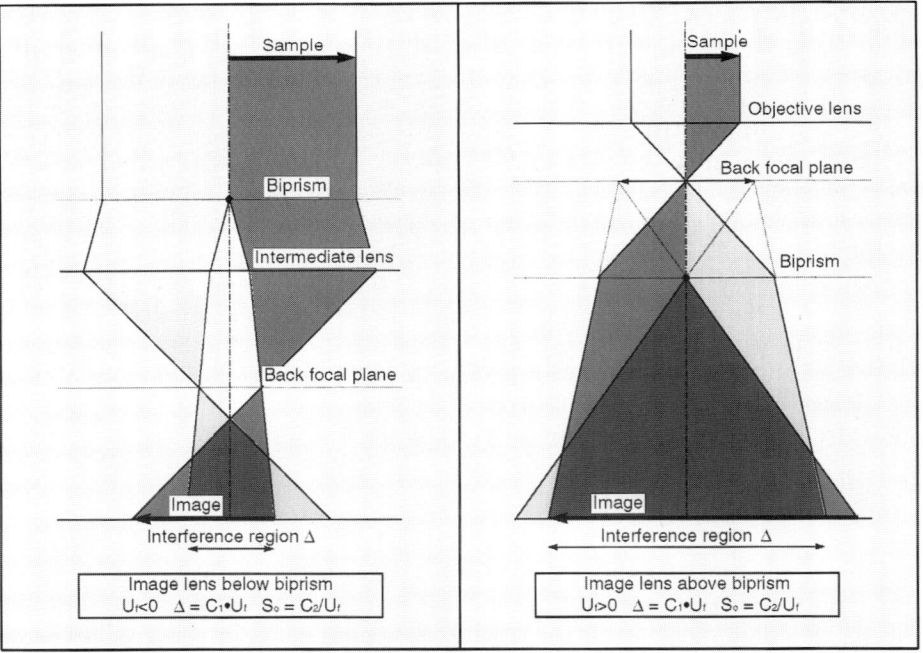

Figure 1. Two possible ray diagrams for taking off-axis holograms.

Table 1
Lens conditions

C1(A)	C2(A)	Obj(A)	I1(A)	I2(A)	P1(A)	P2(A)	U_f(V)	$\Delta(\mu m)$	n
> 1.0	0	0	0.87	3.0	3.0	4.5	-30	5.36	100
> 1.0	0	6.85	0.86	3.0	3.0	4.5	-16.3	0.96	94
> 1.0	0	6.60	0.87	3.0	3.0	4.5	-16.8	0.53	93

2. THEORY OF ELECTROSTATIC FIELDS

The phase shift $\delta\varphi$ of an electron wave influenced by the electric potential V_{el} and the vector potential $\vec{A}$ is given by [3]

$$\delta\varphi = \frac{e}{\hbar} \oint (V_{el}dt - \vec{A}d\vec{r}) \qquad (3)$$

where e is the charge of an electron, $\hbar$ is Planck's constant and the intergration goes over any closed circuit in space time. Landau [4] evaluated this phase shift in a non-relativistic approximation to

$$\delta\varphi(x,y) = \frac{\pi}{\lambda E} \int V_{el}(x,y,z)\, dz \qquad (4)$$

where E and λ are the accelerating voltage and the electron wavelength respectively. The integral is taken along an electron trajectory, which in the following simulation is assumed as a straight line parallel to the z-axis. The microscope gives the integral limits by the crossover of the condensor lens and by the first image plane.

Since the diameter of dielectric polystyrene latex spheres is well known, we used these as samples to determine the magnification of the microscope and the interference width of the holograms for selected microscope data. Holographic investigations on charged latex microspheres were first published by Matteucci et al. [5]. The high symmetry of spheres simplifies the simulations of the phase distribution and the evaluation of the number of electric charges on the sample. In the experiment the sphere is usually placed on a carbon holey foil or on the bar of an empty copper grid. In both cases the carriers are large, conducting, and grounded planes. This allows us to use the charge image method [6] as a model. The electric potential of two oppositely charged points is not changed by a grounded conducting plane with infinite extend centered between the points and perpendicular to a line connecting the points (Figure 2). Removing one charged point does not change the potential on the opposite side of this plane.

Since the potential of one charged point is given by $V_{el} = q/(4\pi\epsilon_o r)$, the potential of a sphere with radius r_s, which can be assumed to be a point at the distance r_s from the carbon foil (x-y-plane), is given by

$$V_{el}(x,y,z) = \frac{q}{4\pi\epsilon_o}\left(\frac{1}{\sqrt{x^2+y^2+(z-r_s)^2}} - \frac{1}{\sqrt{x^2+y^2+(z+r_s)^2}}\right) \qquad (5)$$

where the center of the sphere is on the z-axis. In the case of a sphere on a copper grid (y-z-plane) with its center at a distance r_s from this plane and on the x-axis, the potential is also given by equation (5), but we have to exchange the x-component with the z-component. In order to evaluate the phase distribution we integrate equation (4) using equation (5) for the electric potential. In the case of a sphere on a carbon foil, the phase shift at points with $x^2 + y^2 > r_s^2$ follows as

$$\varphi(x,y) = \frac{\pi}{\lambda\,E} \frac{q}{4\pi\epsilon_o} \ln \frac{r_s + \sqrt{r_s^2 + x^2 + y^2}}{-r_s + \sqrt{r_s^2 + x^2 + y^2}} \tag{6}$$

and in the case of a sphere on a copper grid, the phase shift at points with $(x-r_s)^2 + y^2 > r_s^2$ and $x > 0$ is given by

$$\varphi(x,y) = \frac{\pi}{\lambda\,E} \frac{q}{4\pi\epsilon_o} \ln \frac{(x+r_s)^2 + y^2}{(x-r_s)^2 + y^2} \tag{7}$$

Equiphase lines are equipotential lines of the projected potential. For reasons of symmetry in the case of a sphere on a carbon foil, the field lines are perpendicular to the phase lines. In the case of a sphere on a copper grid, this is valid only as an approximation.

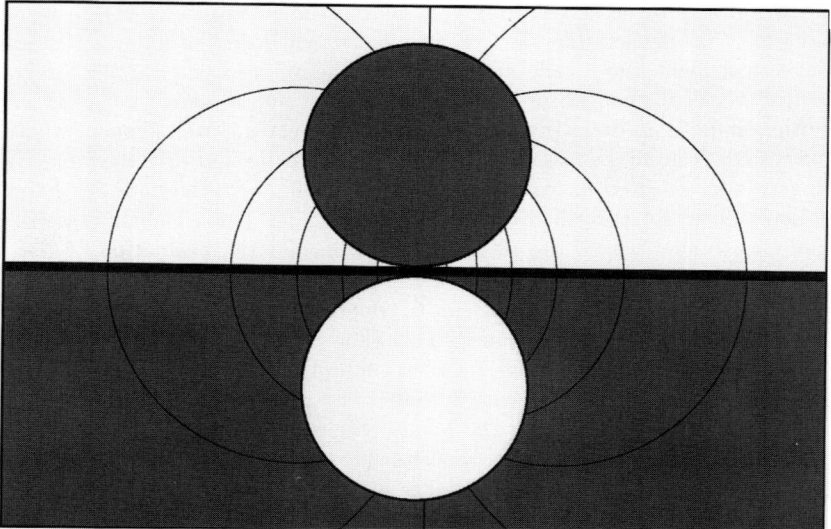

Figure 2. Field lines between charged point and grounded plane are identical to field lines between two oppositely charged points.

2.1. Influence of the field on the reference wave

Since the electric field of small charged particles in the microscope is long ranged with respect to the beam diameter, the reference wave is no longer a plane wave. Figure 3, left, shows an electrically charged latex sphere and an interference pattern at a distance of about twice the sphere diameter. The reconstructed phase distribution reveals an electric field at this distance. As we move the biprism towards the sphere, the reference wave is more strongly influenced by the field.

In order to simulate the phase distribution or to evaluate the number of charges on the particle we have to take into account this effect. To do this we first and most importantly have to assume the shape of the wave front, which is very difficult to find unless we use a sample with a high symmetry. Second, we have to know the distance from the specimen area of interest in the image plane to the centerline of the interference pattern (Figure 3, left). This distance is found adding half the interference width to the distance from the specimen to the centerline of the interference pattern. Third, we have to know the interference width.

Figure 4, left, shows a possibility to precisely measure this width. The biprism fiber is placed below the center of the sphere. If no voltage is applied on the fiber two different rays leaving one point of the sample intersect in the image plane following the laws of optics. The sphere is imaged by the objective lens into one sphere. However, if a voltage

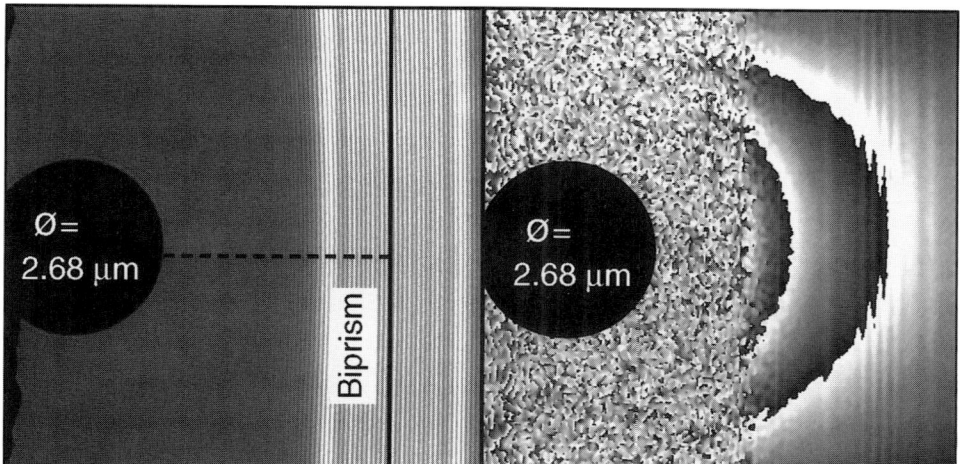

Figure 3. A charged sphere influences the interference pattern (left) at a distance of about twice of the diameter of the sphere, as seen in the phase distribution (right). In order to correctly interpret the measured date we should not neglect the influence of the field on the reference wave.

is applied on the biprism fiber then the rays on the left side of the fiber are deflected by the electric field around the biprism in such a way that they do not intersect in the image plane with the rays on the right side of the fiber. Rays on the right (left) side of the fiber do intersect in the image plane with each other. This means, that one point of the sample is now imaged into two points in the image plane. Since the deflection angle of the beams at the fiber depends on the applied voltage there is exactly one voltage at which the two spheres in the image are adjoining. Then the distance from the object wave to the superposed reference wave which is the interference width is the precisely known sphere diameter. A higher fiber voltage separates the spheres, a lower voltage overlaps them. Figure 4, right, shows the measurement of the interference width. The two spheres with a diameter of $2.68 \mu m$ are adjoining at a fiber voltage of $-15V$. Since the interference width depends only on the applied fiber voltage according to equation (1), we now know all widths for this magnification.

In order to simulate the phase distribution of a sphere on a foil or a grid measured with a non-plane reference wave we have to replace x and y in equations (6) and (7) by the coordinates of the reference wave x_r and y_r and subtract the resulting phase from the object phase. Since we know now the phase difference $\delta\varphi$ between two points (x_1, y_1) and (x_2, y_2) corrected for a non-plane reference wave, the evaluation of the only missing parameter, the amount of charge q on the sphere, is straightforward.

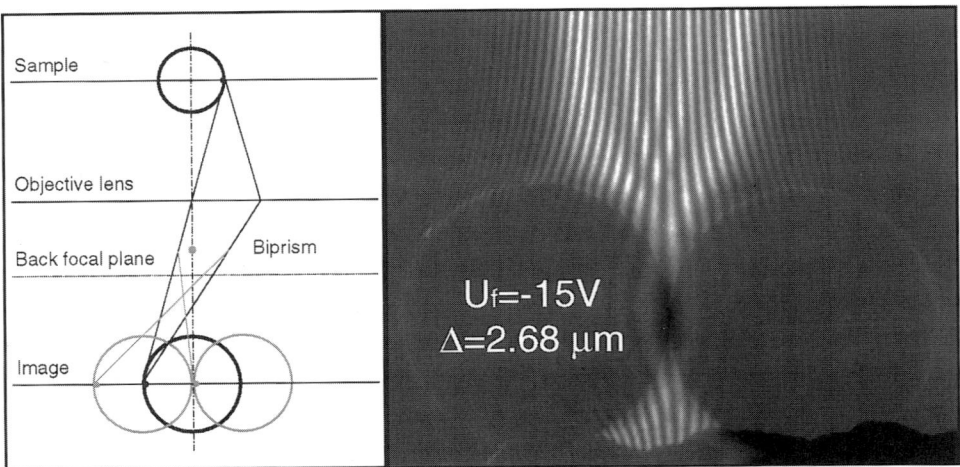

Figure 4. Left: To measure the interference width we place the biprism fiber below the center of the sphere. The deflection of some beams to the wrong side of the fiber results in two spheres in the image plane. Right: At one specific fiber voltage the two spheres are adjoining indicating the sphere diameter is width of the interference pattern.

2.2. Simulations

A plane reference wave results in a phase distribution of concentric equiphase lines around a sphere on a foil. Taking into account the influence of the field on the reference wave results in a disturbed symmetry seen in Figure 5, left side. The image at top shows the simulation in good agreement to the measurement seen on bottom. The noisy phase lines around the sphere are due to a reference wave traveling through the carbon foil which decreases the fringe contrast. The right side shows at top a simulation of a sphere on a copper grid and on bottom the measurement in good agreement too. Since the field lines are nearly perpendicular to the phase lines, the left images suggest field lines ending at the grounded plane.

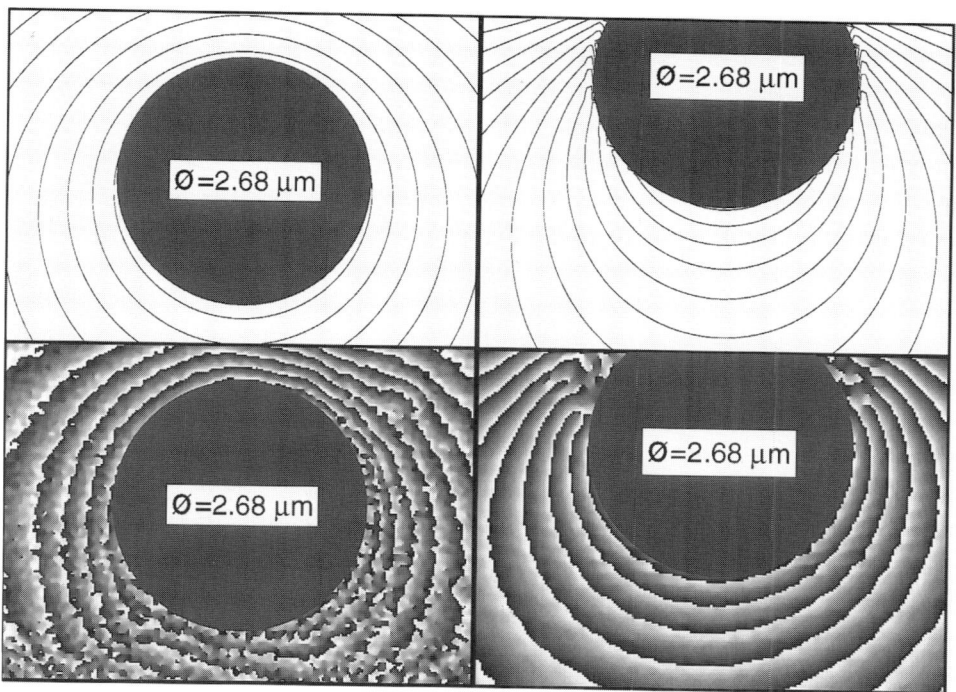

Figure 5. Left (Right) shows at top a simulation of a charged sphere on a carbon foil (copper grid) in good agreement to the measurement on bottom.

3. EXPERIMENTAL RESULTS

A higher intensity of the electron beam charges the sample more than a lower intensity, as seen by different phase lines on the left side of Figure 6. On bottom the electron intensity was about 14 times higher than at top resulting in the absolute value of charge of about 16 electrons at top and 35 electrons on bottom. The interference width in both images is exactely the same. Changing this width also changes the phase lines due to a different reference wave. The broader interference width in the right side image on bottom results in more phase lines than in the image at top. Both images were taken with a constant beam intensity positively charging the spheres by the same absolute value of about 26 electrons. Assuming a plane reference wave gives only 20 electrons at the top image. The charging additionally depends on the supporting layer. Most spheres on a carbon foil show a higher amount of charge than spheres on a grid.

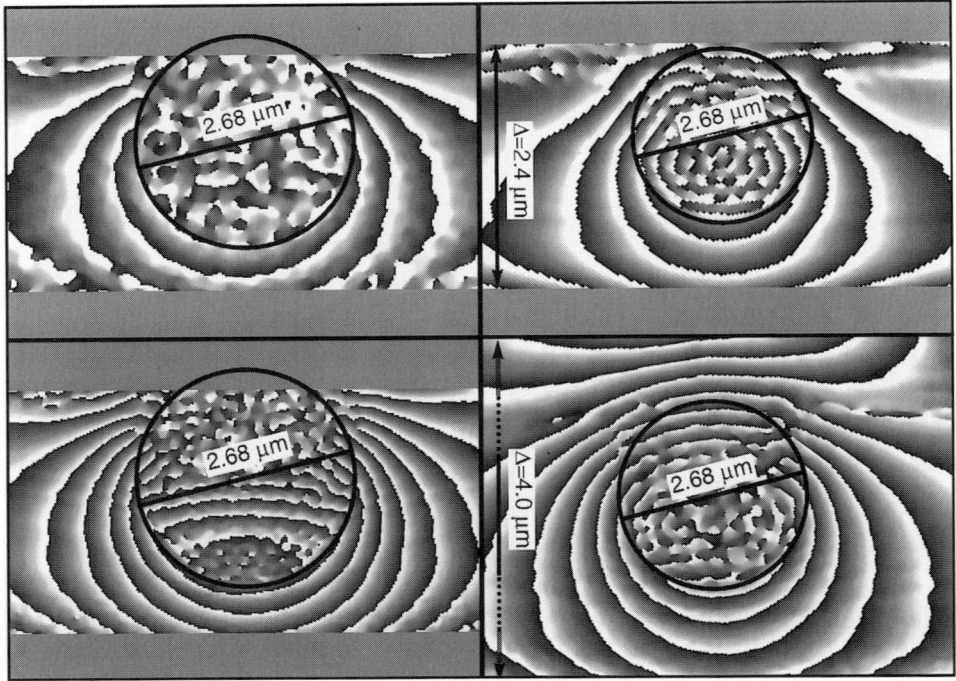

Figure 6. Left (Right): The different number of phase lines is due to a lower intensity of the electron beam (a smaller interference width) at top.

4. ARTEFACTS

Any difference in the reconstructed image wave from the object wave is an artefact. The artefacts most often encountered are the aberrations of the electron lenses which limit the resolution of the microscope. The only artefacts discussed here are the Fresnel fringes produced by the biprism fiber and the fringe branching of hologram fringes.

4.1. Fresnel fringes

If we use a thick sample impervious to the electron beam we still see the electrons of the reference wave in the image but there are no interference fringes because the superposed second wave is missing. The fringes we still see inside the sample with a highly coherent beam are Fresnel fringes caused by diffraction at the edge of the biprism fiber.

These Fresnel fringes are seen in the hologram in Figure 7 inside a thick sphere. Outside the sphere we see both the Fresnel fringes and the interference fringes. Since these fringes are not part of the object wave the structures inside the sphere seen in the reconstructed phase distribution in Figure 7, middle, are artefacts. This is not always obvious. It is very tempting to interpret the lines inside the sphere in the reconstructed phase distribution in Figure 6, bottom right, as phase lines due to the electric field.

The phase distribution of a $482nm$ sphere in Figure 7, right side, shows the inner potential as concentric circles superposed to the Fresnel fringes seen as noisy straight lines. Only the simple geometry of this sample allows us to see these as separate and different effects. Samples with more complex structure mix up both parts in such a way that it becomes nearly impossible to get reliable informations on the potential distribution of the sample. Evaluating the mean inner potential of latex using equation (4) and a phase shift of 7π in the center of the $482nm$ sphere yields $7.9V$. This reconstruction clearly shows that $482nm$ thick latex can be coherently imaged by $200kV$ electrons.

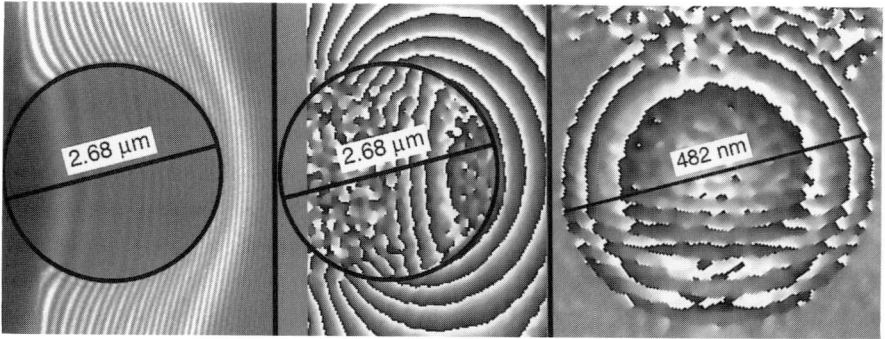

Figure 7. Left: Hologram with Fresnel fringes inside a $2.68\mu m$ latex sphere. Middle: Structures in reconstructed phase are artefacts due to Fresnel fringes. Right: Inner structure caused by Fresnel fringes and mean inner potential of a $482nm$ sphere.

4.2. Fringe branching

In some holograms we detect fringes which seem to branch as seen in the hologram of a nickel tip shown in Figure 8, right, [7] marked with an arrow. In the hologram of a CrO_2-needle in Figure 8, left, the fringes left of the branching fringe are bent to the left, the fringes right of it are bent to the right. Note that the fringe contrast along the border line to the right of the marked fringe is poor compared to the fringes in vacuum or inside the sample. The corresponding phase distribution shows a phase jump of π at the point of the branching represented by a colour jump from black to white.

A possible explanation for this effect is an object charged by the electron beam. This may be modelled by a capacitor consisting of two thin parallel lines, one charged positive, the other one charged negative. Figure 9 shows a simulated interference pattern using this model. The fringe spacing inside the sample is much smaller than outside. If the spacing outside is at a resolution limit, the spacing within the specimen cannot be detected. Then the poor resolution results in a decreased fringe contrast inside the capacitor and in an artificial fringe branching at its ends. The decreased fringe contrast and the fringe branching may cause misinterpretations of the reconstruced amplitude and the phase as well.

In order to avoid the fringe branching there are two parameters which are easy to vary in the microscope. First, a different fringe spacing is achieved by a different fiber voltage, and second, the position of the biprism fiber relative to the sample may be varied using a rotatable biprism. Rotating the biprism fiber also changes the fringe spacing inside the sample.

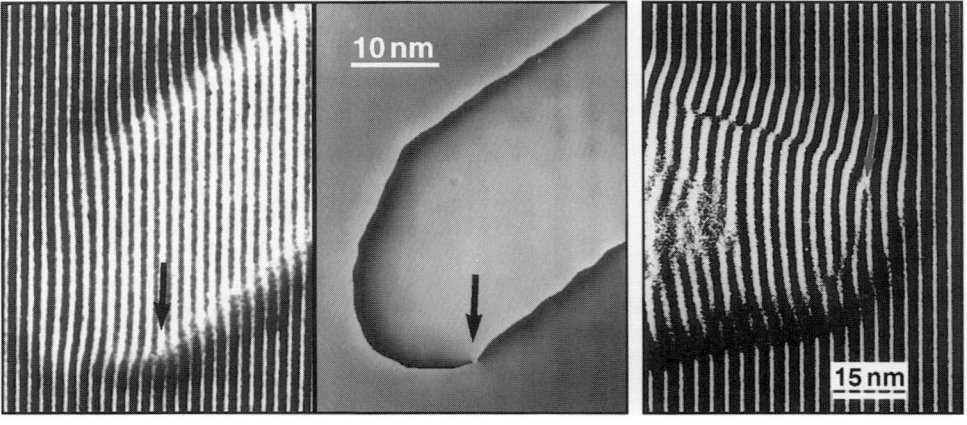

Figure 8. Holograms with fringe branching. The branching fringe is marked with an arrow. Left: CrO_2-needle and reconstructed phase. Right: Nickel tip.

Figure 9, right, shows at top the biprism fiber placed at an angle of about 45° with respect to the capacitor and on bottom it is placed nearly parallel to the sample. In the first case the spacing inside and outside is nearly the same avoiding the branching. In the second case the fringe branching still remaining is caused by a fringe spacing being too large simulating Figure 8, right. Choosing the suitable parameters should always eliminate the artificial fringe branching. With respect to the branching in magnetic samples Tonomura [8] stated 'It should be remembered that such a dislocationlike pattern should never appear for any divergence-free flux field'.

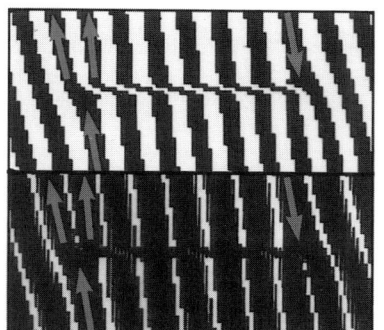 

Figure 9. Simulation of fringe branching. The fringe branching disappears when using a high fringe contrast and/or the right orientation of the fiber with respect to the sample. Bottom right simulates the nickel tip in Figure 8.

5. REFERENCES

1. G.Möllenstedt and H.Düker, Zeitschrift für Physik, Vol.145 (1956) 377.
2. R.Buhl, Zeitschrift für Physik, Vol.155 (1959) 395.
3. Y.Aharonov and D.Bohm, Phys.Rev., Vol.115,No.3 (1959) 485.
4. L.D.Landau and E.M.Lifshits, Quantum Mechanics: Non Relativistic Theory, Pergamon, Oxford, 1965.
5. G.Matteucci, G.F.Missiroli, J.W.Chen and G.Pozzi, Appl.Phys.Lett., Vol.52,No.3 (1988) 176.
6. J.W.Chen, G.Matteucci, A.Migliori, G.F.Missiroli, E.Nichelatti, G.Pozzi, and M.Vanzi, Phys.Rev., A40 (1989) 3136.
7. B.Frost, Dissertation, University of Tübingen,Germany (1993).
8. **A.Fukuhara, K.Shinagawa, A.Tonomura and H.Fujiwara, Phys. Rev., B27 (1983) 1839.**

This work was performed at the Oak Ridge National Laboratory, sponsored by the Directed R&D Program of Oak Ridge National Laboratory, managed for the DOE by Martin Marietta Energy Systems, Inc. under contract DE-AC05-84OR21400 and supported by an appointment to the Oak Ridge National Laboratory Postdoctoral Research Program administered by the Oak Ridge Institute for Science and Education.

Quantitative applications of off-axis electron holography

David J. Smith[a,b], W.J. de Ruijter[a,c], M. Gajdardziska-Josifovska[a,d], M.R. McCartney[a] and J.K. Weiss[a,c]

[a]Center for Solid State Science, Arizona State University, Tempe, Arizona 85287

[b]Department of Physics and Astronomy, Arizona State University, Tempe, Arizona 85287

[c]EMiSPEC Systems Inc., 2409 S. Rural Rd., Suite D, Tempe, Arizona 85282.

[d]Department of Physics and Laboratory of Surface Studies, University of Wisconsin-Milwaukee, P.O. Box 413, Milwaukee, Wisconsin 53201.

Quantitative recording with a slow-scan CCD camera has facilitated several novel applications of off-axis electron holography. The mean inner potential of several crystalline materials, namely Si, MgO, GaAs and PbS, has been determined with an accuracy of about 1%. Heterogeneous interfaces have been analyzed to obtain information about layer width and interface abruptness. The reconstructed, energy-filtered amplitude signal has been used to extract the inelastic mean-free-path of MgO and Si, and to obtain compositional images that are independent of thickness. The potential distributions across Si/Si p-n junctions have been directly observed with nanometer-scale spatial resolution.

1. INTRODUCTION

The technique of electron holography was originally proposed as a means of overcoming the resolution-limiting spherical aberration of the objective lens [1], since it provides access to phase and amplitude information that is normally inaccessible to conventional TEM imaging. Historically, off-line processing of photographic negatives to extract the desired holographic information has been laborious and time-consuming. Nowadays, with the advent of the slow-scan charge-coupled-device (CCD) camera [2,3], off-axis electron holograms can be recorded digitally with high linearity over a wide dynamic range, something which is difficult to achieve photographically. It is then possible to extract reliable quantitative information about the sample under examination, including specimen thickness, mean inner potential and potentials of magnetic and electrostatic fields. Care must still be taken to differentiate the effects of intrinsic fields and to avoid complications arising from dynamical diffraction which can be substantial for zone-axes projections of crystalline samples. In this paper, we begin by briefly discussing some experimental factors which might potentially limit the accuracy of phase and amplitude determination. We then describe recent holographic studies in our laboratory in which we have taken advantage of digital recording to obtain precise and useful structural and compositional details about several important types of materials.

2. QUANTIFICATION

In order to quantify reliably and accurately the amplitude and phase information contained in a digitally recorded off-axis electron hologram, in particular their relationship to physical characteristics of the specimen, it is first necessary to correct for geometric distortions introduced by the microscope imaging system, as well as for fiber optic shear in the recording device [4]. These effects result in local or extended shifts in interference fringe positions that are liable to be misinterpreted as due to phase shifts caused by the sample. For example, phase shifts due to the projector lens system may occur that can be on the order of π across a typical field of view so that phase measurements will be seriously compromised in accuracy unless proper correction is done. Fortunately, this correction is simply accomplished with a CCD detector having a fixed position relative to the imaging system (unlike the case with photographic film). As demonstrated in Fig.1, a reference hologram from vacuum should be systematically recorded with the specimen removed and the reconstructed distorted phase can then be subtracted from that of the original image wave.

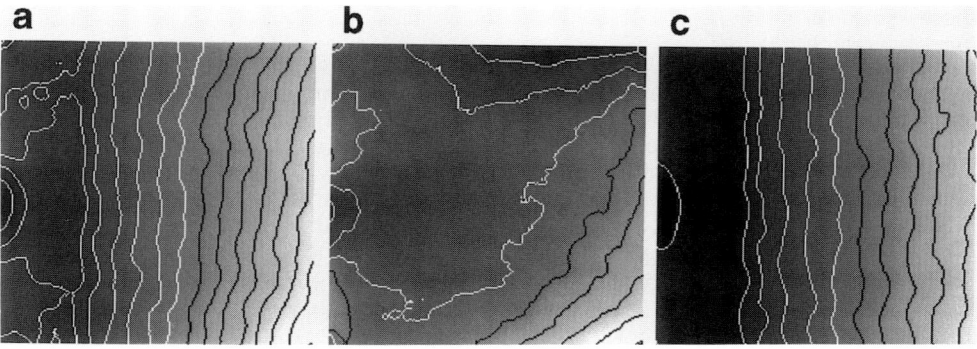

Figure 1. Demonstration of geometric distortion correction for Philips 400ST-FEG equipped with Gatan 679 slow-scan CCD camera. Sample: GaAs cleaved crystal wedge. (a) initial phase image with $\pi/4$ equiphase contour lines; (b) phase image of vacuum (sample removed) with $\pi/8$ equiphase contours showing distortions due to imaging system; (c) distortion-corrected phase image with $\pi/4$ equiphase contours.

For the highest possible accuracy in phase determination, careful attention must also be given to phase unwrapping and flattening of the vacuum phase. The former process is necessary to confirm the occurrence of any 2π phase discontinuities, and can be carried out routinely using special phase-unwrapping algorithms [5]. The latter problem arises from the fact that the position of the sideband resulting from conventional Fourier processing of the hologram can only be located to within one pixel [6]. The subsequent reconstructed phase image contains an additional phase due to the incorrect centering which is in the form of a tilted plane: removal is again easily accomplished with a simple processing routine [5]. Finally, reduction in statistical errors can be reduced by area averaging but at the obvious expense of spatial resolution.

3. SURVEY OF RESULTS

In addition to aberration correction to overcome resolution limits [7], off-axis electron holography has been applied to a wide variety of materials problems, especially those having intrinsic magnetic and electric fields where particular advantage can be taken of phase information that is unavailable to other techniques [8]. We have chosen to concentrate our efforts on quantitative studies at moderate resolution, mostly involving samples without electric or magnetic fields. For such cases, and in the absence of dynamical diffraction effects, the relative phase change (Φ) of the electron wave passing through the sample can be related to the mean inner potential (V_0) of the material, through the expression [9]

$$\Phi = C_e V_0 t \tag{1}$$

where C_e is a constant depending only on the incident electron beam energy and t is the sample thickness in the beam direction. For a sample of known geometry, such as a cleaved crystal wedge, this relationship provides a direct means for determination of V_o although some dynamical correction of the phase is usually still necessary. Additionally, if a thickness profile of the sample can be inferred, profiles across selected interfaces that are dependent on composition can be determined with high spatial resolution. Finally, by reference to the energy-filtered amplitude image, an alternative method becomes available for determining the mean-free-path for inelastic scattering, and it also becomes possible to obtain a composition image that is independent of thickness. We have utilized the amplitude image in the process of observing the potential distribution across an Si/Si p-n junction.

3.1. Measurement of mean inner potential.

The mean inner potential of a solid effectively represents the volume average of the atomic potential, with a value typically between about -5 and -30V, depending on composition, bonding and structure [9]. Accurate knowledge of V_o is important in applications involving low-energy-electron diffraction (LEED) and reflection high-energy electron diffraction (RHEED) but methods for V_o measurement have relative errors reported to be in the range of 2.5% to about 10% [10]. We have recently undertaken off-axis electron holographic studies of crystal wedges of known angle, and hence known thickness, using quantitative digital recording and processing [11,12]. Table 1 summarizes the V_o values that were obtained from plots of phase vs. thickness for crystals of Si, MgO, GaAs and PbS before and after correction for dynamical diffraction. Experimentally, dynamical interactions were minimized by deliberately tilting the respective crystal away from the zone axis, with further reference to the Kikuchi line pattern to ensure that non-systematic reflections were also avoided. By taking proper account of contributions from dynamical diffraction by means of Bloch wave calculations, and with careful evaluation of experimental incertainties, we have achieved measurement accuracies of about 1% which is a factor of about three better than achieved previously by other methods. Nevertheless, significant discrepancies exist (i.e. beyond the calculated experimental errors) between our holographic results and some previous RHEED studies. Further work is needed to fully understand the source(s) of these differences, since they might provide insight into the contributions of surface states to the measured inner potential. It is also of considerable interest to determine whether crystal doping will lead to measurable changes in V_o values.

Table 1
Values of mean inner potential obtained from phase images of electron holograms before and after dynamical correction.

Material	Wedge angle	V_o [V] measured	V_o [V] corrected
Si	109.5°	9.11±0.5%	9.26±0.5%
MgO	90°	12.78±0.6%	13.01±0.6%
GaAs	90°	14.13±1.2%	14.53±1.2%
PbS	90°	16.35±0.7%	17.19±0.7%

3.2. Observation of interfaces.

Interfaces between similar or dissimilar materials have an important role to play in determining macroscopic properties. A knowledge of interfacial structure, particularly local roughness and compositional gradients, is often needed in order to rationalize physical behavior. In our recent studies of interfaces, we have exploited the phase sensitivity of the off-axis electron hologram to provide direct information about compositional changes across several types of heterogeneous junctions [13]. In the absence of strong diffraction effects, and assuming that the specimen thickness does not change appreciably across the interface, then the phase profile extracted from an off-axis electron hologram can be interpreted directly in terms of local variations in the mean inner potential and hence the composition. In practice, sample thickness will increase steadily away from the specimen edge but it was our experience, as demonstrated in Fig. 2, which shows phase profiles from an Mo/Si multilayer film, that a reasonable thickness profile could often be estimated. The thickness-corrected

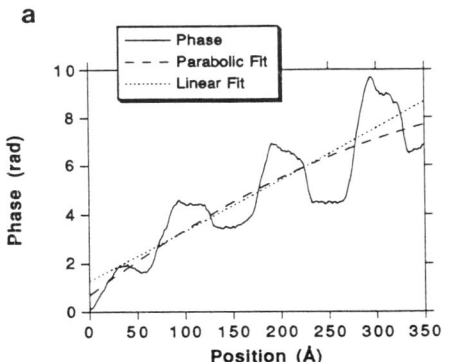
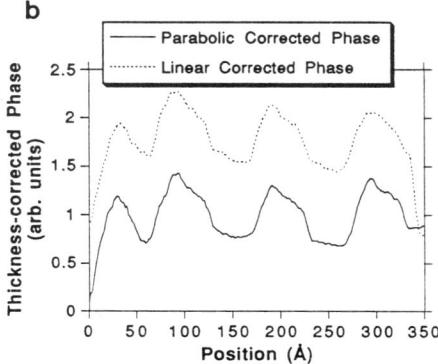

Figure 2. One-dimensional profile of reconstructed phase image from Mo/Si multilayer film showing uncorrected (a) and thickness-corrected (b) profile, as extracted from a ten-pixel-wide area averaged parallel to interface.

profile was then much more easily interpreted to provide the desired compositional information. For example, with the parabolic thickness correction in Fig. 2b, the Mo-Si and Si-Mo interfacial regions were measured to have widths of 12 Å and 6Å respectively, with precisions estimated to be on the order of 2 Å [12]. These values agreed closely with qualitative estimates made previously using high-resolution electron micrographs [14].

Our observations were necessarily restricted to interfaces in the edge-on geometry, and dynamical effects could be ignored since the layered structures in our study were either amorphous or consisting of small, randomly oriented grains. Spatial averaging along each boundary, if reasonably assumed to have one-dimensional composition profile, was also employed to enhance the precision with which the phase determination could be made, and hence the accuracy with which the compositional variations could be determined. An interesting example is featured in Fig.3 which shows the result of processing the off-axis electron hologram of a Si_3N_4/Si_3N_4 grain boundary. The boundary is effectively invisible in the reconstructed amplitude image but it is clearly visible in the phase image. Measurement of the extracted phase profile, averaged across a 40-pixel-wide area, indicates that the amorphous layer between the two grains has a structural width of about (12 ± 2)Å. Thus, the dimensions of the mean inner potential variations can be directly measured from the phase images, unlike defocussed bright-field or weak-beam dark-field imaging where the layer can be imaged but its width cannot be easily or directly quantified.

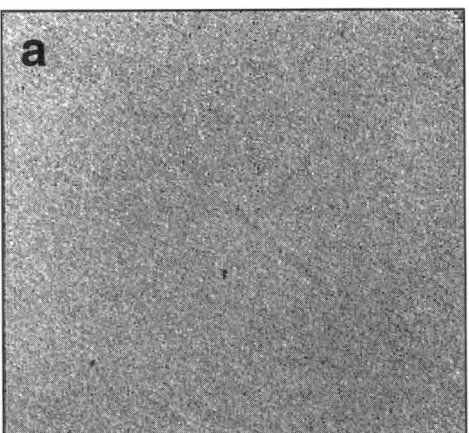
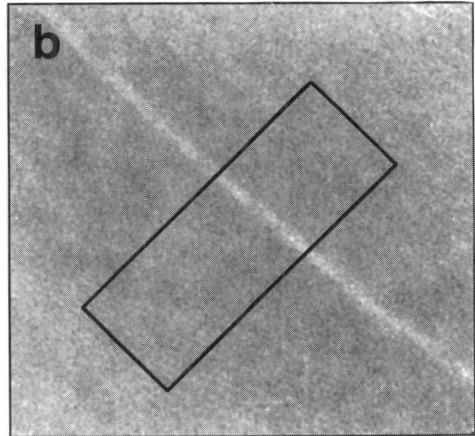

Figure 3. Reconstructed amplitude (a) and phase (b) images from an Si_3N_4/Si_3N_4 grain boundary.

Of course, in these studies of interfaces, it should be appreciated that off-axis electron holography is restricted to thin regions close to the sample edge, diffraction effects must be carefully avoided and specific details of elemental compositions are not provided. Nevertheless, as a technique for studying interfaces, it has the distinct advantages of high spatial resolution and sensitivity, combined with relatively simple processing.

3.3. Determination of t/λ_i.

Interesting and unforeseen developments have grown from the recent realization that the complicated angular distribution of elastically and inelastically scattered electrons contained in an off-axis electron hologram becomes spatially separated as a result of the first Fourier transform (FT) processing step of the holographic reconstruction [15]. Figure 4 shows a line scan taken across the central auto-correlation peak and one sideband of the FT from an MgO crystal wedge hologram, plotted on a logarithmic scale [16]. At the positions of the numerical aperture used for reconstruction, as arrowed, the amplitude of the sideband has dropped by three orders of magnitude. Most of the inelastic scattering from the sample is concentrated close to the central peak so that the sideband used for reconstruction can effectively be regarded as energy-filtered. The corresponding reconstructed amplitude image, for many years almost neglected by the electron holographic community, has some interesting and entirely novel applications [12, 16, 17].

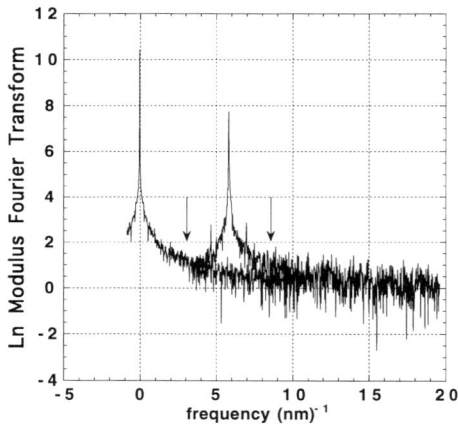

Figure 4. Logarithmic plot showing modulus of the FT from an MgO hologram. The arrows correspond to positions of numerical aperture used for reconstruction.

In electron-energy-loss spectroscopy (EELS), the materials scientist is often interested in the scattering parameter t/λ_i where t is the thickness and λ_i is the mean-free-path for inelastic scattering. Analysis of the various intensity contributions to a hologram [16] indicates that normalization of the reconstructed image amplitude, by reference to the vacuum signal, followed by taking the natural logarithm, gives another image that explicitly displays the spatial distribution of t/λ_i. In the event that the thickness is known, for example by reference to a crystal wedge of known cleavage angle, then λ_i can be determined absolutely. The method is illustrated in Figure 5 which shows averaged thickness profiles from t/λ_i images for single crystal wedges of MgO and Si. Despite the overall noisy appearance, which is related to noise present in the amplitude image, analysis gives values for the bulk mean-free-path for MgO and Si, respectively, of 71±5nm and 92±7nm [16]. These compare favorably with calculated and experimental values from EELS. Further analysis [17] shows that, by dividing the phase image by the logarithm of the normalized amplitude, the thickness dependence can be completely eliminated, leading to a composition image that is independent of specimen topography. The method should thus be particularly useful when off-axis electron holography is applied to specimens of unknown topography.

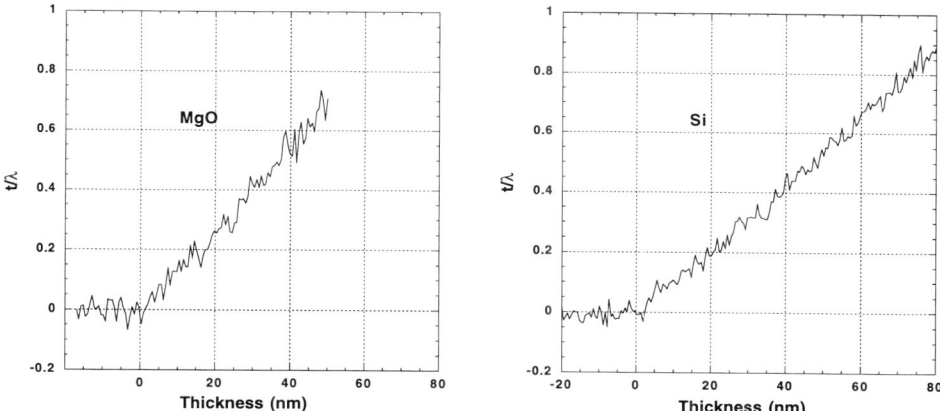

Figure 5. Averaged profiles from t/λ images plotted versus thickness from off-axis holograms of MgO and Si.

3.4 Observation of p-n junctions.

Determination of the two-dimensional potential distributions within semiconductor devices has long been a goal of device engineers. We have used [18,19] off-axis electron holography to observe a $2 \times 10^{18}/cm^3$ p- and n-doped Si/Si p-n junction. By utilizing digital recording and processing, and the method for thickness determination described in the preceding section, we have successfully extracted two-dimensional maps of the depletion region potential. For a defect-free region, we measure a relatively abrupt change in potential of $1.5 \pm 0.2V$ across a lateral distance of 20nm. Our initial results confirm the great promise of off-axis electron holography as a technique for imaging electric potentials but further work is needed to establish whether the sensitivity and spatial resolution are adequate to cope with the next generations of device structures, and also to establish whether these maps can be correlated with dopant profiles.

4. PERSPECTIVE

For many years, electron holography has been a technique mainly of interest to a few specialized laboratories, but it has recently experienced a major revival because of the ready availability of FEG microscopes and the widespread realization that it provides access to much useful structural information. Coupled with very recent developments in digital recording and data processing, it has become possible to extract valuable quantitative data. Our measurements of mean inner potential and its variation across heterogeneous interfaces, and our utilization of the energy-filtered amplitude image to obtain composition images and electrostatic potential distributions, typify the sorts of quantitative applications that are likely to become widespread over the next few years.

The work at Arizona State University was conducted at the Center for High Resolution Electron Microscopy supported by the National Science Foundation under Grant Nos. DMR- 89-13384 and 91-15680.

REFERENCES

1. D. Gabor, Proc. Roy. Soc. A197 (1949) 45.
2. P.M. Epperson, J.V. Sweedler, M.B. Denton, G.R. Sims, T.W. McGurnin and R.S. Aikens, Optical Engineering, 26 (1987) 715.
3. J.C.H. Spence and J.M. Zuo, Rev. Sci. Instrum., 59 (1988) 2102.
4. W.J. de Ruijter, M. Gajdardziska-Josifovska, M.R. McCartney, R. Sharma, D.J. Smith and J.K. Weiss, Scanning Microscopy Supplement 6 (1992) 347.
5. W.J. de Ruijter and J.K. Weiss, Ultramicroscopy 50 (1993) 269.
6. E. Voelkl, L.F. Allard, A.K. Datye and B. Frost, Ultramicroscopy (1995) in press.
7. A. Orchowski, W.D. Rau and H. Lichte, Phys. Rev. Letts. 74 (1995) 399.
8. A. Tonomura, Adv. Phys. 41 (1992) 59.
9. L. Reimer, Transmission Electron Microscopy, 2nd. Ed. (Springer, Berlin, 1989).
10. J.C.H. Spence, Acta Cryst. A49 (1993) 231.
11. M. Gajdardziska-Josifovska, M.R. McCartney, W.J. de Ruijter, D.J. Smith, J.K. Weiss and J.M. Zuo, Ultramicroscopy 50 (1993) 285.
12. M. Gajdardziska-Josifovska, MSA Bulletin 24 (1994) 507.
13. J.K. Weiss, W.J. de Ruijter, M. Gajdardziska-Josifovska, M.R. McCartney and D.J. Smith, Ultramicroscopy 50 (1993) 301.
14. A.K. Petford-Long, M.B. Stearns, C.-H. Chang, S.R. Nutt, D.G. Stearns, N.M. Ceglio and A.M. Hawryluk, J. Appl. Phys. 61 (1987) 1422.
15. A. Harscher, F. Lenz and H. Lichte, Electron holography provides zero-loss images, Proc. 10th. Eur. Cong. Electron Microscopy, last minute brochure, Granada, Spain, p.351 (1992).
16. M.R. McCartney and M. Gajdardziska-Josifovska, Ultramicroscopy 53 (1994) 283.
17. M. Gajdardziska-Josifovska and M.R. McCartney, Ultramicroscopy 53 (1994) 290.
18. M.R. McCartney, D.J. Smith, R. Hull, J.C. Bean, E. Voelkl and B. Frost, Appl. Phys. Letts. 65 (1994) 2603.
19. M.R. McCartney, B. Frost, R. Hull, M.R. Scheinfein, D.J. Smith and E. Voelkl, these proceedings.

Electron holography of p-n junctions

M. R. McCartney[a], B. Frost[b], R. Hull[c], M.R. Scheinfein[a,d], David J. Smith[a,d] and E. Voelkl[b]

[a]Center for Solid State Science, Arizona State University, Tempe, Arizona 85287-1704, U.S.A.

[b]Oak Ridge National Laboratory, Oak Ridge, Tennessee 37831.

[c]Department of Materials Science, University of Virginia, Thornton Hall, Charlottesville, Virginia 22903-2442.

[d]Department of Physics and Astronomy, Arizona State University, Tempe, Arizona 85287-1504.

The potential distribution across a $2 \times 10^{18}/cm^3$ p- and n-doped Si/Si p-n junction has been observed using off-axis electron holography. With digital image recording and processing, and a novel method for thickness determination, we have extracted two-dimensional maps of the depletion region potential. Relatively abrupt changes of potential in the range 1.0-1.5V have been measured across lateral distances of 20-30nm. These values are reasonably consistent with expected Si junction parameters and thus establish the promise of the technique for measuring potential distributions across device junctions and interfaces. The relatively poor signal quality of the reconstructed amplitude image currently limits the available precision of the method. Finally, we report our initial efforts to prepare high-brightness nanotips *in situ* within the electron microscope. These latter experiments are motivated by our desire to improve the spatial coherence characteristics of the electron source.

1. INTRODUCTION

Extraction of the phase image from an electron hologram allows for direct imaging of the electric and/or magnetic potential as a result of phase changes on the incident wavefront caused as the electron wave passes through the sample. For example, the distribution of magnetic lines of force recorded on magnetic tape have been observed [1], and the dynamic motion of individual fluxons in vacuum outside superconducting Pb has been imaged [2]. Quantitative micromagnetic information is now available at nanometer-scale resolution using far-out-of-focus holography implemented in a scanning transmission instrument [3]. Earlier low-resolution investigations of p-n junctions have successfully revealed changes in electrostatic fringing fields associated with changes in sample bias [4,5].

Two-dimensional imaging and measurement of potential distributions within devices is a long-sought-after goal of the semiconductor industry. In principle, electron holography provides this capability and initial results are very promising [6]. It is important, however, to establish whether electron holography has the requisite sensitivity and resolution to image the potential distributions at technologically relevant doping levels. It is equally important to determine whether practical problems such as thickness variations or artefacts due to fringing fields or defects will restrict the usefulness of the technique in mapping potential distributions. This paper presents a theoretical and experimental analysis of the potential distribution for a specific p-n junction, with particular emphasis on the signal-to-noise requirements needed for meaningful mapping. In addition, we describe preliminary results of the *in-situ* preparation of a nano-tip for use as an electron source, in view of its potential for greatly enhancing the coherence of the electron beam and thereby improving the signal-to-noise ratio of recorded holograms.

2. IMAGING p-n JUNCTIONS

2.1 Sample design and field of view

The next generation of semiconductor devices will be built around a sub-0.25-micron design rule. At these dimensions, dopant concentration levels will change by three orders of magnitude over 150nm with corresponding potential drops of 3-4V near the source/drain region of MOSFET devices. With these dimensions and doping concentrations in mind, a highly-doped homo-epitaxial MBE-grown Si/Si p-n junction was prepared as a test structure. For this study, spatial resolution was not an inherent limiting factor, but rather the problem was to maximize the field of view on the sample in order to image the entire potential variation while still including some vacuum at the sample edge so that an absolute, rather than relative, measurement could be made. The off-axis electron holograms were acquired digitally with a 1024x1024 pixel Gatan 679 slow-scan CCD camera. For side-band extraction, 256x256 pixels were used to map the region. i.e., the resolution was limited to less than 1/256 of the desired field of view. The field of view for holographic imaging is, in turn, determined by the overlap region of the electron wavefront which depends on the location of the biprism and the applied voltage, and on the electron optics of the microscope. For the work reported here, the holography was carried out at Oak Ridge National Laboratory on an Hitachi HF-2000 transmission electron microscope equipped with a field-emission gun and an electrostatic biprism located near the second image plane.

The p-n junctions were grown by MBE on n+ Si (100) substrates. Doping was achieved by low-energy (3000eV) implantation of BF_2 (p-type) and As_2 (n-type) during deposition. A 100nm-thick Si buffer layer was first deposited with n-type doping ~$2\times10^{18}/cm^3$, with a deposition temperature ramped from 750°C to 550°C. The next layer was deposited at 550°C, and consisted of a 50nm-thick Si layer also doped n-type (~$2\times10^{18}/cm^3$). The final Si layer had a thickness of 30nm, and was doped p-type (~$2\times10^{18}/cm^3$). Cross-sectional samples of the Si/Si junctions suitable for electron microscopy/holography were prepared by mechanical polishing, dimpling and Ar ion-milling using standard procedures.

2.2 Potential distribution at abrupt p-n junctions

The p-n junction configuration called the step junction is characterized by a constant n-type dopant density that changes with position in a stepwise fashion to a constant p-type dopant density. Using the depletion approximation, we assume that the mobile carrier concentration is completely depleted on either side of the junction and that the mobile majority-carrier densities abruptly become equal to the respective dopant concentrations at the edges of the depletion region. The charge density is therefore zero everywhere except in the depletion region where it takes on the value of the ionized dopant concentrations. In this case, Poisson's equation can be solved analytically to give the expected potential variation, depletion region width and electric field [7]. For the doping level of the sample reported here, $2 \times 10^{18}/cm^3$, we have:

$$\phi_i = \frac{kT}{q} \ln\left(\frac{N_a N_d}{n_i^2}\right) = 0.97 V \qquad (1)$$

where ϕ_i is the potential drop across the unbiased junction, N_a, N_d are the acceptor and donor dopant concentrations and n_i is the number of carriers in undoped Si.

The depletion region width, in the absence of any applied voltage, is given by

$$x_d = x_n + x_p = [\, 2\, \varepsilon_{Si}\, (\, 1/N_a + 1/N_d\,)\, \phi_i/q\,]^{1/2} = 36 nm \qquad (2)$$

where x_d is the total depletion layer width and ε_{Si} is the relative permittivity of Si.

The electric field strength is a maximum at the interface and should vary linearly to the edge of the depletion region:

$$E_{max} = 2\, \phi_i / x_d = 5.6 \times 10^7 V/m \qquad (3)$$

Schematic drawings of the doping profile and charge distribution in the depletion region for a symmetric step p-n junction are shown in Fig. 1. The resulting electric field and potential distribution for an unbiased junction are also shown. The potential exhibits parabolic curvature of opposite sense on either side of the interface and remains flat outside the depletion region in the absence of any net charge.

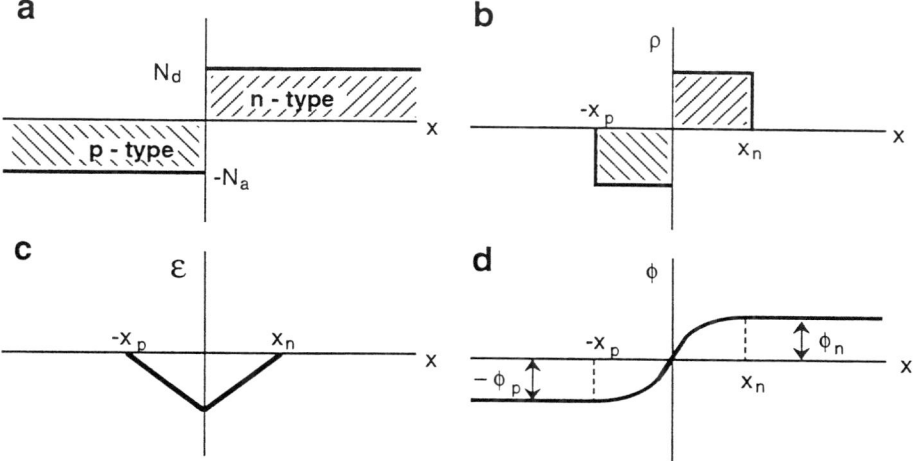

Figure 1. Schematic illustration of symmetric p-n junction: (a) net dopant profile; (b) charge distribution; (c) electric field; (d) potential distribution.

2.3 Phase imaging for electrostatic potentials

In a non-magnetic sample, and in the absence of strong diffracting conditions, any structure in the phase image results from thickness variations and local changes in the projected electrostatic potential of the sample, V(r), as obtained from the electron-optical path difference [8]

$$\phi(r) = \frac{2\pi}{\lambda}\{n(r)-1\}t(r)$$

$$\phi(r) = \frac{2\pi}{\lambda E}\frac{E+E_0}{E+2E_0}V(r)t(r) = C_E V(r)t(r) \tag{4}$$

where λ is the electron wavelength, n is the index of refraction, E_o is the electron rest energy, E is the relativistically- corrected incident electron energy and C_E is an energy-dependent constant which equals 0.009245 rad V^{-1} nm^{-1} for 100keV electrons. The reconstructed phase image thus represents a simple product of the local potential and the local specimen thickness.

Our immediate problem is to remove the thickness dependence from this expression, so that the local variations in the electrostatic potential can be determined. Moreover, since phase changes across the depletion region of a p-n junction may be subtle, the phase images must be carefully processed to remove any changes caused by distortions in the imaging and recording process, as well as thickness variations.

We have recently shown [9] that the relative thickness of the sample at any point in a hologram can be determined using the expression

$$t/\lambda_i = \ln(I_{tot}/I_{zero}) = -2\ln(A_o/A_r) = -2\ln(Amp_n) \tag{5}$$

where t is the thickness, λ_i is the bulk inelastic mean-free-path, and A_o and A_r are the energy-filtered (zero-loss) object wave and reference wave amplitudes, respectively. Thus, Amp_n represents the amplitude image as normalized by the amplitude image of a vacuum hologram acquired under identical illumination conditions. In addition, this vacuum hologram is used to correct for geometric distortions of the imaging system and for non-isoplanicity of the incident wavefront [10]. The normalization and phase corrections may be conveniently carried out in one step by division of the complex sample image by the complex vacuum image. Once the thickness distribution has thus been determined then the phase image can be suitably corrected by division by the thickness image.

Figure 2 shows an original hologram of the p-n junction, plus the corresponding fast Fourier transform, the corrected phase image and the thickness image. The sample, which had been prepared in Arizona, had been partially broken in transit to Tennessee. The lower edge of the crystal, running parallel to the junction (marked on the hologram) was the original ion-thinned edge. The side running approximately perpendicular to the junction is the result of the sample breaking. Also visible in the upper part of the hologram is a faint outline of the lower edge of the sample which arises from the second image formed in the overlap of the images on either side of the biprism. The phase image, Fig. 2c, shows a generally increasing phase from lower right to upper left, following the expected increase with thickness but with some additional contrast visible at the junction. A brighter region near the lower edge of the sample is consistent with slight charging of the sample under electron irradiation, suggesting that the sample was not well-grounded and that the junction was therefore not shorted by the holder. The thickness image, Fig. 2d, has several anomalous contrast features. The lower, p-side, of the interface appears to be thicker than the n-side. In addition, considerable contrast is visible around defects in the sample. These defects were most likely to have been formed as a result of the nearby crystal fracture.

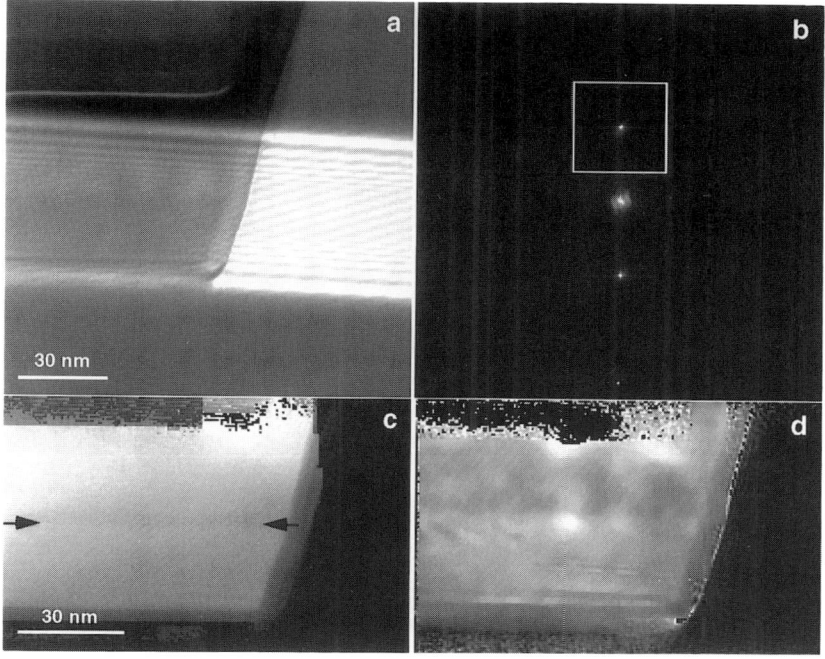

Figure 2. (a) Hologram of Si/Si p-n junction with vacuum at left; (b) Fourier transform of hologram with sideband outlined; (c) phase image showing contrast at junction (arrowed); (d) thickness image showing anomalous contrast at defects and interface.

Another image of the same sample, with an accompanying gray-scale marker calibrated in volts, is shown in Fig.3. This particular image was produced by division of the phase image by the thickness image and the wavelength-dependent proportionality constant. It shows a drop in voltage across the junction with extremes in voltage near the defects. Fig. 4 shows a single-pixel line scan taken perpendicular to the junction at the position arrowed in Fig. 3. This scan shows an overall voltage drop across the whole field of view of approximately 2V centered around a value of about 9 V at the junction. This latter value is in good agreement with the 9.3V measurement of the mean-inner-potential measurement for Si [1]. In the region closer to the junction we measure an averaged voltage difference of 1.5 ± 0.2V across a lateral dimension of 20nm: for an initial experiment, these values are promisingly close to those calculated above for an abrupt junction with doping of $\sim 10^{18}/cm^3$.

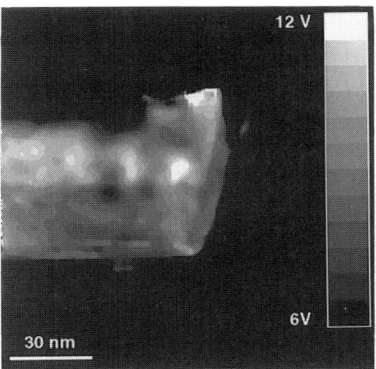
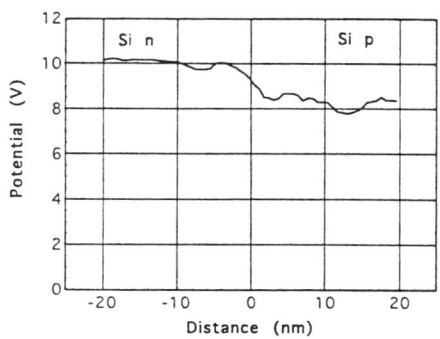

Figure 3. Potential distribution centered around 9.3V mean inner potential of Si with gray-scale values from 6V (black) to 12 (white).

Figure 4. Line profile across p-n junction showing voltage drop of 1.5±0.2V over lateral dimension of about 20nm.

2.4 Discussion of p-n junction results

Evaluation of the experimental results suggests that the electric field is greater than expected and more localized to the region of the interface. However, there are several experimental factors, including sample charging and differences in surface oxide thickness, which may have contributed to the discrepancy. Further controlled experiments with cleaved samples, different dopant distributions and passivated surfaces are needed before the technique is established as a viable method for characterizing p-n junctions. Finally, the accuracy of the technique is currently limited by the inherently poor signal-to-noise ratio characteristics of the amplitude image. Efforts to improve spatial coherence of the source should lead to improved measurement precision.

3. SIGNAL-TO-NOISE REQUIREMENTS

Practical problems associated with the signal-to-noise ratio of the thickness image potentially limit the usefulness of the off-axis holography technique for voltage measurements, primarily because quantitative measurements of potential distributions from phase shifts require prior accurate knowledge of the thickness.

As a representative example of the uncertainties and attainable precision in the measurement of slowly varying potential distributions, we consider the specific case of the p-n junction studied above. In the absence of strong diffracting conditions and magnetic fields, we have, from equation (4)

$$V = \phi / C_E t \qquad (6)$$

The uncertainty, dV, in the potential is thus given by

$$dV/V = d\phi/\phi - dt/t \qquad (7)$$

For Si at 100keV, with a mean inner potential of about 10V and a sample thickness of 50nm, ϕ is about 4.2 rad. Moreover, given that the total potential drop across the junction should be on the order of 1V, the measurement of V should have an uncertainty of no more than 0.5V (and preferably considerably less).

Several experimental variables control the quality of the recorded holographic fringe pattern. The most important is the contrast of the holographic fringes, defined by $\gamma = (I_{max} - I_{min})/(I_{max} + I_{min})$: a generous assumption is that the contrast should be about 0.3. The detection quantum efficiency of the CCD detector at the frequency of the carrier fringes, $DQE(g_h)$, is also important: a reasonable estimate for the camera used in this experiment would be $DQE(g_h)$ of 0.6 [10].

In the event that the thickness is known or constant, then the variance of the phase estimation becomes [11,12]:

$$\text{var}[\hat{\phi}] \ (= [\,d\phi]^2\,) = 14 / (DQE(g_h) \ \gamma^2 \ N_t) \tag{8}$$

where $\hat{\phi}$ is the estimator for the phase measurement and N_t is the total number of counts on the detector in the absence of significant dark current. Taking $d\phi = 0.25$, then we require $N_t = 4.1 \times 10^3$. By taking a sideband with an area that corresponds to 1/16th of the total number of pixels in the hologram, we effectively average the image over a 4x4 pixel area and thus reduce the required total number of counts to $N_t/16 \sim 260$, which would represent a very reasonable 1-2 sec recording at typical exposure levels.

For the particular case when the precision of the voltage measurement depends primarily on the uncertainty in the thickness measurement, the uncertainty can be expressed, using equation (5), as

$$dt/t = -2\lambda_i / (dA_o / A_o - dA_r / A_r) \ t \tag{9}$$

From equation (5), we have $A_o = 0.762 A_r$ for $t = 50$nm and $\lambda_i = 92$nm, and $dA_o = \text{sqrt}(A_o)$ since the noise in the recorded hologram is simple shot noise. With these values, we then calculate that

$dt/t = 4.28 \lambda_i / (dA_r / A_r) \ t$ giving a value for the precision dA_r / A_r of 6.35×10^{-3}

In order to calculate the signal required for this precision in amplitude, the point spread function of the detector and the contrast of the holographic fringes must be taken into account , for example by using the expression [10]

$$\text{var} \{ [\hat{A}]^2 / A^2 \} \ (= [\,dA_r / A_r]^2) = 2 / (DQE(g_h) \ \gamma^2 \ N_t) \tag{10}$$

where $\hat{A}$ is the estimator of the amplitude and A is the mean amplitude. Substitution leads to the value required for N_t of 9.2×10^5 electrons/pixel. With effective averaging as a result of sideband extraction, the required number of counts is reduced to $N_t = 5.7 \times 10^4$ el / px. With a reasonable flux of 200 electrons/pixel/sec, this requirement would necessitate a 5 min exposure. In practice, because of likely sample drift, it will be advisable to average further over the thickness image, for example by using a 5x5 pixel median filter, yielding $N_t / 25 \times 16 = 2000$ el /px as a rough estimate.

We note qualitative agreement of this statistical analysis with the recording of Fig. 2 when the original hologram was acquired with a 16 sec exposure and contained an average signal of 4000 electrons/pixel. The thickness image was smoothed using a

5x5 pixel median filter so that the resulting voltage map had a pixel-to-pixel precision of about 0.5V.

While it would clearly be easier when measuring electrostatic potentials to avoid having to calculate and correct for thickness effects, it is often not possible to produce samples of known or constant thickness. It is also important to note that, in cases where one is tempted to assume that the thickness doesn't change by very much, for the example shown here, 0.5V measurement precision requires that the change in thickness over the measurement area must be less than 2.5 nm, i.e., less than 5 unit cells of Si.

In lieu of the ideal specimen preparation method, what is really needed for enhanced imaging of potential distributions are improved detectors, more stable specimen holders and stages, and most importantly, increased contrast in the holographic fringes since this factor enters into the precision equation as a squared term.

4. EXPERIMENTATION WITH NANOTIPS

Field-emission electron sources are generally considered as essential for obtaining an electron beam of sufficient coherence for electron holography applications. Nevertheless, there has been much recent interest in the further development of nanotips as electron sources for transmission electron microscopes [14, 15], in particular because their enhanced brightness offers the prospect of increased spatial coherence for holographic applications and greater current for microanalysis. It would therefore appear to be of interest to describe our preliminary attempts to prepare nanotips for a 100keV transmission electron microscope from commercially available single crystal tungsten wires using *in situ* Ne-ion etching.

The starting point for our experiments were sharp, <310>- and <111>-oriented tungsten tips provided by FEI. Under normal UHV operating conditions, these tips typically started emitting at extraction voltages in the range of 600-1200V. The gun chamber of the electron microscope was modified by the addition of a leak valve for admission of the Ne gas and a residual gas analyzer was added to monitor the Ne partial pressure, which was maintained in the range of 10^{-5}-10^{-6} Torr during etching. The process of tip sharpening uses field-emitted electrons from the tip to ionize the Ne ions which then follow field lines back to the tip. As sputtering continues, the tip is sharpened and the field emitted current increases, as does the sputtering rate. In order to stabilize the process, the total field emitted current is limited to about 5 microamperes by gradually reducing the extraction voltage as sputtering proceeds. Eventually, a point is reached when no further decrease in extraction voltage occurs. The residual Ne gas is evacuated from the gun chamber with a turbomolecular pump and evaluation of the emission characteristics of the sharpened tip then takes place.

A special connection was made to the external terminals of the electron gun to enable the emission behavior of the sharpened tips to be monitored without any accelerating voltage being applied. Figure 5 compares the I-V curves, in the form of a Fowler-Nordheim plot, for a <111>-oriented tip, before and after sharpening. Interpretation of the slope of the latter implied that the effective radius of curvature of the sharpened tip was 140±10Å. Electron beams were formed using extraction voltages as low as 150V, although about 300V was necessary to provide usable current for recording holograms. Figure 6 shows the interference fringes of a vacuum hologram recorded using a <111>-oriented tip with a 300-V extraction voltage. Preliminary observations indicated reduced flicker compared to that normally

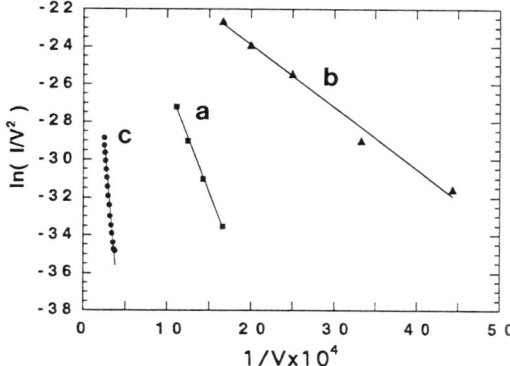

Figure 5. Fowler Nordheim plots for a <111>-oriented tungsten tip a) before and b) after sharpening. For comparison, the curve for a normal field emission tip (c) is also shown.

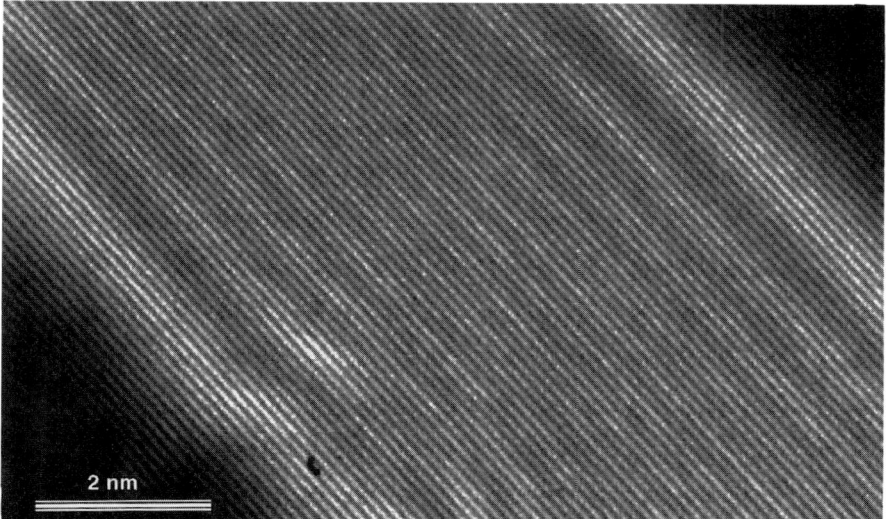

Figure 6. Electron hologram recorded using sharpened nanotip source with extraction voltage of 300V.

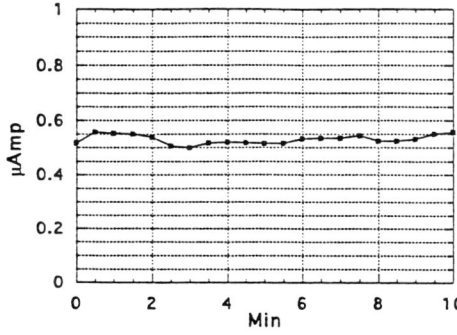

Figure 7. Plot of emission current vs. time for a sharpened <310>-oriented tungsten tip with an extraction voltage of 430V.

observed in the thermally-assisted field-emission mode using extraction voltages of 3-4keV. As shown in Fig.7, measurements of emission stability over time showed that the current remained constant to within about ±5% over 10-minute periods of observation.

The gun-electrode geometry was far from optimized for operation with such low extraction voltages, and it was not possible to make any detailed measurements of the gun brightness. Nevertheless, the success of these initial trials with sharpened nanotips suggests that it should be worthwhile to persevere with further experiments.

Our work in the Center for High Resolution Electron Microscopy at Arizona State University was supported by the National Science Foundation Grant No. DMR-91-15680. Experiments at Oak Ridge National Laboratory were sponsored in part by the Department of Energy under Contract DE-AC05-84OR-21400 managed by Martin Marietta. We are grateful to Dr. J.C. Bean for providing the Si/Si p-n junctions and we thank A.A. Higgs, K. Weiss and J. C. Wheatley for their assistance with implementing the nanotip experiments.

REFERENCES

1. N. Osakabe, K. Yoshida, T. Matsuta, H. Tanaka, T. Okuwaki, J. Endo, H. Fujiwara and A. Tonomura, Appl. Phys. Letts. 42 (1983) 746.
2. T. Matsuda, A. Fukuhara, T. Yoshida, S. Hasegawa, A. Tonomura and Q. Ru, Phys. Rev. Letts. 66 (1991) 457.
3. M. Mankos, M. R. Scheinfein and J. M. Cowley, J. Appl. Phys. 75 (1994) 7418
4. S. Frabboni, G. Matteucci, G. Pozzi and M. Vanzi, Phys. Rev. Letts. 55 (1985) 2196.
5. S. Frabboni, G. Matteucci and G. Pozzi, Ultramicroscopy 23 (1987) 29.
6. M.R. McCartney, D.J. Smith, R. Hull, J.C. Bean, E. Voelkl and B. Frost, Appl. Phys. Letts. 65 (1994) 2603.
7. S.M. Sze, Physics of Semiconductor Devices (Wiley, New York, 1981).
8. L. Reimer, Transmission Electron Microascopy, 2nd. Ed. (Springer, Berlin, 1989).
9. M.R. McCartney and M. Gajdardziska-Josifovska, Ultramicroscopy 53 (1994) 283.
10. W.J. de Ruijter and J.K. Weiss, Ultramicroscopy 50 (1993) 269.
11. M. Gajdardziska-Josifovska, M. R. McCartney, W.J. de Ruijter, D.J. Smith, J.K. Weiss and J.M. Zuo, Ultramicroscopy 50 (1993) 285.
12. J.F. Walkup and J.W. Goodman, J. Opt. Soc. Am. 63 (1993) 399.
13. W.J. de Ruijter, Ph. D. thesis, University of Delft.(1992)
14. H.W. Fink, Phys. Scr. 38 (1988) 260.
15. W. Qian, M.R. Scheinfein and J.C.H. Spence, Appl. Phys. Letts. 62 (1993) 315.

Electron Holography of Heterogeneous Catalysts

A. K. Datye[a], D. S. Kalakkad[a], E. Völkl[b] and L. F. Allard[b]

[a]Center for Microengineered Ceramics and Department of Chemical and Nuclear Engineering, University of New Mexico, Albuquerque, NM 87131, USA

[b]High Temperature Materials Laboratory, Oak Ridge National Laboratory, Oak Ridge, TN 37381, USA.

Abstract

Electron holography has been performed in a coherent beam TEM to characterize Pd metal particles, 5-10 nm in diameter, which are of interest as heterogeneous catalysts. The Pd metal particles were supported on a silica microsphere model support and subjected to oxidation-reduction treatments. The particle morphology was found to depend on pretreatment method. Under some conditions, we obtained single crystal Pd metal particles with a central contrast feature suggestive of an internal void. The presence of internal voids within the single crystal Pd particles was confirmed by electron holography. Particle morphologies were deduced from phase images, obtained from holographic reconstruction, which are sensitive to particle thickness. The ability to determine particle shapes and surface facets should be of great interest as a method for the characterization of heterogeneous catalysts.

1. INTRODUCTION

Heterogeneous catalysts are of great commercial importance both in the synthesis and processing of materials as well as in developing alternative sources of energy and minimizing the impact on the environment. A noteworthy example of the latter application is the use of supported Pd for catalytic combustion where the lower combustion temperatures help avoid the generation of NO_x byproducts [1]. The active phase in such heterogeneous catalysts consists of nanometer-sized metal or oxide particles dispersed within a stable, high surface area oxide support matrix. The oxide support serves to provide thermal stability as well as to act as a promoter for the catalytic reactions. One of the ways in which a support promotes catalyst activity is by facilitating the creation of catalytically active sites having the desired properties [2]. The catalyst support and pretreatment may alter the morphology of the nanometer-sized particles, thereby affecting the activity and selectivity of chemical reactions.

Even though the effect of catalyst supports and pretreatment on catalyst activity and selectivity have been well documented, the underlying mechanisms are still the subject of much debate. For example, when Pd metal catalysts are used for methane oxidation, it is observed that the catalytic activity increases with time. It is not known what causes this induction period [3]. Electron microscopy provides, in principle, a means to directly visualize the morphology and structure of such nanometer-sized particles of interest to catalytic chemists [4].

While the resolution of conventional transmission microscopes has steadily improved over the years, there are two major limitations in the use of TEM for the study of catalysts. First, the observed image represents a two dimensional projection of a 3-D structure. While the 2-D image may adequately portray the outline of the catalyst phase, the contrast within the image cannot be simply interpreted to yield the 3-D morphology of the object being observed. A second limitation is that the catalytically active phase is located within the tortuous pore structure of a catalyst support, which obscures contrast as well as resolution and makes it difficult to study the surface structure of the nanometer-sized particles of interest. We have circumvented the latter limitation with the use of nonporous oxide particles as model catalyst support [5]. The model supports help to localize the dispersed phase on the external surface of the support, making edge-on views possible that improve contrast and resolution and reveal the surfaces of these particles.

An inherent limitation of the conventional TEM imaging process is that image intensity is recorded but the amplitude and phase of the exit electron wave cannot be independently obtained. The availability of TEMs with coherent electron sources makes it possible to apply the technique of electron holography wherein the phase and amplitude of the exit electron wave can be independently reconstructed. Since electron phase shift is directly proportional to the mean inner potential of the sample (for suitably thin, kinematic scatterers), the method provides a direct means of investigating the three dimensional morphology of heterogeneous catalysts at the nanometer scale. In this paper we report a study of silica supported model Pd catalysts.

2. EXPERIMENTAL

The Pd catalysts were prepared by nonaqueous impregnation from a Pd acetylacetonate precursor. The precursor was decomposed by calcination at 573K in 10% O_2. Subsequently, the catalyst was reduced in flowing H_2. Typically, after the calcination step, the catalyst was flushed with 20 sccm of He and the flow switched to 20 sccm H_2 before ramping the temperature to 573 K. After reduction, the catalysts were cooled to room temperature and passivated by exposure to 1% O_2 in flowing helium. The passivated catalysts were stored in glass vials for examination in the TEM. For some of the experiments, after calcination and exposure to flowing helium, the catalyst temperature was raised to 383 K before switching over to H_2. In this manner, exposure of the sample to H_2 at room temperature, which might have resulted in formation of the Pd hydride phase, could be avoided. Further details are provided elsewhere on the results of oxidation and reduction treatments and the catalytic behavior of the Pd catalysts [6].

For TEM, the sample was supported on a holey carbon film. The grid was simply dipped into the powder and shaken off to remove excess powder. This simple method provides adequate sample on the grid and eliminates the potential hydrocarbon contamination that could be caused if additional solvents are used during sample preparation. The electron holograms were acquired on a Hitachi HF-2000 electron microscope. Images were recorded directly on a Macintosh Quadra computer using Gatan's 1K x 1K slowscan CCD camera and DigitalMicrograph® software. For electron holography, the samples were observed at 1000 KX direct magnification. The electron biprism was maintained at a potential of 110 V which provided interference fringes with a spacing of about 0.1 nm referred to the specimen at a contrast of 25% or better. Holograms were analyzed using a Fourier analysis algorithm implemented in a computer program called Holoworks©, which runs within the DigitalMicrograph software [7]. Immediately after recording the hologram, the sample was moved away to obtain a reference hologram. The reference hologram permits the correction of non-uniformities of the reference wave.

3. RESULTS

The majority of the Pd particles in this sample after initial H_2 reduction contained a central contrast feature suggestive of an internal void. The presence of an internal void was consistent with the light/dark contrast seen when going through focus; however, similar contrast behavior would be expected if the particle were filled with amorphous material. To confirm the internal structure of these particles, we utilized electron holography and an analysis of phase images reconstructed from the electron holograms. Preliminary results have been recently reported in the literature [8] and show that the phase shift in the center of the Pd particle is consistent with an internal void. If the Pd particle contained a phase of lower mean atomic potential, say for example amorphous PdO, the calculated phase shift is quite different from that observed experimentally. These preliminary results encouraged us to continue our work on holography of catalysts, and in this paper we describe an investigation of the shapes of 5-10 nm Pd particles supported on a model support consisting of 200 nm SiO_2 microspheres.

Fig. 1a shows on the top left a bright field HRTEM image of one of the Pd particles obtained after H_2 reduction of the calcined catalyst. The image shows clearly the outline of the internal void and the lattice fringes show that the Pd particle is a single crystal. The lattice fringes seem to be unaffected by the presence of the internal void and no internal defects are observed. As mentioned in the introduction, the HRTEM image does not directly provide information on the 3-D shape of the metal particles. To illustrate this, we have plotted in Fig. 1b, a histogram of image intensity along the line marked A in Fig. 1a. In this histogram, as well as others reported in this paper, going from left to right in the histogram corresponds to moving into the particle from its surface. While the histogram does show the periodicity in contrast caused by the lattice fringes, it does not give any information about the 3-D shape of the Pd particle. As we show below, the phase image derived from the electron hologram of this particle is very useful in this regard.

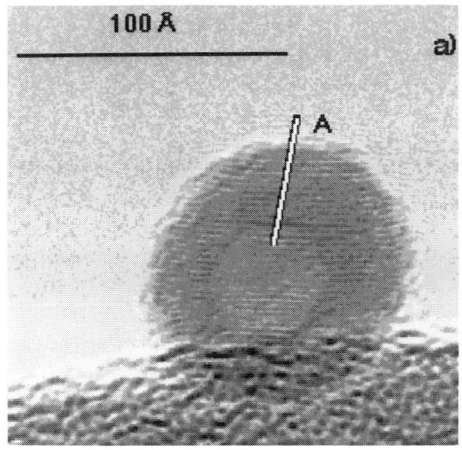

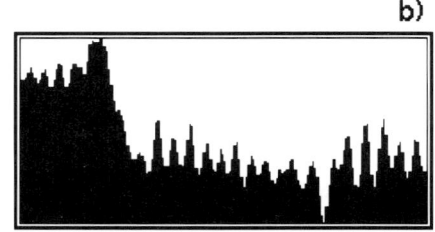

HRTEM image of the Pd particle

Histogram of Image Intensity along the direction A

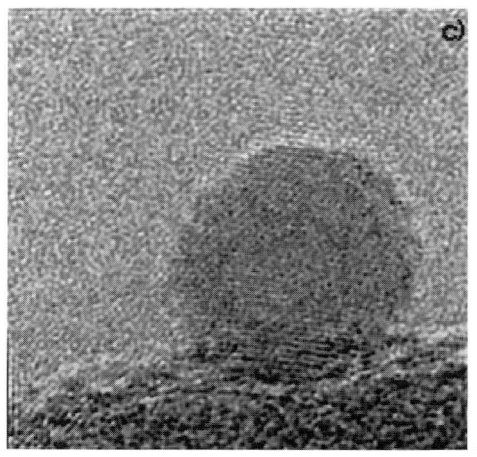

Amplitude image

Phase image

Fig. 1   Analysis of a Pd metal particle having a faceted void

Figs. 1c and 1d show, respectively, the amplitude and phase images derived from the hologram of this particle. Holoworks allows the choice of an appropriate aperture for the Fourier transformations, to de-emphasize the lattice fringe contrast in favor of the morphological details of the particle. Further analysis of the phase image in Fig. 1d is presented on the next page. Fig. 2a shows a histogram derived from the phase image along the line marked A in Fig. 1a. The histogram plots the phase as a function of distance. It is seen that the phase is relatively flat in the vacuum surrounding the metal particle. The value of the phase in the vacuum has been arbitrarily set to 0.85. The phase changes rapidly with distance as the surface of the particle is encountered. These phase shifts are shown more clearly in Fig. 2b with a 3-D model of the metal particle being shown in Fig. 2c.

From the TEM image in Fig. 1a, we can see that the exposed face parallel to the electron beam direction is a (111) surface with a certain degree of surface roughness. The surface roughness may arise in part from electron beam damage caused by observation of the particle at 1000 KX direct magnification. Observation at such high magnifications is necessitated by the need to record the fine interference fringes which are spaced ≈ 0.1 nm apart (in the image) under our conditions. Since, the optimal illumination condition for recording the image in Fig. 1a is quite different from that for recording the holograms, the time spent in adjusting the imaging conditions contributes to unnecessary beam exposure. These experiments were performed before computer control of the microscope had been fully implemented. Subsequently, with direct computer control, it is now possible to preset the imaging conditions and jump back and forth from hologram recording to image acquisition. This should reduce the amount of beam damage and the possibility of introducing artifacts, in future work.

The phase of the electrons with increasing distance from the particle surface agrees with that expected from the 3-D model of the particle in Fig. 2c. The phase first changes rapidly over a distance of approximately 0.5 nm which would correspond to the surface roughness on the exposed (111) surface of the particle. After that point, the phase shift changes more gradually corresponding to the gradual change in thickness as we move into the particle. Eventually, as the void within the particle is reached, there is a reversal in the direction of the phase. This reversal is caused by the fact that the electron beam encounters fewer Pd atoms as it passes through the internal void. After that the phase is nearly constant with distance, suggesting that the internal and external facets are similarly inclined. The preliminary analysis is therefore consistent with a cubooctahedral particle having a pronounced (111) facet.

Fig. 3 shows a particle of Pd from a catalyst sample that was reduced in flowing H_2 after the oxidized catalyst was flushed in helium, heated to 383 K and then exposed to flowing H_2. The temperature was then raised to 573 K to complete the reduction process. A high resolution TEM image of this particle is shown in Fig. 3a, and the corresponding reconstructed phase image in Fig. 3b. The phase along the line marked in Fig. 3b is shown in the histogram in Fig. 3c. The phase profile in Fig. 3c is consistent with a dense particle without the voids

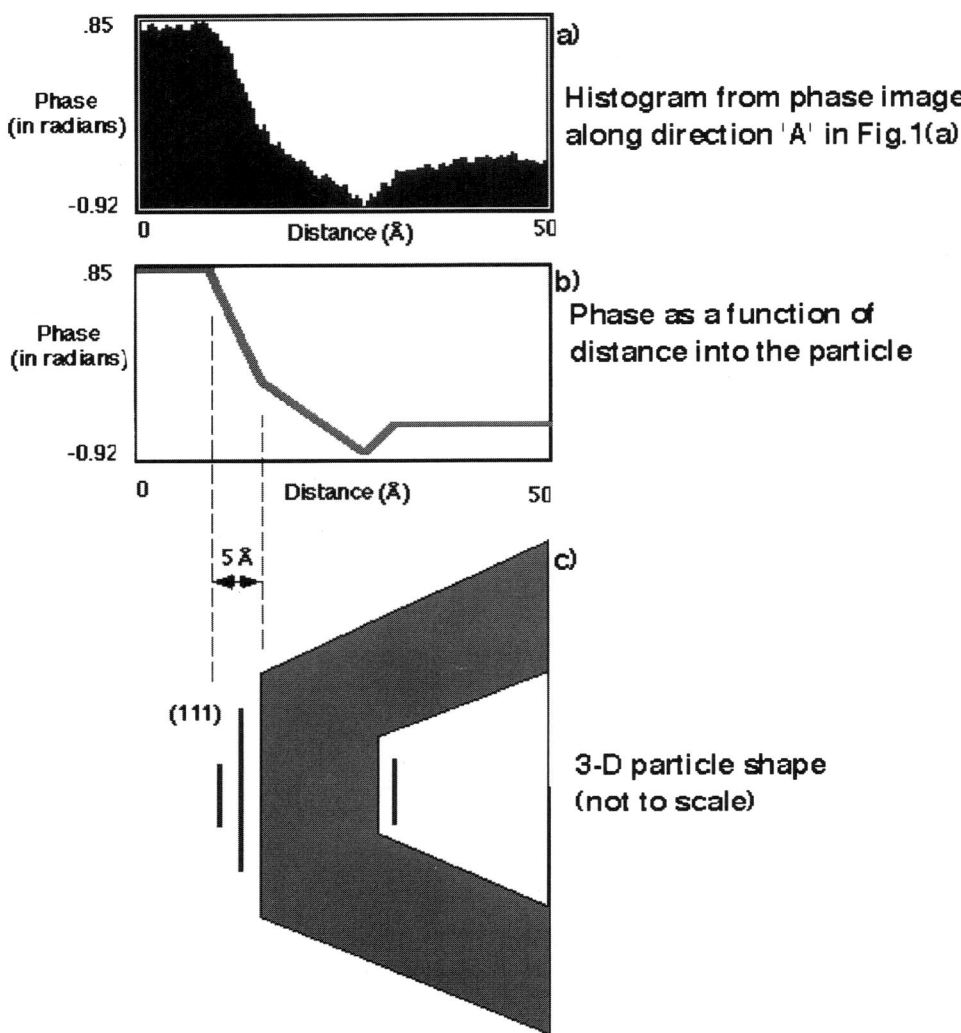

Fig.2 Particle Shape Determined by Electron Holography

Electron holography of heterogeneous catalysts

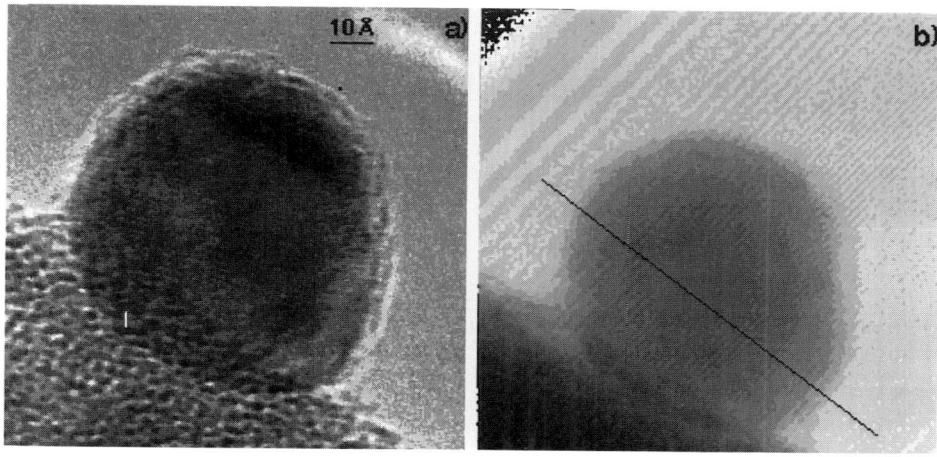

High Resolution TEM image   Reconstructed phase image

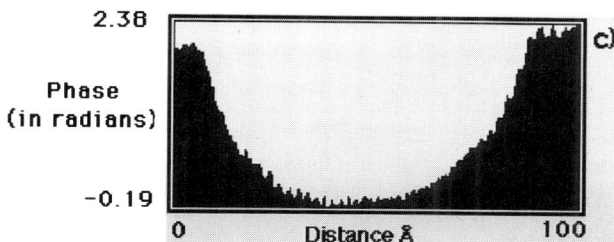

Phase profile along the line marked in Fig. 3(b)

Fig. 3   Analysis of a Pd particle reduced at 300°C without exposure to H_2 at room temperature

seen in Fig. 1 and 2. However, the profile shows some deviation from a smooth profile on the right hand side. This aspect is investigated in more detail in Fig. 4. The computed profile corresponds to a spherical particle. The calculated phase is proportional to the height of a spherical particle. As seen in Fig. 4, the match is very good, except for the region on the right. This region is expanded further in Fig. 5 and we examine two different line traces within the same particle. Fig. 5a shows a histogram along the line marked A, and 5c shows the histogram along line B. The phase shift with distance is linear at A but shows pronounced curvature along B. We interpret this to mean that we are dealing with a surface facet along A while the particle is more rounded along B. These results imply that the shape of nanometer-sized particles is revealed very well in phase images obtained by reconstruction of off-axis holograms.

4. CONCLUSIONS

We have demonstrated that electron holograms of nanometer-sized Pd metal particles can be reconstructed, and the phase image utilized to reveal the morphology of these particles. The phase shift depends linearly on the particle thickness under conditions of kinematic scattering, as occur with nanometer sized particles. Therefore, in effect the phase image provides a thickness map of the sample. As shown in this paper, the thickness map yields information on internal structure of the particle as well as its three-dimensional morphology. The phase images can be analyzed to reveal the presence of surface facets and the overall shape of the particle. The overall shape is of interest to heterogeneous catalysis since it is the coordination of surface atoms that determines the types of catalytic sites available and catalyst activity and selectivity.

ACKNOWLEDGMENTS

This research was supported in part by grant 26856-AC5 from the American Chemical Society, Petroleum Research Fund and by the Laboratory Directed Research and Development Program of Oak Ridge National Laboratory. AD was supported also via the High Temperature Materials Laboratory (HTML) Faculty Fellowship program. The HTML is supported by the U.S. Department of Energy, Assistant Secretary for Energy Efficiency and Renewable Energy, Office of Transportation Technologies, under contract DE-AC05-84OR21400 managed by Martin Marietta Energy Systems, Inc.

REFERENCES

1 L. D. Pfefferle and W. C. Pfefferle, Catal. Rev. Sci. Engg., 29 (1987) 219.
2 M. Boudart and G. D. Mariadassou, Kinetics of Heterogeneous Catalytic Reactions, Princeton University Press, 1984, pg 201.
3 F. H. Ribeiro, M. Chow, and R. A. Dalla Betta, J. Catal., 146 (1994) 537.
4 A. K. Datye and D. J. Smith, Catal. Rev. Sci. Engg., 34 (1992) 129.
5 A. K. Datye, A. D. Logan and N. J. Long, J. Catal., 109 (1988) 76.
6 D. S. Kalakkad, I. Y. Chan, L. F. Allard and A. K. Datye, submitted.
7 E. Völkl, these proceedings.
8 L. F. Allard, E. Völkl, D. S. Kalakkad and A. K. Datye, J. Mater. Sci. **29** (1994) 5612.

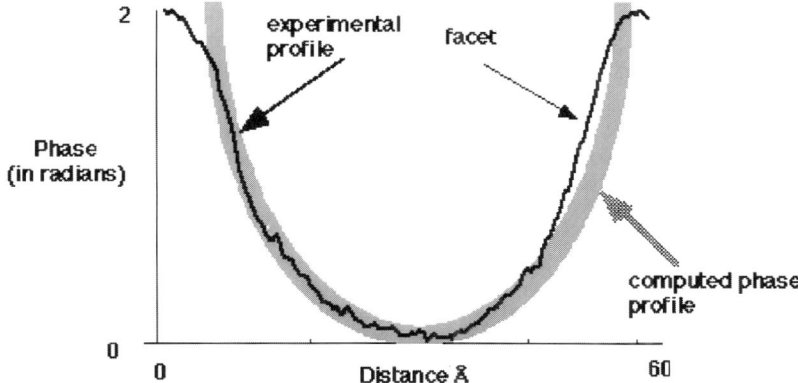

Fig. 4   Comparison between the phase profile obtained from the Pd particle shown in Fig. 3(a) and that computed for a spherical particle

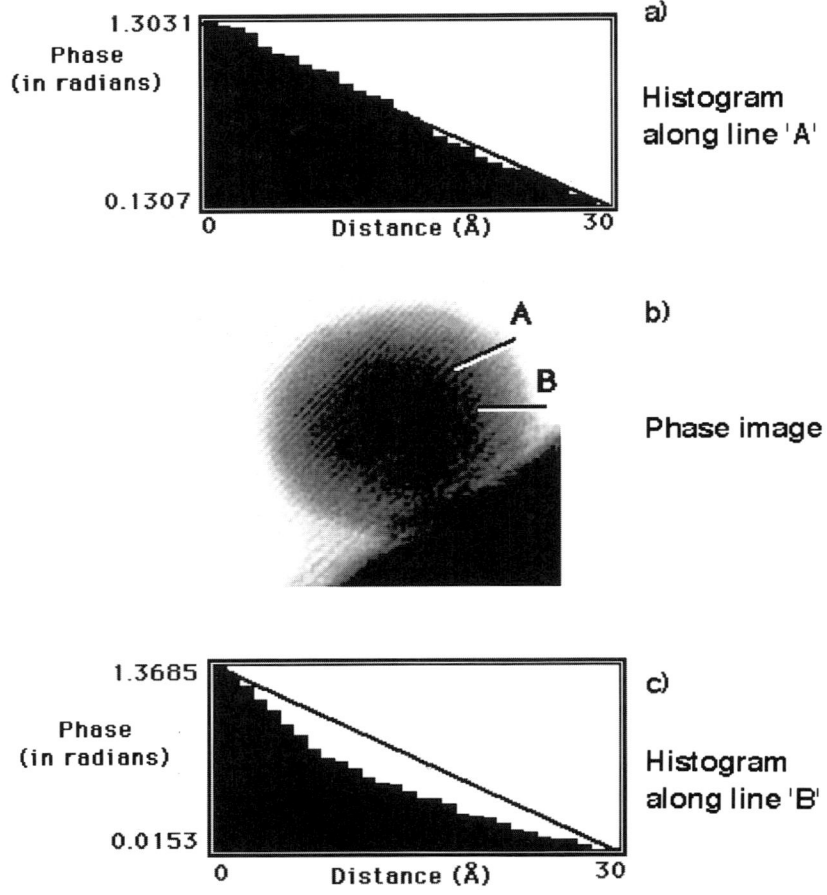

Fig. 5   Phase profiles of the Pd particle in Fig. 3(a) obtained from a surface facet region (A) and a rounded region (B)

Electron Holography in Materials Science

Xiwei Lin, V. Ravikumar, R. Rodrigues, N. Wilcox and Vinayak P. Dravid

Department of Materials Science & Engineering
Northwestern University, Evanston, IL 60208, USA

ABSTRACT

The ability of electron holography and interferometry to extract phase information from the specimen exit wave has potentially considerable, but rather specific, applications in materials science. In this paper, we present two sets of applications of holography in nanoscale phenomena in materials science.

In the first case, we have performed holography experiments on the all-carbon tubular graphitic structures (buckytubes). The hollow nature of the tubules determined from holography matches well with the computed thickness profiles. Electron holography appears to be quite sensitive to defects in these materials, and hints at discriminating different bonding types and molecular coordination.

Holography of crystalline interfaces, however, suffers from many problems, for example - of diffraction effects, asymmetric exit wave function across interfaces and that of thickness variations. Despite these problems, some encouraging results are obtained for grain boundaries in $SrTiO_3$. It is shown that electron holography is sensitive to changes in mean inner potential at grain boundary core and to the boundary electrical charge in Mn- doped (electrically active) specimens.

1. INTRODUCTION

In recent years materials science has been dominated by materials, phenomena and processes which occur at a nanometer scale. Electron microscopy, in its various forms, has contributed immensely to our understanding of nanoscale phenomena in solids. The "amplitude" dominated electron microscopy has served us well over several decades, now is the time to extend it to "phase" information which has become practical to extract with electron interferometry and holography.

There are a number of materials science areas which can make use of phase change information. Phase change of the electron wave may occur due to a differential change in mean inner potential (MIP), thickness, diffraction/channeling effects, electrostatic and magnetic fields. [1-4] With the acquisition of a cFEG TEM (Hitachi HF-2000) a few years ago, we have been exploring the techniques of electron interferometry and holography to assess its utility in solving some

---

Work Supported by U.S. DOE, Grant No.: DE-FG02-92ER45475

outstanding materials science problems.[5,6] This contribution concerns with our results on buckytubes and grain boundaries in $SrTiO_3$ model electroceramics.

1.1. Buckytubes

Theoretical studies on monolayer and multilayer buckytubes indicate that their electronic structure is intimately connected to number of sheets in the tubes.[7,8] The change in curvature in the buckytubes and their termination at a tip are also attributed to insertion of 5-fold or 7-fold member rings which accommodate the change in curvature.[9,10] These buckytubes serve as a classic weak-phase object of well-defined geometry to assess the sensitivity of holography to subtle phase changes.[6] It is also expected that changes in bonding at critical points (curvatures) and defect content of the buckytubes may change the phase of transmitted electrons which could be imaged using holography.

1.2. Grain Boundaries (GBs) in $SrTiO_3$

Grain boundaries in electroceramics are electrically active and are responsible for the useful electronic properties of polycrystals, such as varistor, capacitor, PTCR behavior.[11,12] The electrical activity of GBs is often brought about by segregation of aliovalent species. It has been proposed that segregation and processing treatments form a net charge at GBs (+ve or -ve) which is compensated by a space charge which extends about 5 - 20 nm across the GB. It would be most useful if one can visualize the space charge directly. We have made some calculations to see if holography would be sensitive to the presence and extent of space charge across GBs. Most recently, we have resorted to tilt bicrystals of electroceramics to minimize differential phase change problems due to diffraction effects.[13,14] We have selected $\Sigma = 5$, 13 pure and doped $SrTiO_3$ GBs. The zone axis for all these bicrystals is [001] and the specimen thickness on either side of the GB as well as the GB region is considered to be the same. Under these conditions, we expect that the differential phase shift across GBs would represent a phase change at the GB region. The phase change at GB region may arise due to all, reduction in density (or mean inner potential), strain and electrostatic potential (in doped GBs).

2. EXPERIMENTAL

All electron holography experiments were conducted using a Hitachi HF-2000 cold field emission gun TEM (cFEG TEM) operated at 200 kV, which is equipped with a Hitachi rotatable electrostatic biprism. The cFEG TEM is also equipped with an Oxford ultra-thin-window (UTW) x-ray detector/analyser, a Gatan 666 parallel EELS spectrometer and EL/P acquisition/analysis software.

Holograms are recorded on TEM negative films (adjusted for exposure nonlinearity), followed by digitization with a microdensitometer and subsequent processing with a HP workstation equipped with Semper suite of programs. Typically, a 2048 x 2048 image is extracted after digitization and a 1024 x 1024 area is then selected for Fourier transform. A 256 x 256 size is subsequently extracted from an appropriate side-band for holographic reconstruction.

3. RESULTS AND DISCUSSION

3.1. Buckytubes

The buckytubes have a (predominantly) cylindrical cross-section from which through-thickness can be readily calculated. Such thickness plots can be compared to profile scans of the phase images. Figure 1A is a hologram of a cone-shape buckytube of 5 graphite layers obtained with biprism fringe spacing of 0.09 nm, with reconstructed phase(Fig. 1B) and amplitude images (Fig. 1C). The amplitude image shows very little contrast and does not contain graphite (002) fringes since the aperture used for reconstruction from side band omitted the (002) reflections. Lack of contrast in amplitude image indicates that these all-carbon molecular structures are close to a weak-phase object. A phase image line scan of the buckytube is in good agreement with the calculated thickness profile (see Figure 2). The phase change at the center of the tube is ~ 0.6 for 10 layers of graphite separated by ~ 0.345 nm.

The tapering ends of the buckytubes, i.e. change in curvature, must be accommodated by a carbon network other than the hexagonal geometry as in planar graphite. It has been shown that positive curvature can be brought about by inserting pentagons in an otherwise hexagonal network. Such defects are characterized by disclinations (or dispirations if screw axis and/or glide plane operations is (are) utilized) involving lattice strain and change in local bonding. Figure 3A is a hologram of such a positively "bent" buckytube. The reconstructed phase image (3B), unlike an HREM image, shows considerable change in contrast at the defect area. We suggest that this change reflects change in mean inner potential of the defect area brought about by change in local bonding and coordination.

3.2. Grain Boundaries (GBs) in $SrTiO_3$ Electroceramics

We have selected symmetrical tilt-bicrystals of $SrTiO_3$ as model systems since the phase shift across GBs due to diffraction/orientation effects would be identical in both adjacent crystals. The resultant phase shift at GB core would then be due to a combination of: mean inner potential, charge, thickess and lattice strain. The undoped GBs are likely to be electrically neutral and are not expected to exhibit significant phase shift due to GB charge or space-charge. Thickness could be different at the GB due to differential thinning by IBT, but was minimized using careful thinning procedures. Lattice strain due to dislocations and change (reduction) in local coordination and bonding can together be considered as change in mean inner potential.

Figure 4 is a high resolution electron hologram of a 24° symmetrical tilt-GB in $SrTiO_3$. The biprism fringe spacing is about 0.08 nm and the phase contrast image of the crystals is seen underneath the biprism fringe pattern. No visible fringe bending or termination could be seen. The reconstructed phase image (Figure 5) showed very little overall contrast indicating that thickness change across the field of view was minimal. A careful examination of the phase image indicates a subtle but noticeable change at the GB core. This change becomes visible in the phase profile obtained by averaging the phase image along the GB plane, as shown in the inset of Figure 5. The relative phase change calculated from

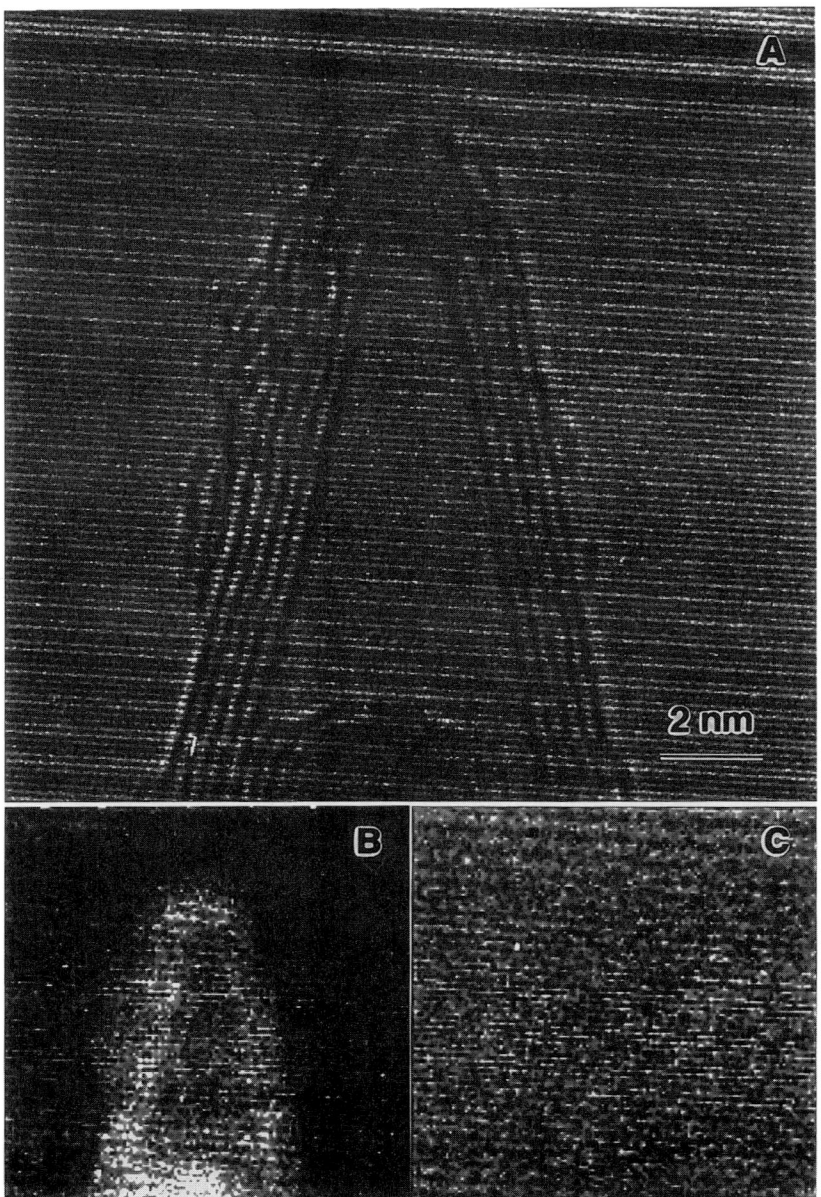

Fig. 1: (A) Hologram of a cone-shaped buckytube, with the extracted phase (B) and amplitude (C).

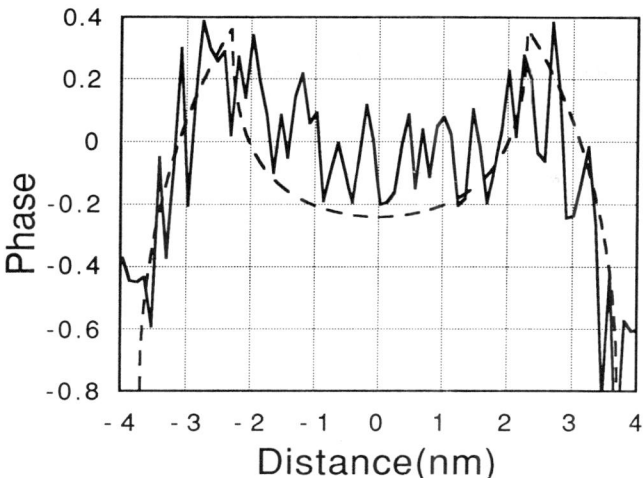

Fig. 2: A line scan of the phase image across the buckytube (solid curve), with the calculated thickness profile (dashed curve)

this profile is ~0.15, which is approximately $2\pi/40$. This change is attributed to a change in local mean inner potential at the GB, since neither charge or thickness are expected to change at the GB core in these specimens.

Mn-doped grain boundaries, however, showed more interesting behavior. It has been shown[15] that Mn preferentially and tightly segregates at GBs in $SrTiO_3$ and the valence of Mn at the GB core is a combination of +2 and +3 states. Since Mn+2 and/or Mn+3 replace Ti, the resultant defect should have a net negative charge. Thus, Mn-doped GBs should exhibit a negative GB charge which is compensated by a positive space charge region which may contain holes and oxygen vacancies. We indeed appear to have electron holography evidence for the presence of charged interfaces. Figure 6 is an electron hologram of a Mn-doped GB in $SrTiO_3$. The biprism fringes are clearly shifted towards the vacuum in the vicinity of the GB. The shift of the biprism fringes towards the vacuum is clear indication that the GB is negatively charged. The details of these observations along with analysis of reconstructed phase images will be discussed elsewhere. [16]

4. SUMMARY

Electron holography certainly has considerable potential for solving some of the outstanding problems in nanoscale phenomena in materials science. The materials science of nanostructured materials would have immediate pay-offs

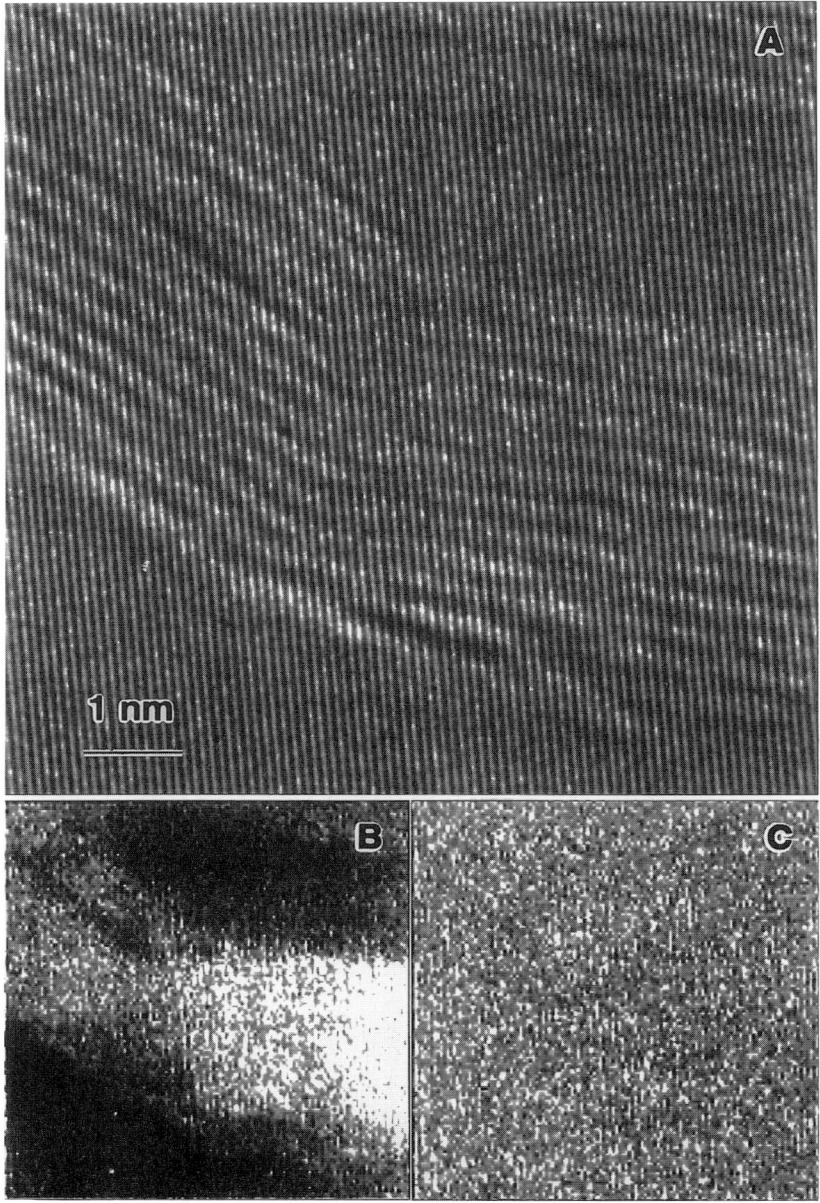

Fig. 3: (A) Hologram of a positively "bent" buckytube, with the extracted phase (B) and amplitude (C).

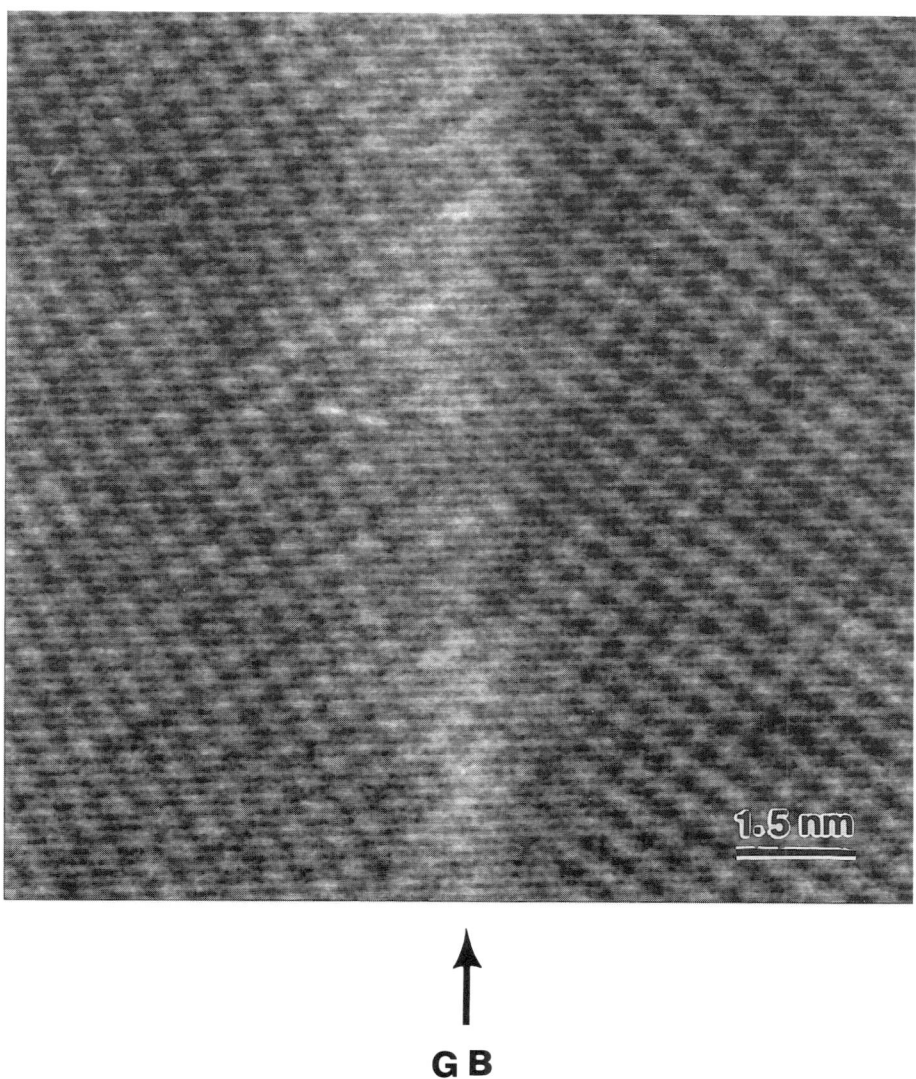

GB

Fig. 4: High resolution hologram of a 24° symmetrical tilt grain boundary in strontium titanate. The fringe spacing is about 0.08 nm.

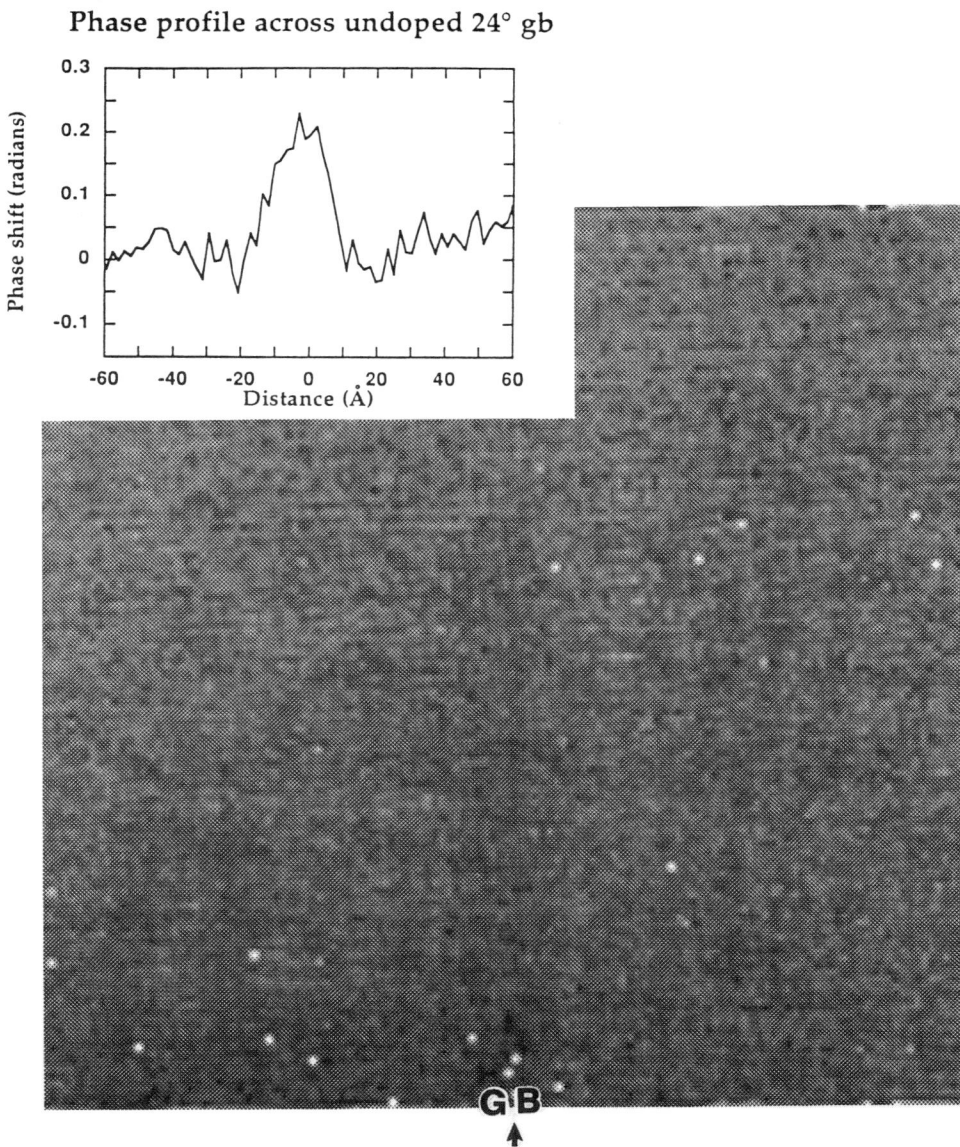

Fig. 5: The reconstructed phase image of the hologram in Fig. 4. The line average of the phase profile across the grain boundary is given as an inset.

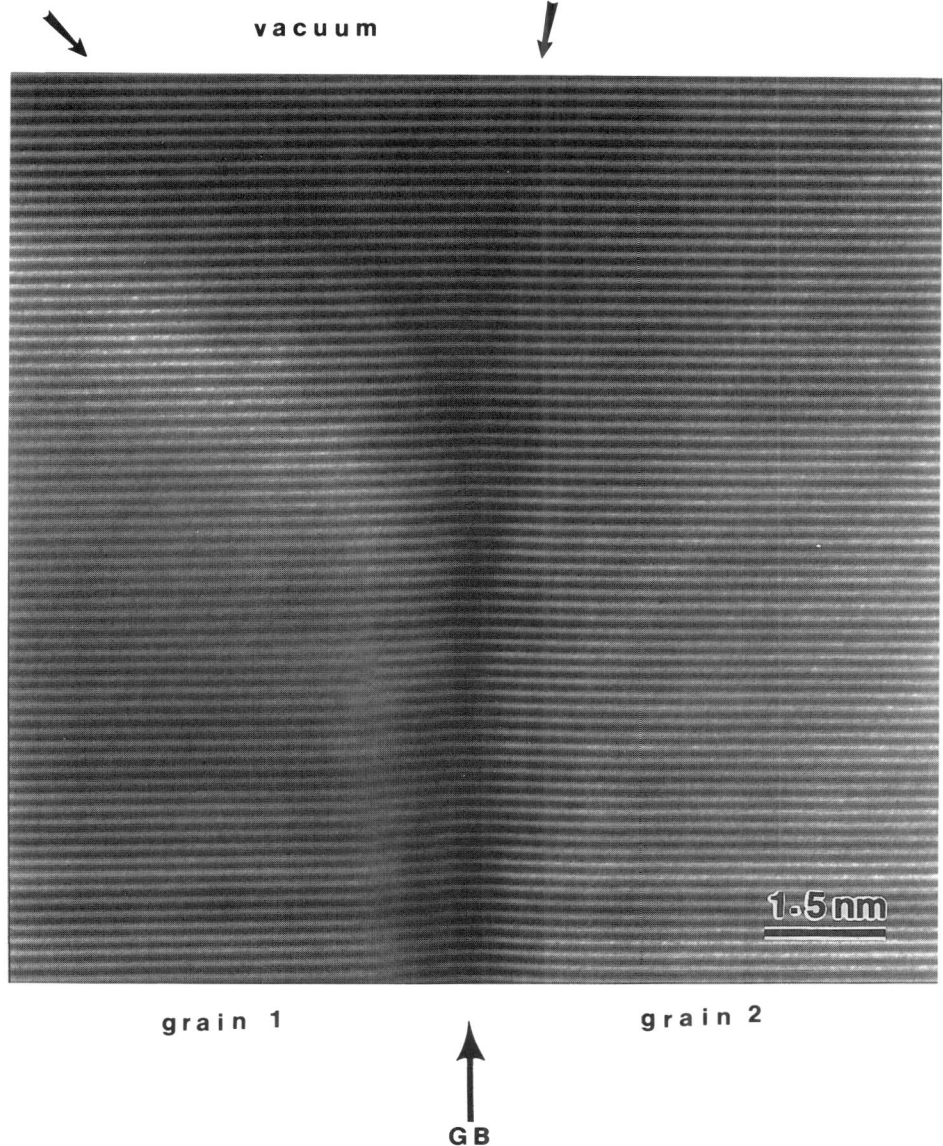

Fig. 6: Hologram of a charged grain boundary in Mn-doped strontium titanate.

with electron holography as demonstrated above for buckytubes. However, applications of electron holography in strongly diffracting systems such as crystal defects and interfaces, are faced with problems of separation of diffraction, channeling and thickness variation effects which superimpose one another. The progress and fate of electron holography in such applications would hinge critically on experimental and thoretical/modeling efforts. However, there are indications that useful information such as change in mean inner potential, charge at interfaces and the associated space charge could be imaged and perhaps quantified for the first time.

REFERENCES:

1. A. Tonomura, Adv. in Physics, Vol. 41 (1992) 59.
2. H. Lichte, P. Kessler, F. Lenz and W.-D. Rau, Ultramicroscopy, 52 (1993) 575.
3. M.R. McCartney and M. Gajdardziska-Josifovska, Ultramicroscopy, 53 (1994) 283.
4. M. Mankos, M.R. Scheinfein and J.M. Cowley, J. Appl. Phys., 75 (1994) 7418.
5. V.P. Dravid, X. Lin, V. Ravikumar, R. Rodrigues and N. Wilcox, Proc. 52nd MSA Mtg., Vol. 52 (1994) 542.
6. X. Lin and V.P. Dravid, Proc. 52nd MSA Mtg., Vol. 52 (1994) 764.
7. N. Hamada, S. Sawada and A. Oshiyama, Phys. Rev. Lett., 68 (1991) 1579.
8. R. Saito, M. Fujita, G. Dresselhaus and M.S. Dresselhaus, Phys. Rev. B, 46 (1992) 1804.
9. S. Iijima, T. Ichihashi and Y. Ando, Nature, 356 (1992) 776.
10. S. Amelinckx, X.B. Zhang, D. Bernaerts, X.F. Zhang, V. Ivanov and J.B. Nagy, Science, 265 (1994) 635.
11. N. Yamaoka, M. Masuyama and M.Fukai, Bull. Amer. Ceram. Soc., 62 (1983) 698.
12. P. Gaucher, R.L. Perrier and J.P. Ganne, Adv. Ceram. Mater., 3 (1988) 273.
13. V. Ravikumar and V.P. Dravid, Ultramicroscopy, 52 (1993) 557.
14. M.M. McGibbon, N.D. Browning, M.F. Chisholm, A.J. McGibbon, S.J. Pennycook, V. Ravikumar and V.P. Dravid, Science, 266 (1994) 102.
15. N. Wilcox, V. Ravikumar, R. Rodrigues, V.P. Dravid, M. Vollmann, R. Waser, K.K. Soni and M. Adrieans, submitted, Solid State Ionics, September 1994.
16. V. Ravikumar, R. Rodrigues and V.P. Dravid, in preparation.

Electron Holography Applied to the Study of Fullerene Materials

L. F. Allard, E. Völkl, S. Subramoney[a] and R. S. Ruoff[b]

Oak Ridge National Laboratory, Oak Ridge, TN, USA 37831-6064
[a]Dupont Experimental Station, Wilmington, DE 19880-0228
[b]SRI International, Menlo Park, CA 94025

Electron holography permits the precise determination of aspects of the morphology of nanometer-sized particulates, because the pure phase information in the image wave can be reconstructed and displayed as a phase intensity image. Profiles of intensity over the image are, in particularly advantageous cases, directly related to particle morphology, since the phase intensity for materials that are essentially kinematic scatterers and are of homogeneous composition is directly related to thickness. Holograms at high resolution can also be used for many other analyses of a material's structure; they provide, for example, a method for precise calibration of lattice spacings in nanometer-sized areas. Examples of the use of holography for studies of nanoparticles of novel carbon materials such as giant nested fullerenes and carbon nanotubes are given.

1. INTRODUCTION

The characterization of the structure of novel carbon materials such as nanotubes and fullerene structures produced by a variety of electric discharge techniques has been of great interest in electron microscopy in recent years. The primary tool to date to elucidate the structure of these materials has been high resolution electron microscopy [1], which permits lattice images to be recorded from which certain inferences can be derived regarding the material's crystallography. However, high resolution TEM techniques provide little information, beyond the imaging of the edge profiles of these nanoparticulates, that would lead to unambiguous determinations of overall particle morphologies. Also, again because of the nanoparticle nature and composition of the material,

---

Research sponsored by the Laboratory Directed Research and Development Program of Oak Ridge National Laboratory, managed for the Department of Energy by Martin Marietta Energy Systems, Inc. under contract DE-AC05-84OR21400, and by a Post-doctoral scholar appointment (E.V.) with Oak Ridge Institute of Science and Technology.

high resolution SEM techniques do not yield images from which three dimensional morphologies can be determined, even when immersion lens SEM techniques are used that provide resolutions at better than the 1 nm level [2]. Even if morphological information were able to be derived from ultra-high resolution SEM imaging techniques, no crystallographic information to complement morphological information on nanoscale particles is available from the SEM.

To overcome these limitations, we have utilized electron holography to study the structure of carbon nanotubes and so-called giant nested fullerenes, some of which encapsulate heavy elements such as La and Gd. This paper describes aspects of the morphology of carbon nanotubes elucidated by holography techniques, and aspects of the crystallography of giant nested fullerenes which are determined with great precision by interferometry methods.

2. EXPERIMENTAL METHODS

Carbon nanotubes were created by an electric arc discharge method in an atmosphere of 500 Torr of He [3]. Polyhedral nested-shell carbon nanoparticles are often found as a byproduct of the carbon-arc discharge conditions that create nanotubes [4]. If La_2O_3 or Gd_2O_3 powders are packed into a cylindrical hole drilled into the carbon electrode, internal cavities in these 10-40 nm particles can be filled with these carbides during the discharge process. Both types of materials were selectively collected and separately dispersed onto holey carbon support films for observation in the electron microscope.

Our holograms were acquired digitally using a Hitachi HF-2000 cold FEG-TEM operated at 200 kV. A Möllenstedt biprism [5] positioned between the first and second intermediate lenses was used to form the holograms. The biprism voltage was adjusted to give 0.1 nm or finer fringes (referred to the specimen) having contrast greater than 25% at magnifications of 700kX and 1MX. The voltage on the biprism fiber (a drawn quartz fiber coated with gold with a total diameter of 0.3 µm) was supplied by a Keithley 487 picoammeter/voltage source, which supplies up to 500 volts at a stability of 10^{-6} (about 10 times more stable than a battery). Holograms were recorded using a Gatan 1k x 1k CCD slow scan camera, and phase images were reconstructed and quantitative measurements from the holograms were made using Holoworks© software described elsewhere in this proceedings [6].

3. RESULTS

3.1. Carbon nanotubes

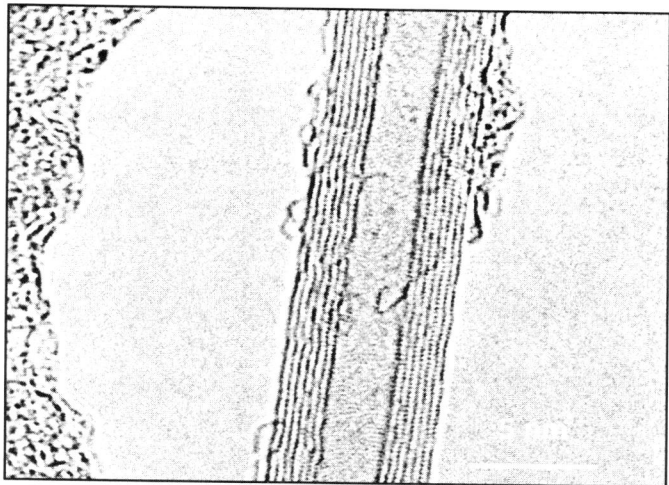

Figure 1. Typical carbon nanotube extending over a hole in the carbon film.

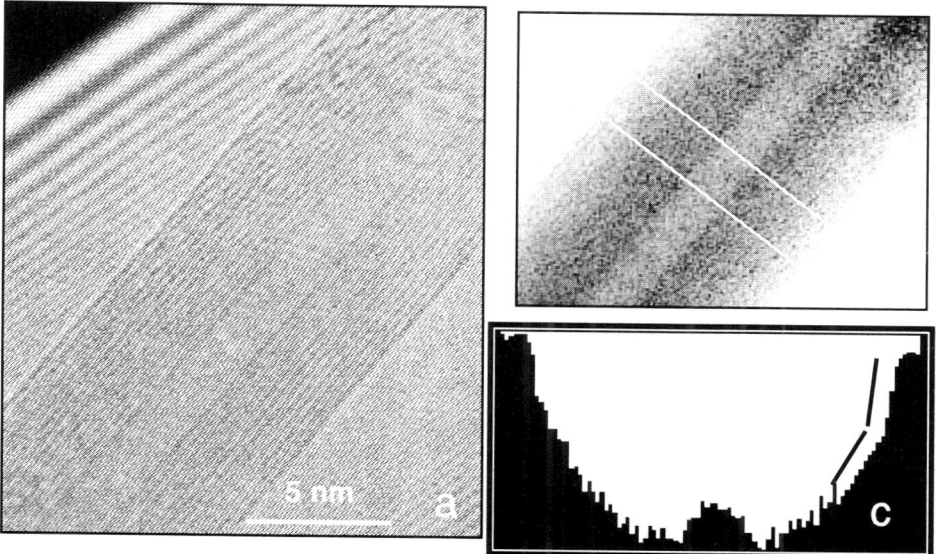

Figure 2. (a) Hologram of a carbon nanotube having 13 nested shells and a 1.6 nm core. (b) Phase image reconstructed from hologram. (c) Phase profile showing some facetting. Central "bump" results from decreased phase change through hollow core of tube.

Figure 1 shows a typical carbon nanotube extending across a hole in the carbon support film. This particle shows 7 nested shells and a core with a diameter of about 2.6 nm. An example of a hologram of a similar tube is shown in Fig. 2a. A reconstructed phase image is shown in Fig. 2b, and the associated phase profile from the outlined area in the phase image is shown in Fig. 2c. This profile shows a generally rounded morphology, but there is a suggestion of facetting associated with the cross-section of the tube, as noted in the profile. The total range of phase change over the tube is less than π, and the presence of the hollow core is evident in the decrease in the phase change due to the decrease in the volume of material traversed by the beam in the core region.

That the phase profile is a good indicator of the thickness of the object has been shown recently by Allard et al. [7], for the case of 10-15 nm diameter nanocrystals of Pd which were shown quantitatively to have hollow cores. Since Pd crystals deviate from kinematic scattering conditions at smaller thicknesses than lower atomic number materials such as carbon nanotubes, it is reasonable that the phase profiles from these materials are indeed directly related to thickness variations and in fact can be expected to accurately indicate thickness at the sub-nanometer level.

Carbon nanotubes are often observed to lie in bundles of two to many tubes.

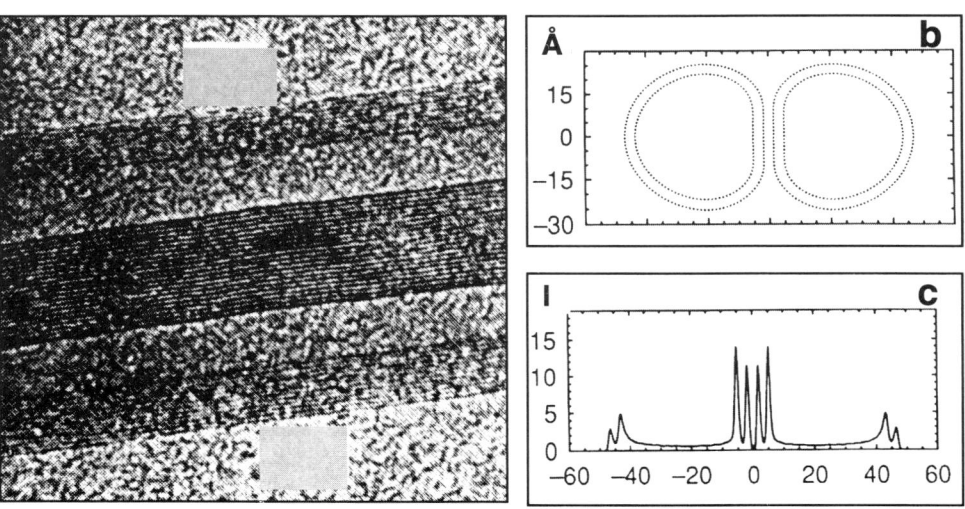

Figure 3. (a) A pair of nanotubes touching each other longitudinally. Contrast in the central region suggests the model of (b), where the thickness of material in which basal planes lie parallel results from deformation of the tubes. (c) shows the expected intensity profile from the model, and this is consistent with the observation. See [4] for detailed discussion.

Recently, Ruoff et al. [4] analyzed high resolution images of the simple case of two nanotubes lying adjacent and touching, oriented with the electron beam parallel to the boundary between the two tubes. Figure 3a is the TEM image of a pair of these tubes, one tube having 6 nested shells and the other having 7 shells. The contrast of the basal planes in the center region is clearly higher than in the outer walls. This observation was suggested to result from the beam traversing a greater thickness of material with lattice planes parallel to the beam, as would be expected if the tube pair were deformed as indicated in the diagram of Fig. 3b. This model shows the simple case of touching nanotubes having only two nested shells each, and the intensity profile computed from the model shown in Fig. 3c correlates to the observed experimental result where the central planes have higher contrast than the outer.

We have observed a similar case of a pair of nanotubes, shown in the hologram of Fig. 4a oriented as in Fig. 3a with the beam essentially parallel to the

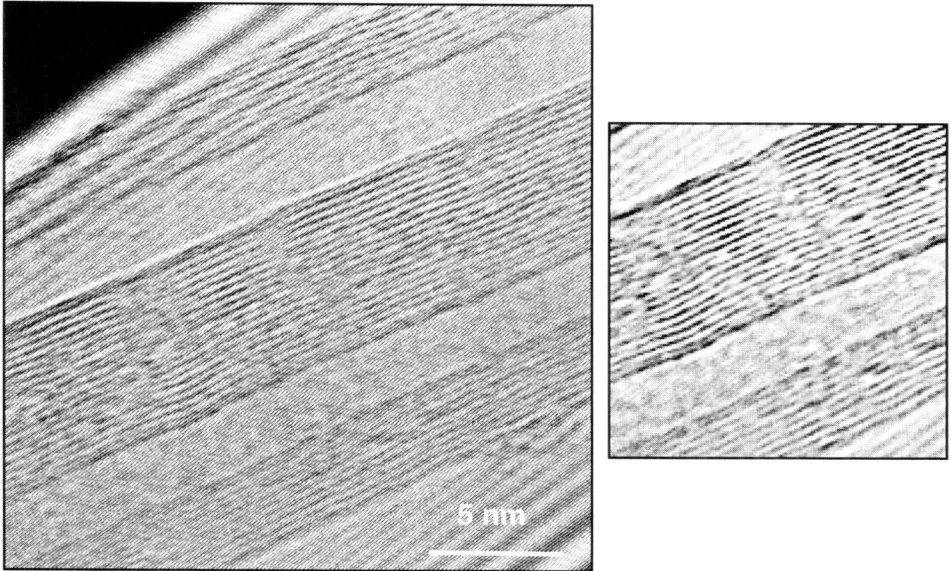

Figure 4. a) Hologram of nanotube pair oriented with the beam parallel to the boundary between tubes. b) Intensity image reconstructed from the autocorrelation region of the FT of the hologram, showing higher contrast in the central region consitent with the observations of Ruoff, et al [4].

boundary between the tubes. Although a direct high resolution TEM image was not recorded for this tube pair, Fig. 4b shows the intensity image obtained by inverse Fourier transform from the autocorrelation region of the FT of the hologram. As in the tube pair shown in Fig. 3a, the central region of this tube

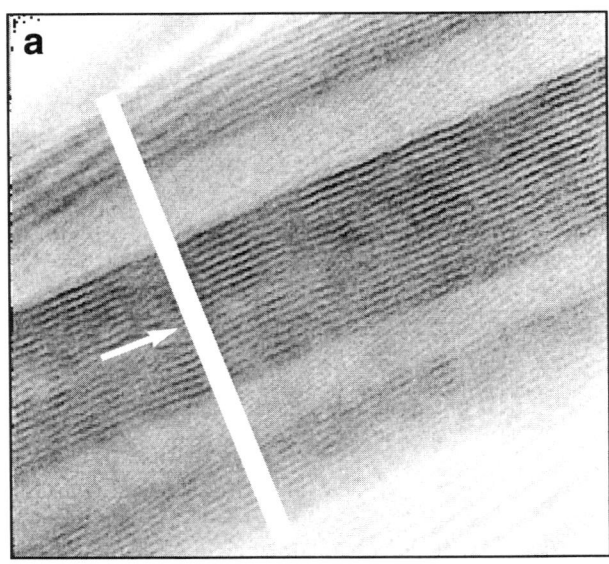

Figure 5. (a) Phase image of double nanotube, not corrected for artefacts using a reference hologram, and with no mask used to remove lattice fringes. (b) Phase profile over the area indicated, showing clear dip in phase change consistent with model proposed by Ruoff, et al.

pair also shows higher contrast than the outer wall regions, suggesting that the tubes are deformed in a similar manner. The phase image (with no corrections made to remove artefacts) from this tube pair is shown in Fig. 5a, and the phase profile across the area outlined is shown in Fig. 5b. There is a clear dip in the center of the phase profile, corresponding to the arrowed location of Fig. 5a. This is consistent with the model structure suggested by Ruoff et al., in which the overall thickness change over the center region would result in the observed change in phase.

The phase image of Fig. 5a was not corrected for artefacts such as Fresnel fringes [6], and neither were contrast effects from the nanotube basal planes

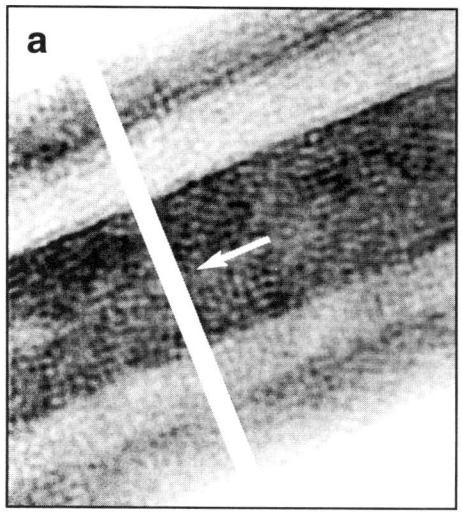 

Figure 6. a) Phase image with lattice fringes eliminated by masking the FT appropriately, and Fresnel fringes minimized by application of reference hologram; b) phase profile showing generally rounded profile, with central phase dip indicating consistency with Ruoff model.

removed. To illustrate the improvement in the interpretation of the phase intensity profile with these effects accommodated, a reference hologram was applied to the original hologram in the initial reconstruction step to minimize Fresnel effects, and the lattice fringes were removed by choosing the appropriate mask in a further processing step. The results of this sequence are shown in Fig. 6, where the phase image of Fig. 6a shows virtually no residual fringe effects. The improved phase profile is shown in Fig. 6b.

3.2. Giant Nested Fullerenes

A typical giant nested fullerene structure encapsulating a Gd compound is shown in Fig. 7. With these particles questions arose regarding the nature of the graphitized carbon structure. "Partially graphitized" carbon has historically been used as a magnification standard for high magnification calibration of the electron microscope [8]. The value commonly given for the basal plane spacing in these materials is 0.34 nm, since pure graphite having a hexagonal structure exhibits an ideal basal plane spacing of 0.3354 nm [9-11] . However, graphite structures can range in the bulk from pure crystalline graphite to the fully turbostratic graphite structure where succeeding planes are randomly oriented [12]. Turbostratic graphite shows no higher order reflections in x-ray diffraction

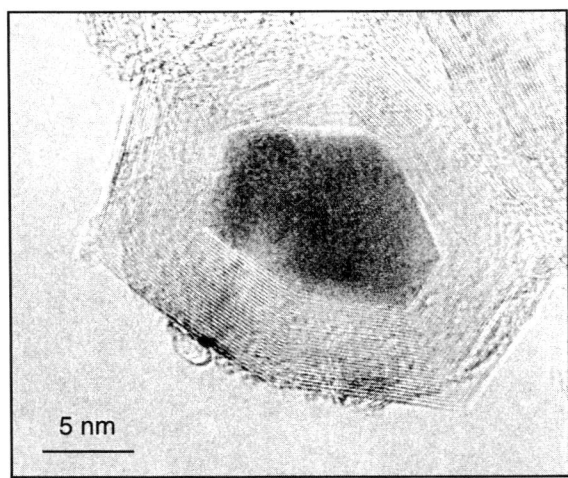

Fig. 7. Typical supergiant fullerene encapsulating a crystal of a heavy metal compound, in this case a Gd-rich compound tentatively identified as GdC_2 (see text for details).

patterns, because there is order in only one direction--along the c-axis. The structure is said to be fully turbostratic when higher order peaks in the x-ray pattern disappear, at which point the basal spacing is typically 0.344 nm or greater [9-11]. Thus the generally accepted value of 0.34 nm for the basal plane spacing in partially graphitized carbon does not guarantee that any individual crystallite has that specific spacing. Without a very accurate calibration of the magnification for an individual image in the TEM, no precise measurement can be made of the basal plane spacings in a given nanocrystal of a graphite-based material, such as partially graphitized carbon or a giant nested fullerene. Electron holography provides a means to determine lattice spacings with precision, as illustrated in the following example.

Figure 8a shows a hologram of a segment of a giant nested fullerene encapsulating a single crystal of a Gd compound (expected to be GdC_2 [3]). The hologram was recorded at 140 V on the biprism and at an instrument magnification of 1MX, with the specimen height adjusted to obtain the OL current required for optimum imaging. A 256^2 FFT of the area outlined is shown in Fig. 9a. In addition to the clear basal plane spacings within this area, a second set of apparent lattice spacings is evident. The reflections associated with both sets of planes are indicated on the FFT. In order to determine these spacings with precision, the distance of the sideband from the center of the autocorrelation was determined using a known crystal structure. A hologram

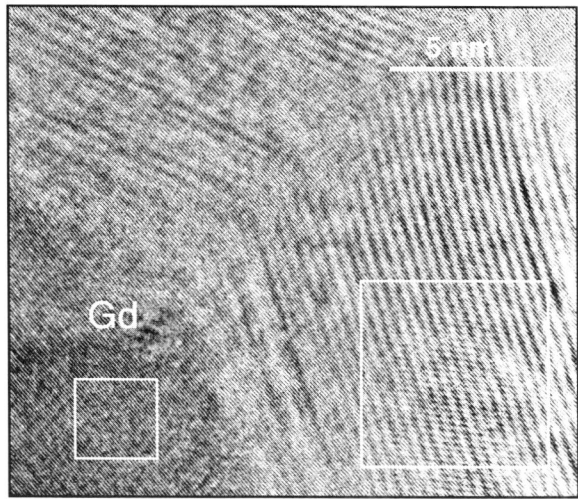

Figure 8. Hologram at 140V showing a segment of giant nested fullerene at the interface with the encapsulated Gd crystal. Areas for FFT analysis are outlined in both regions.

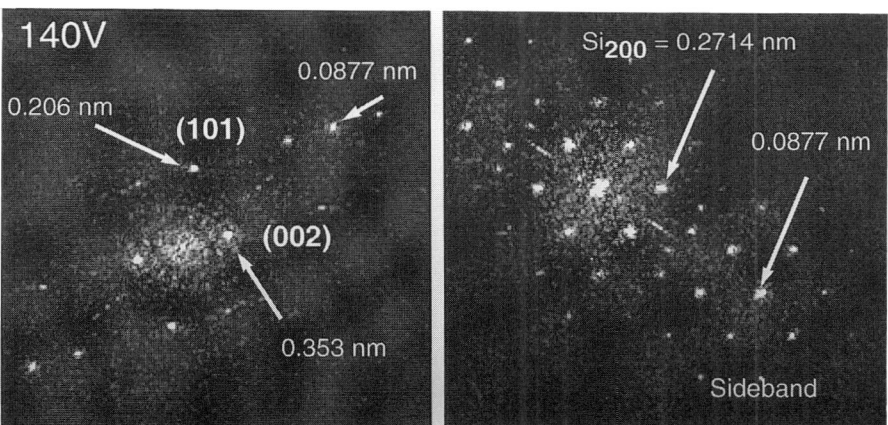

Figure 9. (a) FFT from the graphite area outlined in Fig. 8. Graphite spacings calibrated using the Si <110> zone axis pattern of (b).

was made at 1MX of a Si crystal in a <110> orientation, as shown in Fig. 9b. Care was taken to reproduce the operating conditions obtained for the fullerene hologram. The spacing of the hologram fringes at 140 V was calibrated using the "peak" measuring function provided by Holoworks. With this function, the distance of the center of a sideband was determined in pixels, with an accuracy of better than 0.05 pixels, using methodology developed by de Ruijter, et al. [13]. Using the same process to determine the pixel distance of the known {200} reflection (0.271 nm), the 140 V fringes were thus found to represent a spacing of 0.0877 nm related to the specimen.

The distance of the sideband at 140 V in the fullerene hologram also represents a spacing of 0.0877 nm related to the specimen, and this spacing was used to calibrate the {002} basal reflections as well as the reflections associated with the additional lattice spacings in the fullerene crystal. The basal spacings were thus found to be 0.353 nm and the additional spacings were 0.206 nm. These latter spacings are consistent with the {101} reflections of pure graphite, and are properly positioned, as also determined by the peak program. The basal spacings are, however, significantly larger than the 0.335 nm expected for a pure hexagonal graphite structure and in fact are consistent with spacings found in purely turbostratic structures (i.e. d > 0.344 nm), where higher order reflections are not expected because of the random arrangement of each succeeding basal plane.

The reason for the development of the 0.353 nm basal plane separation is purely geometric. If the supergiant fullerene is considered to comprise a series of nested polyhedra, it results that an odd number of extra hexagons is added at each facet for each neighbor polyhedron, resulting in an ABABAB... stacking of polyhedral basal planes (see [14] for a detailed discussion). The nested polyhedra evidently are conformal (that is, they are similar geometric objects). Reduction of 0.353 (the experimental value) to 0.335 (the value for single crystal graphite) would mean a reduction in the c-c bond length of 5%, from a likely value of 0.1415 nm (the value appropriate for a graphene sheet) to that of 0.1342 nm. This is much too costly in energy, so the covalent bonding network (and the conformal layering) determines the interlayer separation. It is interesting to note that this 0.353 nm interlayer spacing has been observed in BN films deposited onto MgO nano-cubes (smoke particles) [15], along with lattice planes and higher order reflections consistent with a non-turbostratic structure. Since BN is a structural analogue to graphite, these films on MgO mimic the morphologies observed in the encapsulation of carbides by graphite, and it is suggested that a similar geometric argument accounts for the observed BN interlayer spacings.

The calibration of the hologram spacings permits also the measurement of lattice spacings associated with the Gd crystal being encapsulated. Although fringes are not easily visible in the digital image of Fig. 8, the FFT from the 128^2 region outlined on the Gd area showed faint reflections indicating the presence of lattice planes in the image having a spacing of 0.161 nm. This corresponds to the {210} reflections of GdC_2, a result consistent with x-ray observations from bulk materials.

4. CONCLUSIONS

Of the many applications of electron holography, this paper describes two which we have applied to the characterization of fullerene materials. First, because of the small dimensions of the crystals being studied, kinematic scattering conditions apply and the phase images retrieved from holograms can yield phase intensity profiles which directly relate to thickness profiles. This has allowed direct confirmation, for example, of the cross-sectional shape of carbon nanotubes that are parallel and touching. This type of measurement would be difficult to obtain by other methods (e.g. cross-sections prepared by any thin foil preparation technique would probably be ambiguous due to the possibility of sample preparation artefacts). Second, digital processing of holograms provides a method for precise measurement of image dimensions, by using the highly reproducible spacings of hologram carrier fringes as a standard. Giant nested fullerenes, which show lattice images that exibit a precise ordering relationship between basal planes consistent with perfect graphite structure, have been shown to exhibit basal plane spacings that are equivalent to those typically observed in a fully turbostratic graphite structure. This is due to a purely geometric effect. While such a measurement could be made using more standard techniques of magnification calibration, these would always be suspect, whereas the generation of known fringe spacings that can be digitally processed offers a method to obtain more reliable crystallographic data.

REFERENCES

1. S. Iijima, *Nature* **354**, 56-58 (1992).

2. D. J. Smith, M. H. Yao, L. F. Allard and A. K. Datye, *Applied Catalysis*, in press (1994).

3. R. S. Ruoff, D. Lorents, B. Chan, R. Malhotra, and S. Subramoney, *Science* **259** (1993) 346-348.

4. R. S. Ruoff, J. Tersoff, D. C. Lorents, S. Subramoney and B. Chan, *Nature* **364,** August 1993, 514-516.

5. G. Möllenstedt and H. Düker, *Z. Physik*, **145** (1956) 377.

6. E. Völkl, this proceedings.

7. L. F. Allard, E. Völkl, D. Kalakkad and A. K. Datye, *J. Mat. Sci.* **29** (1994) 5612-5614.

8. R. D. Heidenreich, W. M. Hess and L. L. Ban, *J. Appl. Cryst.* **1** (1968) 1-19.

9. W. Ruland, *Acta Cryst.* **18** (1965) 992-996.

10. D. B. Fishbach, "Chemistry and Physics of Carbon", Vol. 7, P. L. Walker, ed., Marcel Dekker (1968) 7.

11. M. Shioya and A. Takaku, *Carbon* **28** No. 1 (1990) 165-168.

12. J. Biscoe and B. E. Warren, *J. Appl. Phys.* **13** (1942) 364.

13. W.J. deRuijter et al., *Scanning Microsc. Supp.* **6** (1992)347.

14. D. C. Lorents, R. S. Ruoff, R. Malhotra and S.Subramoney, *Mol. Mat.* **4** (1994) 15-22.

15. L. F. Allard, A. K. Datye, T. A. Nolan, S. L. Mahan and R. T. Paine, *Ultramicroscopy* **37** (1991) 153-168.

TRANSMISSION ELECTRON HOLOGRAPHY OF POLYMER MICROSTRUCTURE

M. Libera, J. Ott, and Y. C. Wang

Department of Materials Science and Engineering,
Stevens Institute of Technology, Hoboken, NJ 07030

1. Introduction

Electron-optical contrast of multi-phase amorphous polymers is poor because the local average atomic number changes little. Better electron-optical contrast is typically induced by preferentially staining one phase with a heavy element (e.g. Os, Ru). Incident electrons traversing stained regions are Rutherford scattered ($\sigma(\alpha) \sim Z^{3/2}$ to Z^2) outside an objective aperture. This method has been widely used for studies of phase distribution and morphology in block copolymers and blends (1). Identifying appropriate stains for new systems is often difficult, and, ultimately, the information gathered by staining methods is limited by the stain's ability to faithfully label a feature of interest. This latter issue is particularly important at polymer interfaces. Identifying an alternative method to image polymer microstructure would be a significant step forward in this field.

Despite the fact that unstained polymeric specimens provide little modulation of the electron wave amplitude, there should be angle variations in the wave phase. These phase modulations can in principle be recovered by holographic imaging methods. This paper describes work beginning at Stevens to do holographic imaging of polymer microstructure. Our initial focus is on amorphous two-phase polymers, and we first describe preliminary work on the preparation and morphology of solution-cast thin films of styrene-butadiene-styrene (SBS) tri-block copolymer. This paper then outlines the basis for holographic phase contrast in such system - namely differences in mean inner potential. We conclude by describing the construction of an electron biprism built for a Philips CM 20 FEG TEM and preliminary results using it.

2. SBS Morphology

We have been concentrating our research on SBS tri-block copolymers. Thin specimens are prepared by placing one drop of 0.1wt%SBS in toluene onto a polished NaCl crystal (~1cm^2) and allowing the toluene to evaporate. The SBS film is then scored with a razor blade, floated from the NaCl substrate, and pieces of the film are collected on copper grids. SBS is a well studied class of block co-polymer. It has a regular morphology of alternating styrene-rich and butadiene-rich domains in either a rod-like or lamellar sort of morphology. There are slight differences in density ($\rho_{st} \sim$ 1.05 g/cm^3; $\rho_{bu} \sim$ 0.95g/cm^3) which lead to some intrinsic contrast when using an objective aperture, but this is weak. Since both materials are amorphous, strong Bragg peaks are absent, and traditional diffraction contrast is not possible. Figure 1 describes the molecular structure of the styrene and butadiene mers and illustrates the basic problem associated with TEM imaging of these materials. Both consist of carbon and hydrogen. Figure 2a shows a bright-field image from an unstained film. Lamellae can not be resolved here. Figure 2b shows a bright-field image from a similar specimen that has been stained by exposing it to osmium tetroxide vapor which attacks the unsaturated carbon double bond in the butadiene. The butadiene-rich lamellae thus appear dark and the styrene-rich lamellae appear light.

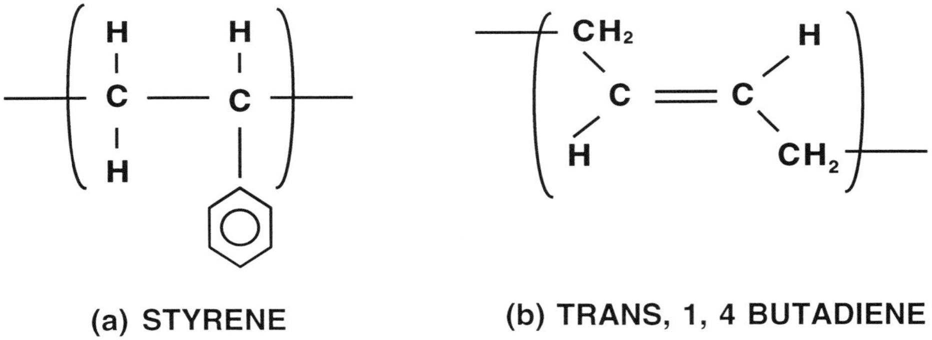

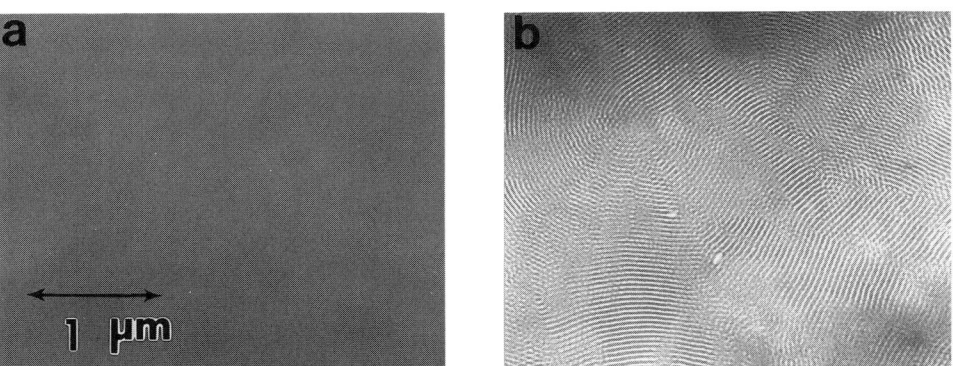

(a) STYRENE   (b) TRANS, 1, 4 BUTADIENE

Figure 1 - Mer structure of (a) styrene and (b) butadiene.

Figure 2 - Bright-field TEM images of solution-cast SBS films: (a) without OsO$_4$ stain; and (b) with OsO$_4$ stain.

3. Origin of phase contrast in polymers

An electron wave (the object wave) is refracted as it passes through a specimen. Relative to an unrefracted wave that passes through vacuum (the reference wave), the difference in phase $\Delta\phi$ due to refraction is given as:

$$\Delta\phi = (2\pi/\lambda)(n-1)t \qquad \qquad ...[1]$$

where n is the electron-optical index of refraction of the specimen and t is the specimen thickness. In the absence of magnetic fields, the electron-optical refractive index is related to the electrostatic potential of the specimen. In crystalline specimens, the electrostatic potential is traditionally represented as a Fourier expansion based on the reciprocal lattice. The first term of that expansion is the average scalar potential and is known as the mean inner potential (Φ_o). While a Fourier representation of the atomic potential is not possible in amorphous materials since there is no long-range structural periodicity, the mean inner potential remains a valid concept. It is simply related to the total electrostatic contributions of all the atoms divided by the volume occupied by those atoms. The refractive index due to the mean inner potential is given as:

$$n = 1 + \frac{|e|\Phi_o}{E} \frac{E_o + E}{2E_o + E} \qquad \qquad ...[2]$$

where E_o is the rest energy of an electron, E is the energy of the fast electron, and |e| is the magnitude of charge carried by an electron. Φ_o ranges between approximately 5V and 30V. Since it is positive, the refractive index through matter is greater than unity as in light optics.

The mean inner potential can be estimated from tabulated values of the atomic scattering factor at zero scattering angle (2,3,4). Results from such an estimate can be misleading, however, because bonding effects are difficult to take into account. In the case of MgO, Φ_o calculated using $f_i(s=0)$ for neutral Mg and neutral O (O'Keeffe and Spence's so-called pro-crystal) one finds Φ_o (MgO neutral)=18.43V. Using f_{O-2} and f_{Mg+2} from recent calculations by Rez et al (5), Φ_o (MgO fully ionized)=12.65V. This latter value compares much more favorably with the value of 13.01V determined experimentally by Gajdardziska et al. (6) using electron holography. One can estimate using the Pauling electronegativity formula that the percent ionicity in MgO is 70.2%. In other words, modelling MgO using fully ionized Mg and O ions is largely correct. To the best of our knowledge, however, the effect of covalent bonding has not yet been addressed in first-principles calculations of the f_i. Discrepancies between the estimated and measured values of Φ_o for Si and GaAs (4), the latter having a Pauling ionicity of 4%, are likely due to these bonding effects.

Because bonding affects the magnitude of the mean inner potential, we expect many polymer systems to give phase contrast measurable by electron holography. Figure 1 shows that while styrene and butadiene are both composed solely of hydrogen and carbon, styrene has a bulky pendant aromatic ring whereas butadiene instead has an unsaturated carbon double bond. Without better information for f_i, we have not yet been able to estimate the effect of these steric and bonding differences will have on the Φ_o characteristic of each of these two polymer phases. For the case of a lamellar morphology, however, we anticipate a characteristic modulation of the specimen potential such as illustrated by figure 3. The extremes of the modulation should correspond closely to the Φ_o values characteristic of single-phase styrene and single-phase butadiene. Using the relation

$$\Phi_o \text{ (Volts)} = \frac{47.878}{\Omega} F^B_{g=0} \qquad ...[3]$$

where Ω is the scattering volume (unit cell in the case of a crystalline material) and $F^B_{g=0}$ is the structure factor calculated from electron-scattering factors determined using the Born-approximation system (3), we have estimated the mean inner potential characteristic of low-density polyethylene (LDPE). LDPE is amorphous and has a density of 0.85g/cm^3. From this, we estimate that the average volume per mer is 54.7A^3. Using $f_{el}^C(0)=2.509A$ (2) and $f_{el}^H(0)=0.529A$ (7) we find Φ_o(LDPE) = 6.24V. This again ignores the effect of bonding on the magnitudes of the f_i. We would expect most polymeric hydrocarbons to have an inner potential of this order.

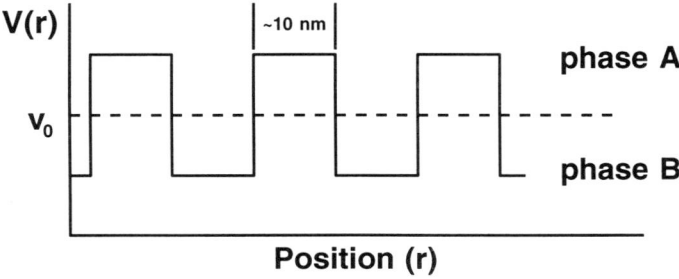

Figure 3 - Schematic illustration of typical fluctuations in specimen potential as a function of position in a two-phase polymer with a lamellar morphology.

4. Biprism Construction

To do holography at Stevens we have built a modified SAD aperture rod for a Philips CM20 FEG TEM. This assembly includes a non-rotatable biprism along with three conventional SAD apertures (figure 4). Since specimen tilt is not critical for these amorphous polymer specimens, relative specimen/filament orientations can be achieved using a tilt-rotate specimen stage. We have used an oxyacetylene flame to draw quartz filaments. Submicron filaments are coated with a layer of evaporated gold. We have successfully biased the filament using either batteries connected in series or a low-ripple DC power supply. We have found that shielded cable and contacts between the power supply and the biprism feedthrough are essential to minimize noise induced by stray fields. We record holograms on film, digitize these using a 600dpi scanner, and manipulate/analyze them using HIP software from the Tennessee group (8) and NIH Image (9). We are currently developing satisfactory reconstruction methods.

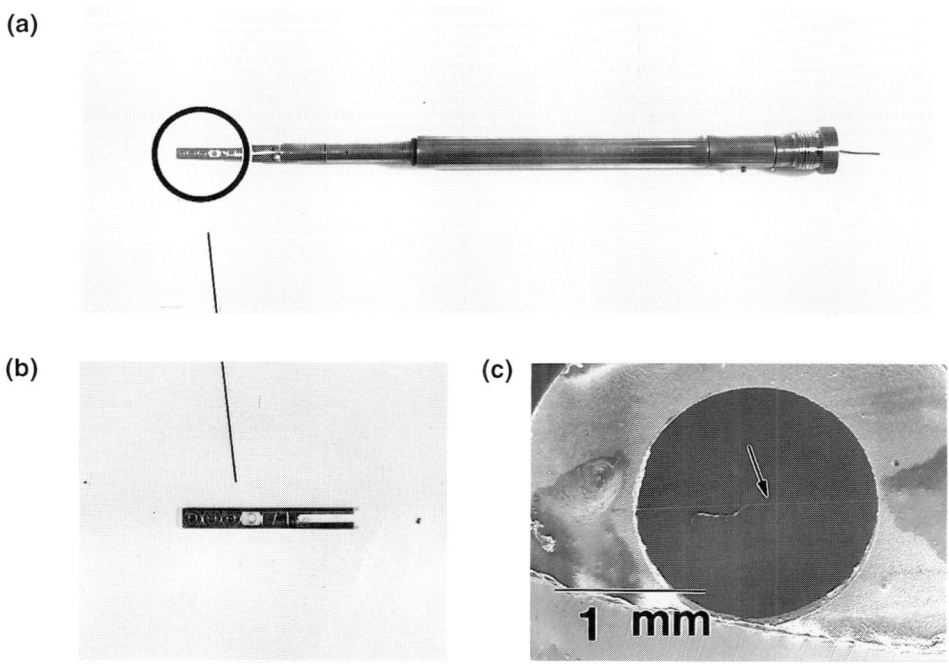

Figure 4 - (a) modified SAD aperture rod with electrical feedthrough; (b) detail of aperture holder with three conventional SAD apertures; and (c) SEM image of the biprism filament (arrow) mounted on its insulating ceramic washer.

5. Results

A typical hologram recorded at Stevens is shown in figure 5. The specimen is graphitized carbon on an amorphous carbon substrate. Fresnel fringes from the biprism are visible as well as holographic interference fringes carrying information characteristic of the graphite, the amorphous carbon, and the vacuum (empty hologram). First results of applying holography to determine the morphology of stained SBS are presented in (10).

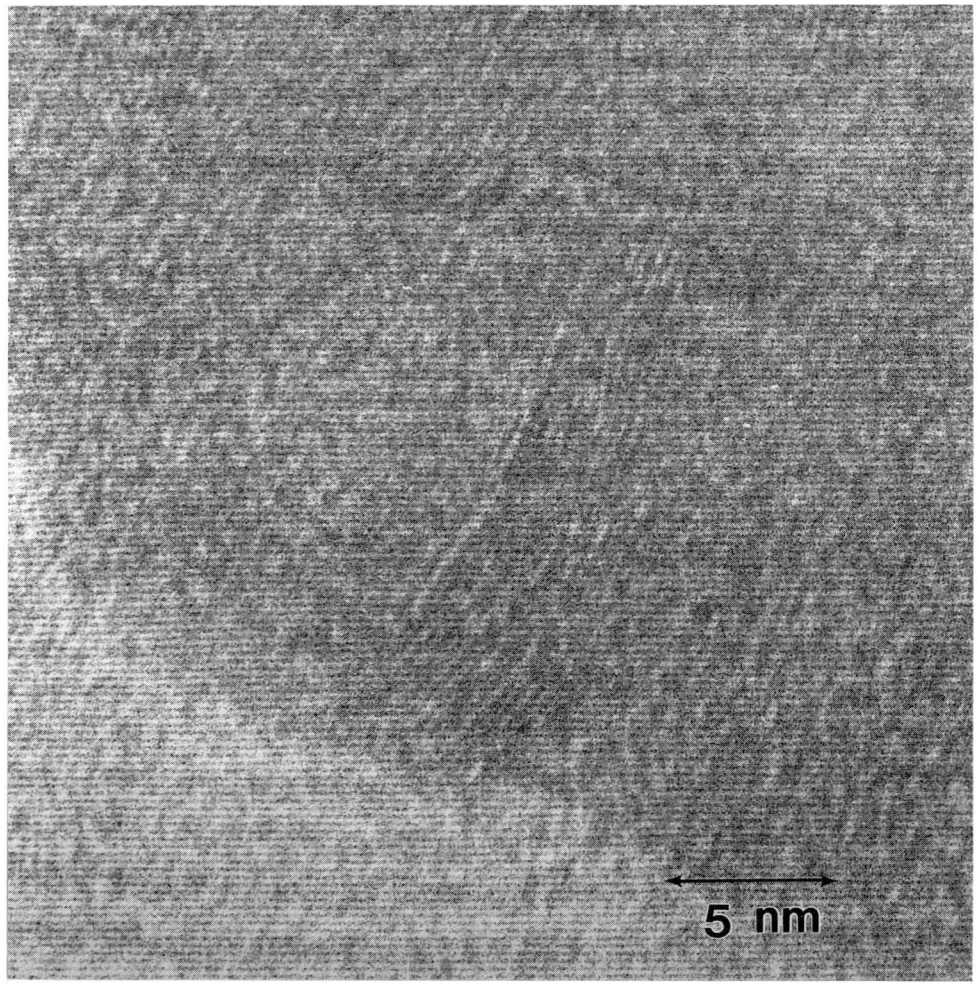

Figure 5 - Electron hologram of graphitized carbon on an amorphous carbon substrate.

6. Key Issues to be addressed for holography of polymers

There are three principal challenges in the application of electron holography to polymers. First, no quantitative information is really known concerning the mean-inner potentials characteristic of engineering polymers. Second, because of their varying elastic properties, traditional methods of ultramicrotomy usually lead to systematic differences in the thickness of different phases. Casting methods should mitigate this effect, since cutting is not involved. Preliminary measurements using AFM and SEM on an SBS film cast on glass indicate that the thickness and average roughness are ~70-100 nm and +/-10 nm or less, respectively. Higher resolution AFM measurements are needed to resolve thickness variations between individual styrene and butadiene lamellae. Third, the familiar problem of radiation damage may be slightly mitigated since holography does not discard as many incident electrons as traditional methods, but low-dose considerations will remain important.

Acknowledgment

This research is supported by the Army Research Office and uses microscopy resources provided by the New Jersey Commission on Science and Technology and the NSF.

References

1. L. Sawyer and D. Grubb, *Polymer Microscopy*, Chapman and Hall, London, 1987.
2. P.A. Doyle and P.S. Turner, Acta Cryst. A24 (1968) 390.
3. J.C.H. Spence and J. M. Zuo, *Electron Microdiffraction*, Plenum, New York, 1992
4. M. A. O'Keefe and J.C.H. Spence, Acta Cryst. A50 (1994) 33.
5. D. Rez, P. Rez and I. Grant, Acta Cryst. A50 (1994) 481.
6. M. Gajdardziska-Josifovska, M.R. McCartney, W.J. de Ruijter, D.J. Smith, J.K. Weiss and J.M. Zuo, Ultramicroscopy 50 (1993) 285.
7. *International Tables for X-ray Crystallography*, Vol. III, The Kynoch Press, England, 1968.
8. D.C. Joy, Y.-S. Zhang, X. Zhang, T. Hashimoto, R.D. Bunn, L.F. Allard and T.A. Nolan, Ultramicroscopy 51 (1993) 1.
9. Image version 1.45, public-domain software (Macintosh) available from the National Institute of Health
10. M. Libera, M. Gajdardziska-Josifovska and M.M. Disko, Proc. 52nd Annual Meeting of MSA (1994) 444.

Electron holographic observation of thin biological filaments

K. Aoyama [a], G. Lai [a and b] and Q. Ru [a]

[a]Tonomura Electron Wavefront Project, ERATO, Research Development Corporation of Japan (JRDC), P.O. Box 5, Hatoyama, Saitama 350-03, Japan

[b]Department of Electrical and Electronic Engineering, Shizuoka University Johoku 3-5-1, Hamamatsu, Shizuoka 432, Japan

Electron holography is applied to observe thin biological filaments. This method has two advantages for observing weak phase objects such as biological materials. One advantage is sufficient contrast of the reconstructed phase image from the electron beam's phase distribution caused by the target molecules. The other advantage is the accuracy of the phase image's shape. This was clearly demonstrated with two observations of filaments. In the first experiment, an unstained bacterial flagellum filament adsorbed on a carbon supporting film is observed with high contrast in the in-focus condition by using the phase-shifting method. Subsequently, a tobacco mosaic virus (TMV) bridged over a hole of supporting film is observed as a cylindrical shape that represents the actual shape of the filament. An exactly in-focus image of a weak-phase object can be observed with sufficient contrast by electron holography, in contrast to ordinary transmission electron microscopic methods.

1. INTRODUCTION

Because biological macromolecules primarily consist of light atoms, the low contrast in observed image is large disadvantage in ordinary transmission electron microscopy (TEM). Staining or shadowing methods have been widely used to increase contrast. However, the contrast obtained by these indirect methods is not due to the biological molecule itself, but heavy atoms. Moreover, an artificial structural change may be introduced in specimen preparation. If we want to observe the actual structure, such pre-treatments should not be used.

There are some methods for easily obtaining sufficient contrast in TEM without staining or shadowing. One method is to make a dark-field image, but because most of irradiated electrons do not contribute to the image contrast, a large number of electrons and a very long exposure time are required to obtain visible contrast. Because of specimen drift, it is difficult in practice to use so long exposure time. Moreover, a large total dose causes large irradiation damage, which is the most serious problem in observing biological materials. Another method is to introduce large defocusing in the image formation, making a phase-contrast image. Defocusing, however, always causes artificial transformations of the image which are hard to correct. Namely, the contrast and the accuracy (or resolution) are trade-off, and it is very difficult to obtain an accurate image with sufficient contrast by using conventional imaging technique.

Electron holography has some important advantages over conventional methods. Since a specimen is observed by the phase distribution of electron waves, not by its intensity contrast, an in-focus image of a weak-phase object can be observed with sufficient contrast. Because the phase distribution is quantitatively measured, the details of the specimen can be correctly interpreted. In this report we show that the image obtained using electron holography reflects the correct profile of the specimen with enough contrast.

2. SPECIMEN PREPARATION

2.1. Preparation of flagellum filament

We selected bacterial flagellum filament for the first observation to demonstrate the ability of electron holography to produce a high contrast image. The filaments were prepared from a wild-type strain of *Salmonella typhimurium* SJW1103. They are only 24 nm in diameter and consisted of a single protein flagellin with a counterclockwise helical form of pitch 2.3 µm and a helical diameter 0.5 µm [1]. Because the filament had a characteristic form, it was easy to identify during the observation and easy to recognize during the demonstration. Therefore, it was a suitable sample for the first observation. The specimens were prepared as follows: 1) For fixing the flagella filaments, a solution containing 0.15 mg/ml flagella, 20 mM Tris-HCl, 150 mM NaCl, and 200 mM $(NH_4)_2SO_4$ (pH=7.7) was mixed with an equal volume of a 20% glutaraldehyde solution. After five minutes the mixture was diluted with nine parts of water. 2) Five microliters of the solution were put on the specimen grid covered with a carbon supporting film which had been hydrophilized by ion sputtering. The flagella filaments were adsorbed on the supporting film by hydrophilic interaction. 3) Excess solution was removed with a filter paper and the grid was washed with water to remove salts and the fixer.

2.2. Preparation of tobacco mosaic virus (TMV)

Tobacco mosaic virus (TMV) was selected for the second observation which required mechanical toughness of the sample filament. TMV is a long rod 18 nm in diameter and 300 nm in length which consists of a single-strand RNA molecule of 6000 nucleotides and 2130 copies of identical protein subunits arranged around a helical RNA core [2]. The specimens were prepared as follows: 1) TMV was fixed to increase resistivity against irradiation damage and mechanical strength. A solution 0.3 mg/ml TMV in water was mixed with an equal volume of 10% glutaraldehyde solution. 2) The solution was then dialyzed against enough volume of pure water for one day to remove the excess fixer. 3) Five microliters of solution were put on the specimen grid covered with a holey carbon film. The holey film had been prepared by ion sputtering [3] which achieved hydrophilization at the same time. 4) The specimen was frozen by liquid nitrogen after removing excess solution. 5) The frozen specimen was put into a vacuum chamber to freeze-dry. Because the solution's (ice's) volume was very small, the grid had to be held a low temperature to keep the solution frozen. After several hours, TMV filament was bridged over the hole.

Freeze-drying is required to bridge thin biological filaments over the hole, because of the weak mechanical strength of the filaments. If we do not use this method, the filaments are drawn to the edge of the hole due to the surface tension of water.

3. OBSERVATION OF THE FLAGELLA FILAMENT

3.1 Phase-shifting method

For observation of the flagella filament, A 200-kV field-emission electron microscope (Hitachi HF-2000) operated at 200kV was used for hologram formation, and the phase-shifting method reported previously [4,5] was used for hologram reconstruction. This method is based on tilting the incident electron beam to shift the initial phase of the hologram and then recording a series of holograms with different initial phases. Because interference is observed on the imaging plane, the tilt of the incident beam does not shift the image position. The optics for obtaining phase shifts by beam tilting are illustrated in Figure 1. The holograms are captured with a TV camera and transferred frame-by-frame to a computer that computes the phase image of the specimen.

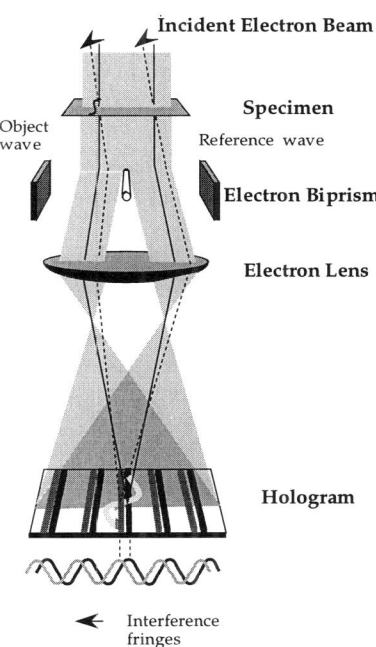

Figure 1   Schematic representation of Phase-shifting electron holography.
  Shadows and Lines indicate electron beams and electron paths corresponding to the maximum fringe intensity, respectively. Because the electron beam tilts, interference fringes from black to gray fringes.

The phase-shifting method has several advantages over the conventional Fourier transform reconstruction method [6]. Since small or thin objects can be reconstructed independently of hologram fringe spacing, flagellum filaments much thinner than the fringe width can be observed. Another advantage is its precise detection of phase distribution. Since a flagellum filament causes a maximum phase distribution of only about $1/10\ \lambda$, a much higher precision; say $1/100\ \lambda$, is required. This precision has been achieved with phase-shifting electron holography.

Because this experiment required a wide interference area to cover the whole filament, an extremely low magnification mode was selected. Low magnification also reduces irradiation damage. To get optimum conditions for this wide interference area, we greatly decreased the power of the objective lens and focused the specimen by using an intermediate lens. To maintain total magnification, the projector lenses were operated at their maximum currents. The resultant interference area was about 5 µm wide and the direct magnification was 1500.

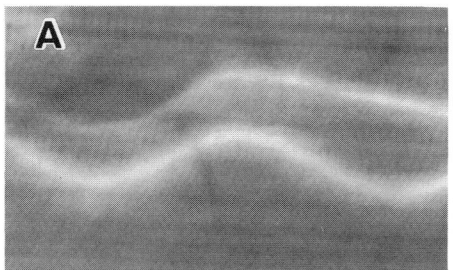
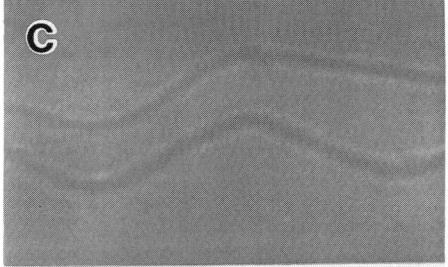
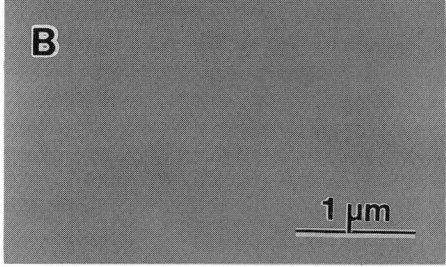

Figure 2
A: The reconstructed in-focus phase image has sufficiently high contrast. image. This is unobtainable with any ordinary TEM method.
B: An in-focus, bright-field TEM image. The contrast is insufficient.
C: TEM image defocused a few micrometers. The image has visible contrast, but artificial fringes are seen at the edge of the filament (Fresnel fringes)

3.2 Ability to produce high contrast images

We were able to observe single flagellum filaments, even though they were only 24 nm in diameter, by using phase-shifting electron holography [7]. Since flagellum filaments have a helical form in solution, they should form a sine curve when they adhere to the carbon supporting film. The reconstructed phase image (Figure 2A) shows the expected sine curve. Figure 2B is an in-focus, bright-field TEM image. The contrast is insufficient. Figure 2C is a TEM image defocused on the order of a μm showing the same area as Figure 2A. This image has visible contrast due to large defocusing, but its details are inaccurate. The width of the filament cannot be determined because artificial fringes are seen at the edge. These results demonstrate that electron holography produced a sufficiently high contrast in-focus image that could never be obtained by ordinary transmission electron microscopic methods.

4 OBSERVATION OF THE TOBACCO MOSAIC VIRUS

4.1 To observe accurate shape

However, the first experiment had some problems. The peak phase shift caused by a single filament was found to be $1/15\ \lambda$, which is smaller than that expected value. The value have been able to estimate about $1/6$ from observation of polystyrene latex which have similar constitution as biological macromolecules and enough mechanical strength. This difference might be due to flattening of the sample or that the real peak thickness might not be detected because of insufficient resolution. Moreover, the flagella filament showed irregular shape, we suppose, due to blots on the supporting film's surface. Additionally, irradiation damage and damage caused by the high vacuum in the electron microscope were presumed to have occurred, since we do not perform counter plan in this observation.

To resolve these problems, three techniques were introduced for the second observation; the Low Dose System (LDS), the cryo (liquid nitrogen) specimen holder and observation without the carbon supporting film. Using the LDS and cryo holder are common methods to reduce irradiation damage for observing biological specimens. However, we did not use them in the first observation because of the restriction mentioned in 4.2 and the complexity of operation.

The supporting film caused three problems:
1) Flattening : Samples are adsorbed on the hydrized carbon supporting film by hydrophilic interaction. This force, however, causes a transformation of the sample. (Figure 3A)
2) Remaining : Biological macromolecules usually require some kind of chemicals and salts in the solution. Those materials remain between sample and supporting film when the specimen dries up. (Figure 3B)
3) Noise (including charge up) : Since an objective wave and a reference wave are independently affected by the supporting film's noise, the supporting film's noise is a more serious problem for electron holographic observation than TEM

observation. If we use an ice embedding technique, flattening and remaining problems do not occur, but noise levels become higher than observation without a supporting film. Moreover, charge up of the supporting film or the embedded ice is a more serious problem for electron holography, as described in 4.4

Observation without supporting film was necessary to observe the actual structure. We achieved the condition using holey carbon film and freeze-drying. Figure 4A shows the TMV filament bridged over the hole.

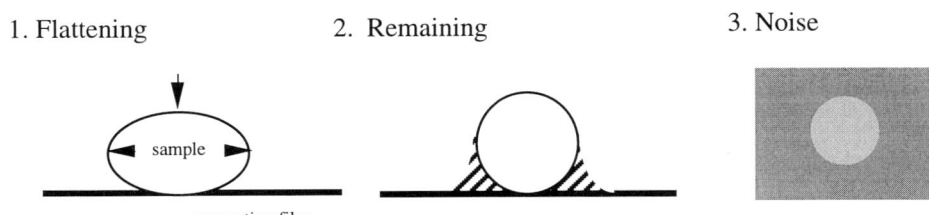

Figure 3   Schematic drawing of problems caused by supporting film

4.2 Observation of the TMV

For observing the correct cylindrical shape of the thin biological filament, a second experiment was performed. In this observation, the electron microscope was operated at 100kV for hologram formation. The hologram was recorded on Kodak 4489 electron microscope film, and conventional Fourier transformation was used for image reconstruction. Electron lenses were operated in normal mode, which is the most user friendly, and the direct magnification was selected at 50,000.

Since a few technical and mechanical problems have not yet been solved, the phase-shifting method was not used. The LDS and the beam tilting system which is necessary for phase-shifting method can not communicate with each other in our electron microscope and this was our most serious problem. Since it has been hard to use these two systems at the same time, we have had to relinquish the phase-shifting method. Moreover, the substantial size of the TV camera detector is seriously deficient for this observation. When the LDS is used for taking image, an operator cannot see the image at real magnification, but only an extremely low magnification for searching. Therefore, a sufficient detector size is required to more easily record images of the TMV. When we use electron microscope film, we can get a huge detector size (approximately 10 x 7 cm : full size) and high resolution (approximately 7 µm : Kodak 4489). Comparatively, the TV camera is only 1 x 1 cm in size and 20 µm in resolution. This size substantially corresponds to only 1/600 (size 1/10 x 1/7 and resolution 1/3 x 1/3) of the film. It is a very large disadvantage to achieve the required observation using the LDS. Since a large number of holograms are required, it is difficult to perform the phase-shifting method by using electron microscope film.

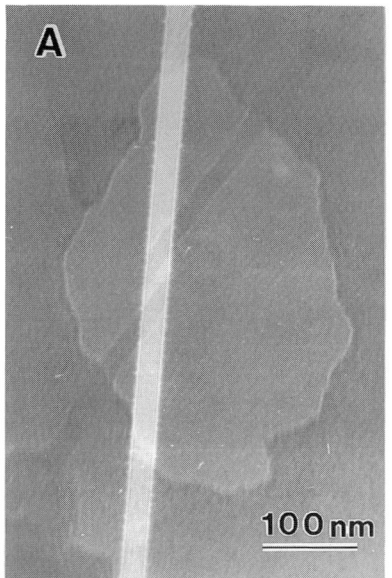 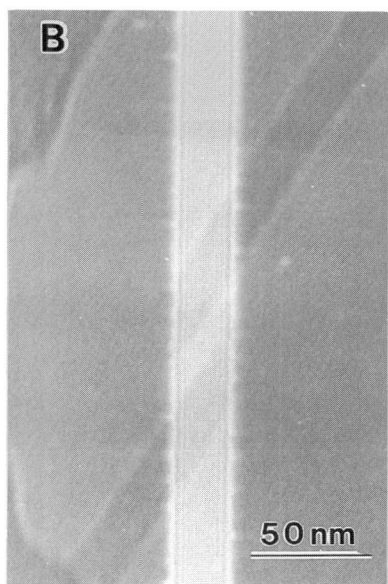

Figure 4   Tobacco mosaic virus observed without supporting film

4.3 Cylindrical shape of the filament

Figure 4A is a TEM image including a hologram. A region overlapped the image of the TMV filament and the interference area is a hologram of the TMV. Figure 4B is a high magnification image around the hologram. Because the image of TMV overlapped from right and left of the interference area, the filament image is looked to make a slide

Figure 5 shows a comparison between a TEM image and a reconstructed phase image. Figure 5A is a TEM image of TMV trimmed from the TEM region of Figure 4B. Figure 5B is a phase image reconstructed from the hologram region of Figure 4B. Figure 5C is an intensity profile of Figure 5A across the filament and Figure 5D is a phase profile of Figure 5B across the filament. The profile of the phase image completely shows a half circular form that can be considered the real projection shape of cylinder-like materials. In comparison, the profile of the TEM image shows an artificial form due to Fresnel contrast which enhances edges of the image. If the filament can be observed at exactly in-focus, the image is not effected by Fresnel contrast. Image contrast, however, can not be detected by such a focus condition in TEM. Therefore, it is impossible for TEM to observe the actual shape with sufficient contrast. The peak phase shift caused by a TMV was found to be 1/7 λ, which almost corresponded with expected value.

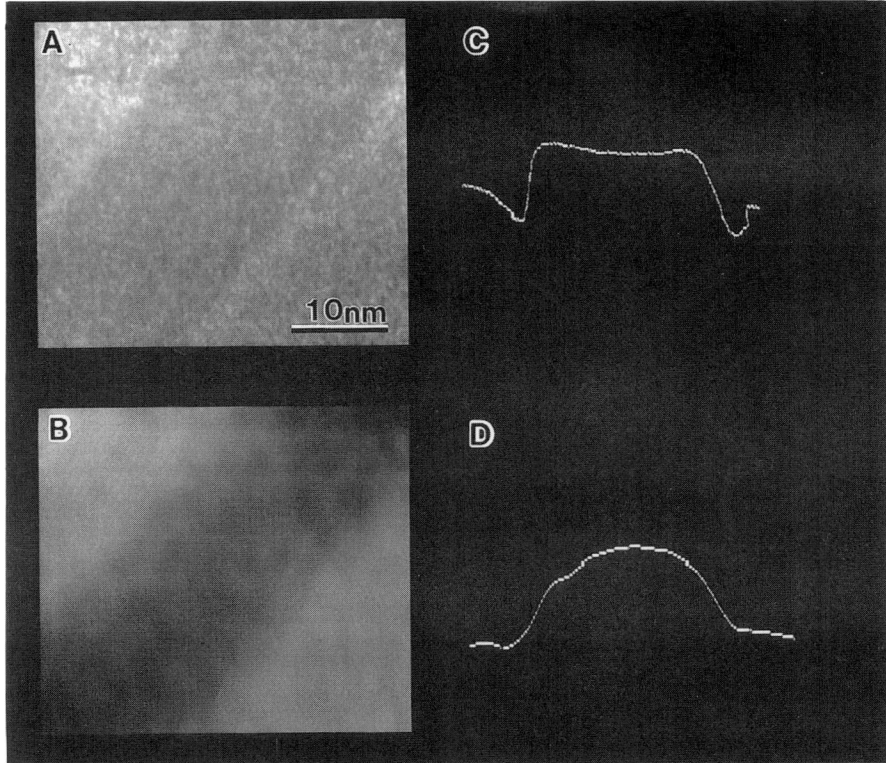

Figure 5   Comparison between TEM image and reconstructed phase image
A) TEM image of tobacco mosaic virus
B) Reconstructed phase image
C) Intensity profile of TEM image across the filament
D) Phase profile of B across the filament

Mechanisms for producing contrasts are different, therefore, a quantitative comparison of the contrast between the TEM image and the electron holographic reconstructed phase image is not so easy by these experiments. However, we could qualitatively demonstrate the high contrast of phase image in the first observation.

4.4 High resolution analysis

With electron holography, it is possible to get a resolution higher than the unique resolution of a microscope by using an aberration correction technique, etc. [8]. However, an effort for such ultra-high resolution analysis is not required for observing biological materials, because irradiation damage limits us to analytical resolution, which is much lower than the instrumental one. For high resolution analysis of biological macromolecules by electron microscopy, two-dimensional crystals of the samples and cryo-microscopic techniques, including ice embedding, are necessary because of irradiation damage.

To achieve a high resolution analysis of biological macromolecules by electron holography, two serious problems must be overcome. One is the narrowness of detectable area. This problem is clearly shown in Figure 3A which shows a wide TEM area and a narrow hologram area. Since it is required to avoid useless irradiation, an operator usually can not see the image before taking a picture and must resort to so-called random shooting. To overcome this problem, the phase shift-method must be applied. Since the resolution of the reconstructed images is independent on hologram fringe spacing, a wide interference area is attainable. When highly precise measurement of phase distribution is required for phase-shifting electron holography, several hundreds of holograms are necessary. However, the holograms can be captured with a TV camera by a video rate of 30 holograms / sec., which is not such a long exposure time. The total dose to the specimen is the same as TEM under ideal conditions. Another problem is charge up of embedded ice due to electron irradiation. Electron holography has a much higher sensitivity than TEM [9]. Usually, this property favors observations of electric fields and magnetic fields and the others. However, in our case, it disturbs the observation. Therefore, the ice embedding technique, which is a very powerful method to resolve problems for high resolution analysis by electron microscopy [10], may not be used. Unfortunately, we currently have no solution to this problem.

5. CONCLUSION

In the first experiment we clearly observed an unstained bacterial flagellum filament adsorbed on a carbon supporting film by the phase-shifting method. We found that electron holography produces a sufficiently high contrast in-focus image that could never be obtained by ordinary transmission electron microscopic methods. However, the experiment had some problems caused by the carbon supporting film. The reconstructed phase image of the flagellum filament showed an artificial shape. We solved the problems by using a holey-carbon film to support the filament. In the second experiment, a biological thin filament, TMV was bridged over a hole from 300 to 1000 nm in diameter, and observed without a supporting film. The reconstructed phase image showed a complete projection of the cylindrical specimen which can be considered the actual projective shape of the TMV filament.

Acknowledgments : The authors thank Dr. Keiichi Namba and Ms. Yuko Mimori, of the International Institute for Advanced Research., Matsushita Ltd., for providing the filaments samples and for their valuable advice.

References

1. S. Asakura, Advan. Biophys. (Japan), 1, (1970) 99-155
2. G. Stubbs, S. Warren and K. Holmes, Nature (London) 267, (1977) 216-221
3. A. Koreeda, J. Electron Microsc. 29 (1980) 61-63
4. Q. Ru, J. Endo,T. Tanji, and A.Tonomura, Optik 1, (1992) 51-55
5. Q. Ru, J. Endo,T. Tanji, and A.Tonomura, Appl. Phys. Lett. 59, 19 (1991) 2372-2374
6. M. Takeda and Q.Ru, Appl. Opt. 24 (1985).3068-3071
7. K. Aoyama, G. Lai and Q. Ru, J. Electron Microsc. 43, (1994) 39-41
8. T.Tanji and K. Ishizuka, Proc. of 51st Ann. Meet. Micros. Soc. Am., Cincinnati (1993) 1084-1085
9. G. Matteucci, G. F. Missiroli, J.W. Chen and G. Pozzi, Appl. Phys. Lett. 52 (1988) 176-178
10. M. Adrian, J. Dubochet, J. Lepault, and A. W. McDowall 1984. Nature, 308 (1984) 32-36

Fraunhofer in-line electron holography of small weak-phase objects

T. Matsumoto[a,b], T. Tanji[a], and A. Tonomura[a,b]

[a]Tonomura Electron Wavefront Project, ERATO, JRDC
[b]Hitachi Advanced Research Laboratory, Hitachi, Ltd.
Hatoyama, Saitama 350-03
Japan

Abstract

We present recent advances in Fraunhofer in-line *transmission* electron holography. These include the reconstruction of the phase image of an undecagold cluster, which is the smallest object ever visualized by the technique. These also include the reconstruction of the phase image of an unstained DNA double helix embedded in an amorphous ice layer, which is the smallest *biological* object ever visualized by the technique. A coherent electron beam from a cold field-emission gun and a proper technique of specimen preparation have enabled us to accomplish these progress.

1. Introduction

The development of a field emission electron gun has made electron holography a practically useful part of electron microscopy [1, 2]. Most of the recent applications of electron holography have been worked out in off-axis configuration using Möllenstedt-type electron biprism. Historically speaking, however, the scheme of holography which Gabor originally proposed was in-line *projection* holography [3]. Unfortunately, in-line projection electron holography could not be put into practice because of the difficulties in obtaining a sufficiently small electron probe in Gabor's days. Instead, in-line *transmission* holography [4] was employed by Haine and Mulvey [5], who obtained the first in-line hologram with electrons in 1952. The last successful application of in-line transmission electron holography was reported by Münch [6] who used a 75-kV field emission gun to obtain reconstructions of gold particles as small as 1 nm in 1975, almost 20 years ago. Those two styles of in-line electron holography have now been revived [7-9]. In this paper, we describe our recent advances in in-line transmission electron holography. Applications of the in-line projection electron holography will be described by others [10, 11] in these proceedings.

2. Principle

The principle of in-line holography is shown in Fig. 1. In recording an in-line hologram, an isolated object in a plane h0 is illuminated by a plane wave (an electron wave in electron holography). The wave scattered by the object and the unscattered plane wave interfere in a recording plane h1. The interference pattern is magnified by electron lenses and recorded on a photographic plate as a hologram. In reconstruction of a hologram, a plane wave illuminates the hologram to produce a primary reconstructed image with an overlapping conjugate image at the plane H2. This conjugate image is one of the disadvantages of in-line holography. But the effect of the conjugate image can be made negligibly small if a hologram is recorded in a Fraunhofer condition [12]. Fraunhofer condition is described by a sufficiently smaller Fresnel number N than one;

$$N \equiv \frac{\pi\sigma^2}{\lambda z} \ll 1, \tag{1}$$

where λ is the electron wavelength, z is the defocus, and σ is radius of the object. The Fresnel number was first introduced in electron holography by Münch [6] to describe the theoretical contrast of an in-line hologram of a weak-amplitude Gaussian object. A theoretical contrast of an in-line hologram of a weak-phase Gaussian object, which is more important for applications of the technique in high resolution, can be similarly described by an analytical formula using Fresnel number [7]. Table 1 lists experimental conditions which were used in successful experiments employing in-line electron holography published so far. The last two rows include our experiments which will be described in this paper. As the Fresnel number becomes smaller, not only the disturbance from conjugate image but also the contrast of hologram is reduced. From the table, it is known that a Fresnel number between 0.01 and 0.05 is a good compromise.

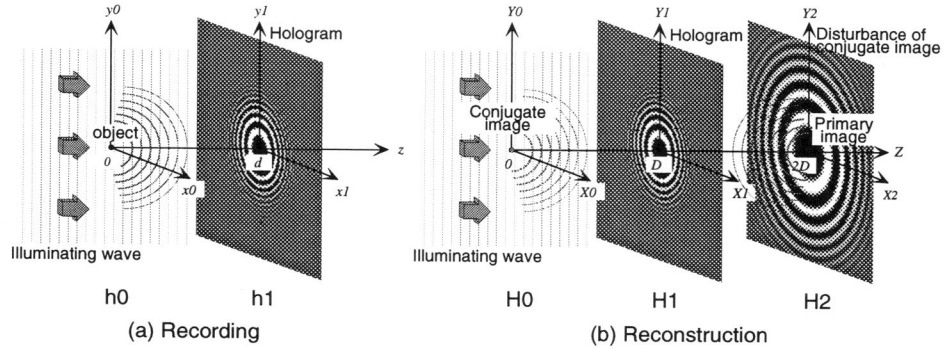

Figure. 1 Principle of in-line holography

Table 1
Published works employing in-line electron holography

Author (year)	particle and diameter	defocus*	acceleration voltage (wavelength)	Fresnel number N	electron source
Tonomura et al. (1968)	gold 10 nm	+2 mm	100 kV (3.7 pm)	0.0106	point filament
Münch (1975)	gold 1 nm	+5 μm	75 kV (4.3 pm)	0.0364	FE
Gallion et al. (1975)	not described 8~30 nm	+280 μm	50 kV (5.4 pm)	0.0335~ 0.471	not described
Troyon et al. (1976)	gold 2 nm	+45 μm	50 kV (5.4 pm)	0.013	FE
Bonnet et al. (1978)	gold not described	-2.1 μm	50 kV (5.4 pm)	not described	FE
Bonhomme and Beorchia (1980)	gold <2 nm	-8 μm	50 kV (5.4 pm)	0.0733	point filament
Present work (1)	gold 0.82 nm	+13.6 μm	200 kV (2.5 pm)	0.016	FE
Present work (2)	DNA 2 nm	+18.8 μm	100 kV (3.7 pm)	0.045	FE

*a positive sign designates an overfocused hologram, while a negative designates an underfocused hologram.

3. Experiments

3.1. Methods

A freshly prepared monoamino-undecagold (Nanoprobes, Inc., Stony Brook, N.Y.) was used as an undecagold specimen. Undecagold particles were prepared on a thin (< 5 nm)carbon film supported by a thick holey carbon film mounted on a copper grid for electron microscopy. The concentrations of the specimens were made as low as possible to prevent aggregation of particles. As a DNA specimen, a plasmid DNA pUC18 (2686 bp) in 10 mM Tris-HCl (pH 8.0), 10 mM $MgCl_2$, and 1 mM EDTA was used. The sample was prepared as a frozen-hydrated specimen as described elsewhere. A model 626 Gatan liquid nitrogen cooling holder was used and operated at a temperature of -165°C. The experimental configuration was shown in Fig. 2 (left for undecagold and right for DNA). A Hitachi HF-2000 field emission electron microscope was used and a minimum-dose procedure was employed to avoid radiation damage. An auxiliary cold-block was used to prevent ice contamination on the specimen during observation. Acceleration voltage was 200 kV for undecagold and 100 kV for DNA. Holograms were recorded on Kodak 4489 or SO-163 electron image films and developed by using Kodak D-19 developer. Holograms were digitized by using Nikon LS-3500 film scanner, transferred onto an Apollo DN-10000, and reconstructed numerically using software coded by one of the authors (T. M.).

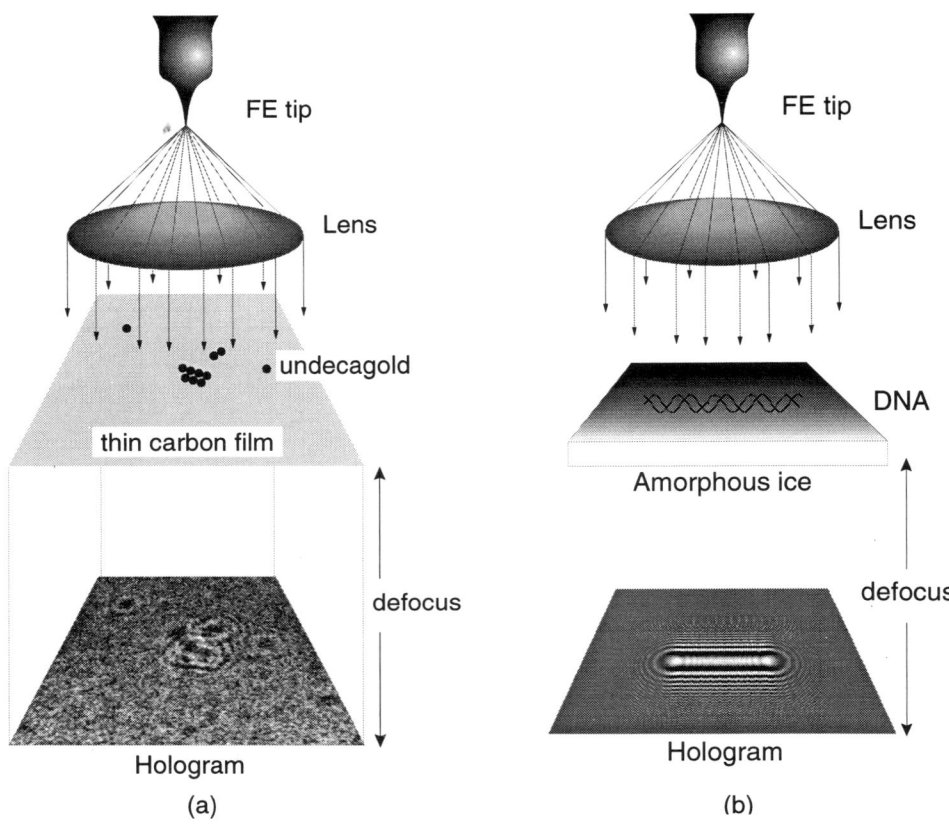

Figure 2. Simplified diagrams of experimental setup for Fraunhofer in-line electron holography of (a) undecagold, and (b) DNA.

3.2 Undecagold

Examples of Fraunhofer in-line electron holograms of undecagold particles are shown in Fig. 3. A hologram recorded with a defocus of 13.6 μm is shown in Fig. 3a, and the one recorded with a defocus of 30 μm is shown in Fig. 3b. The corresponding Fresnel numbers are 0.016 and 0.0074 for Fig. 3a and Fig. 3b, respectively. These are small enough to satisfy the Fraunhofer condition, so that the disturbance of conjugate image is negligible in the reconstructed phase image as shown in Fig. 4a. Fig. 4b shows an enlarged part of the reconstructed phase image. It shows a skewed hexagonal shape of the molecules, which should be compared with a projected structure of undecagold particle determined by X-ray crystallography.

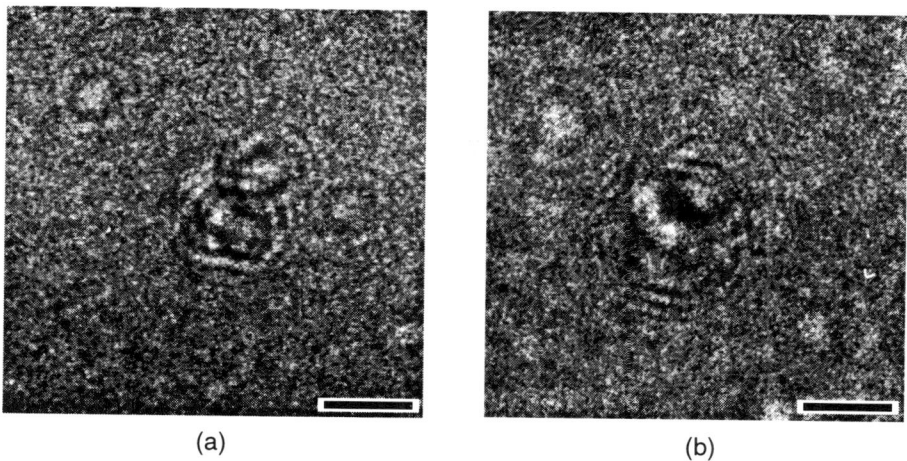

Figure 3. Fraunhofer in-line electron holograms of undecagold particles. Defocus values are (a) 13.6 μm, and (b) 30 μm. Bars represent 20 nm.

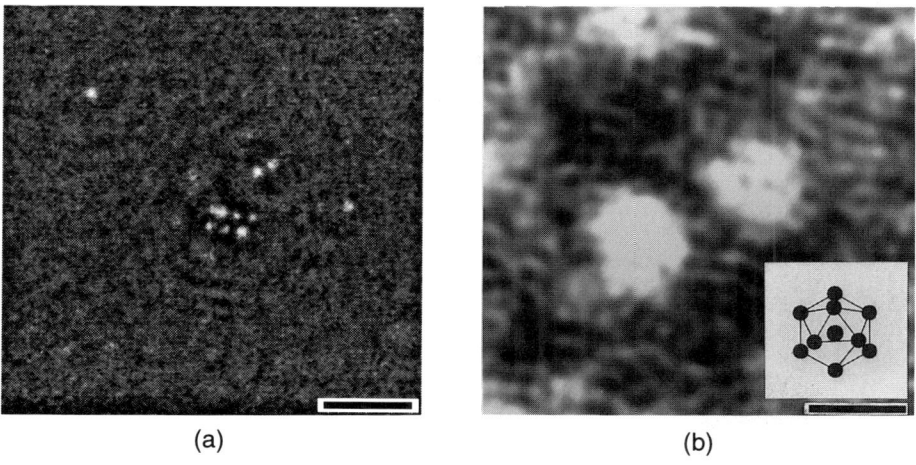

Figure 4. Reconstructed phase images of undecagold particles: (a) large field, and (b) narrow field. The hologram used for the reconstruction is the one shown in Fig. 3a. The corresponding Fresnel number is 0.016, which satisfies a Fraunhofer condition. Hence, disturbance of conjugate image is negligible in the reconstruction. A skewed hexagonal shape of the molecule can be observed, which corresponds to a projected image of the crystallographically determined atomic structure of the molecule (inset). Bar represents 20 nm in (a), and 1 nm in (b).

3.3. Frozen-hydrated DNA double-helix

Fig. 5a shows a computer-simulated phase shift of an electron wave which transmits through a DNA double helix consisting of 6 repeats of CG residues. An acceleration voltage of 100 kV was assumed. It was found that the phase shift is no more than $\lambda/33$ where λ is the electron wavelength (3.7 pm). Fig. 5b shows a computer-simulated in-line hologram of a DNA double helix which consists of 96 repeats of CG residues. A defocus of 20 μm and an acceleration voltage of 100 kV were assumed. Fig. 6 shows two examples of in-line holograms of unstained frozen-hydrated DNA double helix. These were recorded at a calibrated electron-optical magnification of 29,600 with a defocus of 18.5 μm. The defocus was adjusted by mechanically raising the specimen height from a focused position, instead of changing the objective-lens current. The thickest arrow indicates the zero-th order diffraction fringe, the second thickest arrow indicates the first order, and so on. In both of these figures, diffraction fringes up to second order can be observed. By a close inspection, we can observe faint third-order diffraction fringes as indicated with arrowheads in Fig. 6b. The existence of such higher order diffraction fringes proves that these are not blurred images, but holograms.

Fig. 7a shows the numerically reconstructed phase image of a single DNA double helix embedded in amorphous ice layer. Fig. 7b shows the reconstructed amplitude image, and the original hologram is shown in Fig. 7c. A strand as small as 2 nm in diameter is clearly reconstructed in the phase image with high contrast, whereas the reconstructed amplitude image is not clear at all. The diameter of the reconstructed strand appears thicker where indicated by larger arrowheads than where indicated by the smallest arrowhead. This is partly because more numbers of interference fringes are recorded to the lower-left portion of the hologram.

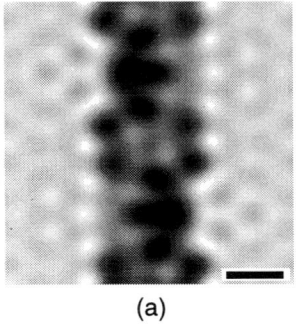
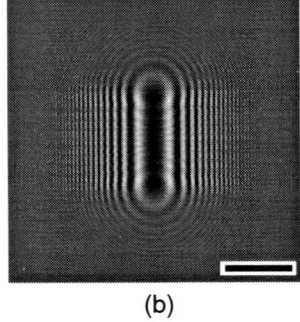

(a)   (b)

Figure 5. (a) Computer-simulated phase image of DNA double helix, and (b) computer-simulated in-line hologram of DNA double helix assuming a defocus of 20 μm and an acceleration voltage of 100 kV. Bar represents (a) 1 nm, and (b) 20 nm.

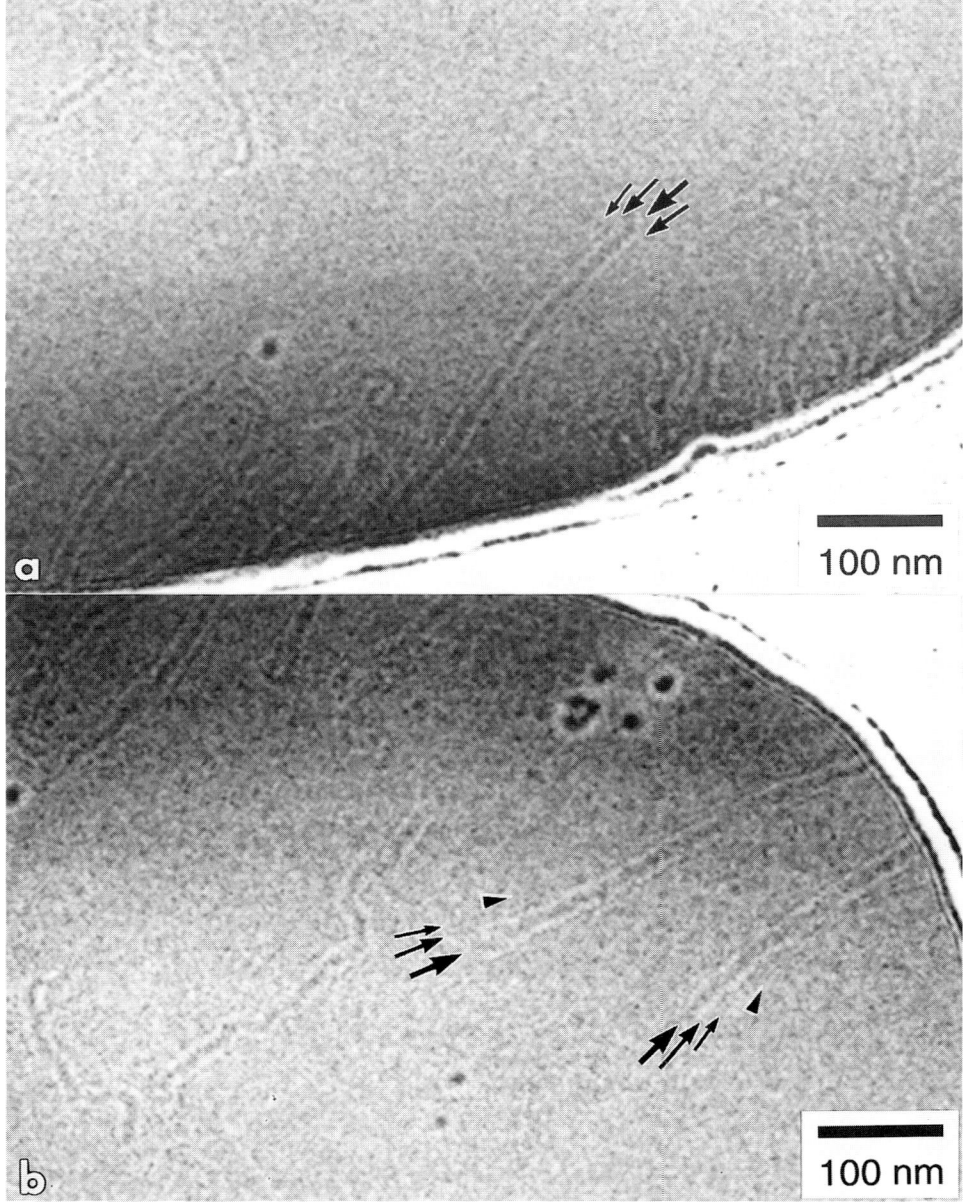

Figure 6. Experimental in-line electron hologram of unstained frozen-hydrated DNA double helices.

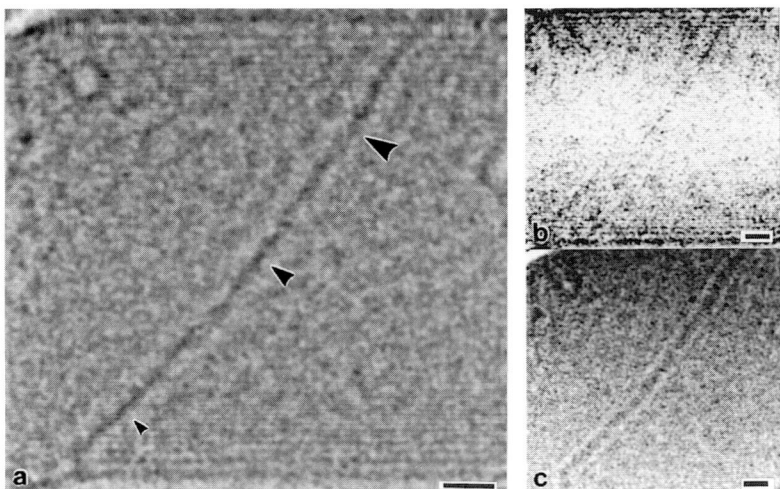

Figure 7. (a) reconstructed phase, (b) amplitude, and (c) original hologram. All bars represent 20 nm.

4. Conclusions

Fraunhofer in-line electron holography has been successfully used to visualize a 0.82-nm gold cluster and unstained DNA double helix in amorphous ice layer taking advantage of phase-contrast enhancement of the technique. This demonstrates the usefulness of in-line electron holography, especially for investigations of extremely weak-phase objects.

References

1. A. Tonomura, Electron Holography, Springer Series in Optical Sciences, Springer Verlag, Berlin Heidelberg 1993.
2. H. Lichte and W. D. Rau, Ultramicroscopy 54 (1994) 310.
3. D. Gabor, Proc. Roy. Soc. London A197 (1949) 454.
4. M. E. Haine and J. Dyson, Nature 166 (1950) 315.
5. M. E. Haine and T. Mulvey, J. Opt. Soc. Am. 42 (1952) 763.
6. J. Münch, Optik 43 (1975) 79.
7. T. Matsumoto, T. Tanji, and A. Tonomura, Ultramicroscopy 54 (1994) 317.
8. H. -W. Fink, W. Stocker, and H. Schmid, Phys. Rev. Lett. 65 (1990) 1204.
9. J. Spence, W. Qian, and A. Melmed, Ultramicroscopy 52 (1993) 473.
10. H. -W. Fink, H. J. Kreuzer, and H. Schmid, in these proceedings.
11. J. Spence, X. Zhang, and W. Qian, in these proceedings.
12. A. Tonomura, et al., Jpn. J. Appl. Phys. 7 (1968) 295 .

State of the Art of Low-Energy Electron Holography

Hans-Werner Fink[a], Hans Juergen Kreuzer[b], Heinz Schmid[a]

[a]IBM Research Division, Zurich Research Laboratory, 8803 Rüschlikon, Switzerland

[b]Dalhousie University, Halifax, Canada

1. INTRODUCTION

In order to carry out electron holography experiments in the geometry envisioned by Gabor [1], one needs a coherent source of electrons, an object that scatters part of the electron wave elastically and a detector to record the hologram. The hologram is the intensity pattern at some screen, brought about by the interference of the scattered object wave with the part of the wave that passed the object unaffected. Early experiments by Haine and Mulvey [2], who used an electron microscope to demonstrate Gabor-type holography, were experimentally extremely difficult because of instrumental instabilities and lack of coherence of the sources available at that time. Today, most electron holography work is done with high-energy electron microscopes that utilize lenses to generate a sufficiently coherent crossover of the electron beam [3]. We would like to review here a recent development in electron holography that uses electrons with low kinetic energy, about three orders of magnitude lower than in conventional electron microscopy, in a microscope without lenses.

2. LOW-ENERGY ELECTRON POINT-SOURCE (LEEPS) MICROSCOPY

The work on projection microscopy with electrons of low kinetic energy started with the invention of an electron source that directly provides a coherent ensemble of electrons. For this electron-point source, emission of electrons is confined to an atomic-sized volume in space, which ensures transversal coherence of the electron ensemble created in this way [4,5]. The principle of projection microscopy itself has a rather long history which started with instruments using light sources that go back to the times of Leonardo da Vinci. What is really new about the work we describe here is that we use, quite analogous to a laser as a photon source, an electron source that directly provides a beam of coherent electrons. This makes apertures or lenses obsolete. Ultimately, these sources are shaped into a single-atom tip created by depositing an individual foreign atom from the gas phase onto the trimer apex of a [111]-oriented tungsten tip, as illustrated in Figure 1.

With this unique source of electrons, a projection microscope was designed that provides high-resolution shadow images greatly magnified without the need for lenses [6]. A schematic of the experimental setup is shown in Figure 2. An approach sequence, demonstrating the increase in magnification by the decreased point-source to sample distance, is shown in Figure 3. Magnifications of the order of one million can readily be achieved

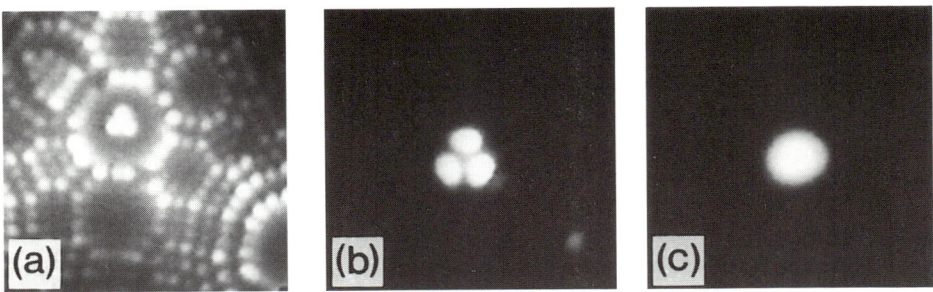

Figure 1. Preparation of a single-atom tip from W [111]-oriented single crystal wires. (a): Field ion image taken with He as imaging gas of a [111]-oriented W tip, (b) FIM pattern of a point-source tip terminated by a W_3 cluster, (c) mono-atomic tip created by depositing a single W atom onto the W_3 tip apex.

Figure 2. Schematic of the LEEPS microscope. *Red*: point-source tip indicates the electron wave in the solid confined to an atomic-sized point where the electrons tunnel into the vacuum to propagate as a spherical wave (*yellow*) representing a coherent ensemble of free electrons. At an object placed in the path of the wave, part of this spherical wave is scattered elastically. This object wave (*blue*) interferes with the yellow reference wave and brings about the hologram (*green*), which is a record of the amplitude and phase of the scattered object wave. (Design is part of the book cover of *Physics with Electron and Ion Point Sources*, World Scientific, Singapore, 1995).

simply by the geometry of the setup without the need for lenses. The intrinsic resolution of this microscope is given by the size of the point source and the wavelength of the electrons.

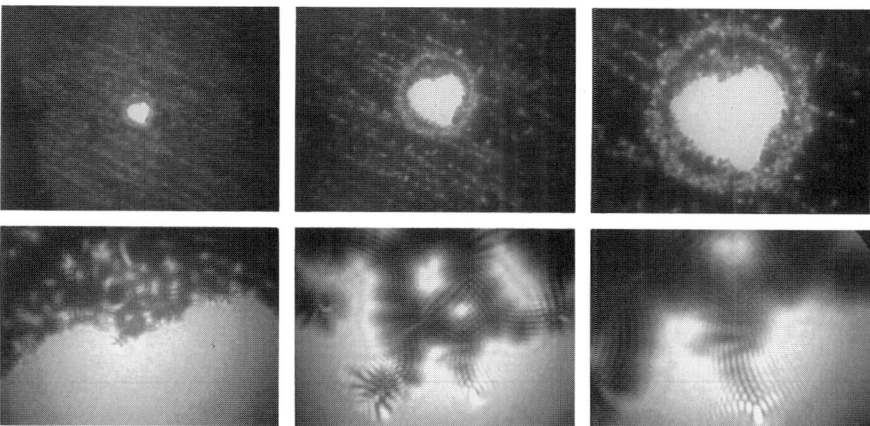

Figure 3. Sequence of LEEPS micrographs in which the point-source to sample distance is decreased from 200 μm to 870 nm for emission voltages between 74 and 34 V. At high magnification, objects at the edge of the Au film are imaged as Gabor-type inline holograms with electrons of only 34 eV kinetic energy.

3. LOW-ENERGY ELECTRON HOLOGRAPHY, EXPERIMENT AND THEORY

Improvements of the instrumental stability soon led to the first demonstration of holography with low-energy electrons. These first experiments of electron holography in a lensless microscope provided high-contrast holograms of carbon fibers about 10 nm in diameter [7]. Fringe separations well below 1 nm had already been obtained at that time, proving the high interference resolution and high contrast of this new holographic microscope. Electron energies between 20 and 300 eV responsible for the high contrast were ensured by the ultimately sharp tip and the close proximity between electron-source and sample. Researchers at the CNRS Laboratory in Marseille were able to compare the holographic information obtained with this new microscope with conventional SEM and TEM images of the same objects [8]. The same group was also able to record holograms with ultra-low-energy electrons down to only 7 eV kinetic energy [9]. This opened up the possibility to penetrate even thicker samples, which might be an important prerequisite to image films or more complex molecular arrangements. However, this will be achieved at the expense of the attainable resolution, which will be limited by the increased de Broglie wavelength of the slow electrons.

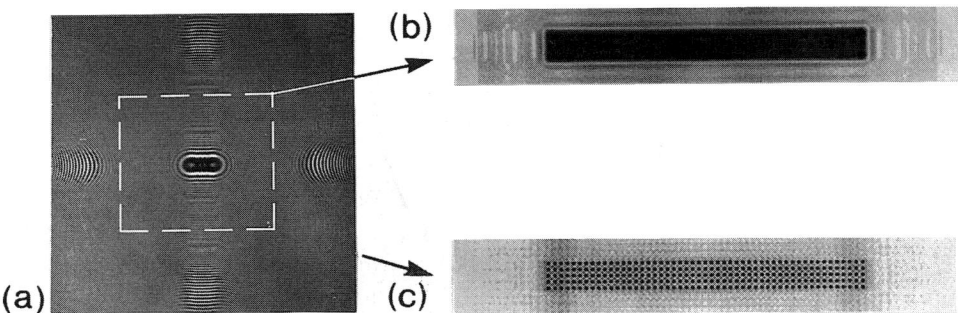

Figure 4. Theory of the LEEPS microscope. (a) Simulated hologram of a C cluster made up of two layers of 5×51 atoms. Reconstructed object wave front from the simulated hologram with an acceptance angle of (b) 38° and of (c) 70°. Only in recording the fine fringe spacings with a high angular acceptance detector is atomic resolution in the reconstruction achieved.

On the theoretical side, the Halifax group was able to simulate the low-energy electron holograms by modeling clusters at which part of the electrons were scattered. The holograms were calculated by taking into account the proper scattering for the specific energies including up to eight partial waves to describe the object wave for each cluster atom. This theory has been successfully tested against experimental results [10]. Soon afterward, the reconstruction of the object wave front from the electron holograms was addressed and included in the theory of the low-energy electron point-source (LEEPS) microscope [11]. The simulated holograms obtained by the cluster calculations were reconstructed in three dimensions. From this theoretical work, predictions of achievable resolution at a given electron energy were obtained. A simulated hologram and the reconstruction of the object wave front are shown in Figure 4. Concerns raised in view of multiple scattering and the twin image problem were also addressed at that time. Chemical specifity was demonstrated and improvements in depth resolution in reconstruction were suggested, e.g. by using a range of electron energies [12] or employing tilted geometries in holographic tomography [13]. The feasibility of tomographic holography with an angular variation as large as 90° has been demonstrated [14].

On the experimental side, the application range of low-energy electron holography has been exploited by experimenting with various organic samples, including preliminary data obtained with unstained DNA molecules [15]. In Figure 5, we show inline holograms of DNA molecules together with the result of a simulated hologram obtained from scattering at carbon atoms that were arranged in a double-helix structure.

One serious concern regarding the imaging of organic molecules with high-energy electron beams is that the molecules are damaged by the radiation with electrons of typically 100 keV. Our preliminary studies indicate that the situation is quite different with our microscope, which uses electron energies of only about 100 eV. Images such as those shown in Figure 5 are stable without any observable changes in the hologram if imaging is done

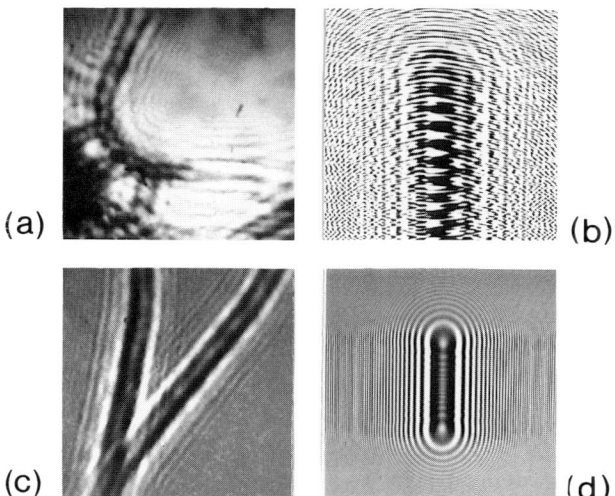

Figure 5. (a) Inline hologram obtained with 100 eV electrons from a sample on which DNA molecules have been deposited from a liquid solution onto a partly transparent carbon film. Under standard imaging conditions with emission current from the source measuring several nanoamperes, no effects of radiation damage on the molecule have been observed. (b) Simulation of a C double-helix hologram computed for 100 eV electrons. (c) Hologram of C nanotubes with 72 eV electrons. (d) Corresponding simulated image.

with total electron currents emitted from the source of the order of nanoamperes. Only if we select extreme imaging conditions of several 10 μA being emitted from the source do we actually observe the slow decomposition of the molecule. However, whether the observed damage is in fact due to the high electron dose or possibly to secondary effects like the creation of ions remains to be examined in detail. In any event, we note that the conditions under which we observe damage are clearly beyond sensible imaging current values. For carbon nanotubes, of which a hologram is also shown in Figure 5, no radiation damage has been observed even at high current imaging.

4. DERIVATIVES OF THE LEEPS MICROSCOPE

If the point source tip is biased positively in the presence of a high partial pressure environment of some gas, the tip becomes a bright gas ion source of atomic dimension. Ions created above a single tip atom are used to perform high-resolution ion projection microscopy. In combination with the LEEPS microscope, images of electric and magnetic fields have been obtained by comparing electron and proton images [16]. Two examples of the complementary information accessible by ion- and electron-projection images are shown in Figure 6. Helium ions have been found suitable to machine small fibers of carbon *in situ* on a nanometer scale.

Thus the ions in this combined microscope provide a way to actually manipulate and change the sample in addition to the nondestructive imaging with the low-energy electrons.

Another effort to influence and manipulate the sample under investigation has been undertaken by incorporating a second tip into the LEEPS microscope. With this tip,

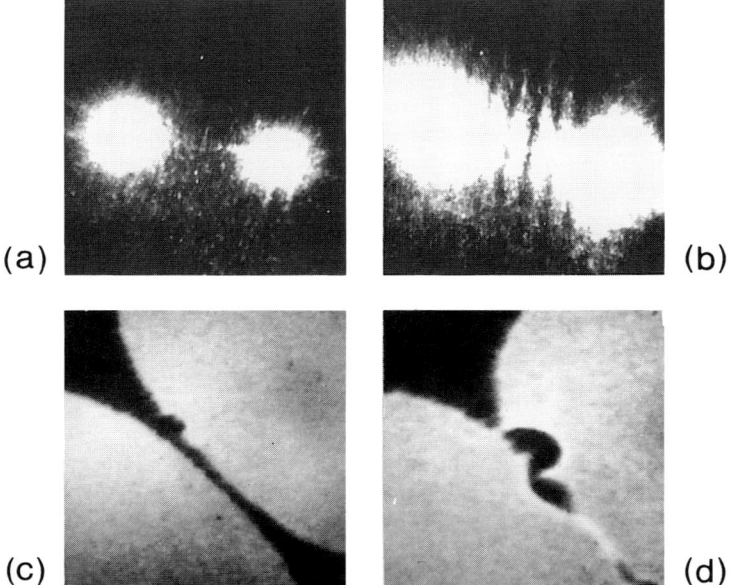

Figure 6. (a) Ion projection image taken with H ions reveals the size of two holes in a thin Au foil to be about 30 nm. (b) In switching to the field emission mode by reversing the polarity at the tip, the coherent electron wave produces the well-known Young interference pattern at the detector plane. (c) Fe clusters deposited onto a network of C fibers are imaged with protons. This reveals the cluster boundaries, from which the size of two adjacent clusters has been determined to be about 40 nm. (d) Electrons experience much greater deviation due to the Lorentz force influencing the electron trajectories by the magnetic field surrounding the Fe cluster.

placed opposite the sample to be investigated, it has become possible to manipulate nanometer-sized objects and to measure currents through individual carbon fibers [17]. A schematic of this microscope is shown in Figure 7. Recent results obtained with carbon nanotubes will be described at the end of the next section.

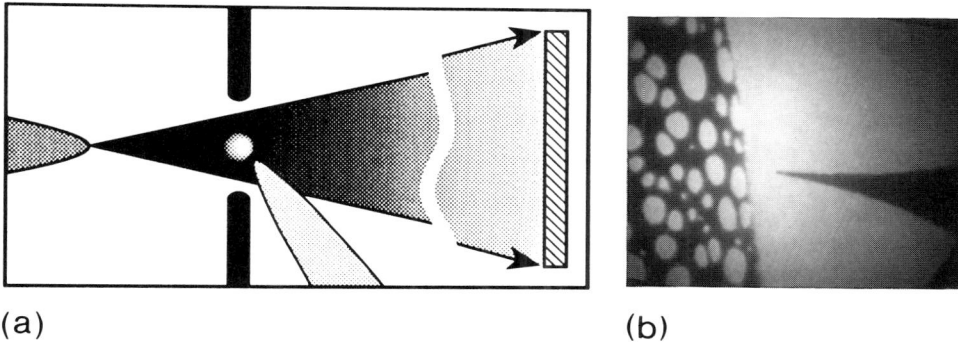

(a)   **(b)**

Figure 7. (a) Schematic of LEEPS microscope with an additional "manipulating tip" incorporated. (b) Projection image of a C foil edge. The manipulating tip is imaged at a lower magnification because it is still far away from the sample.

5. RECENT RESULTS

On the source side, the dependence of the emission properties on the atomic arrangement of the tip apex has been investigated in detail by the Marseille group [18]. They found that the opening of the beam does not depend much on the size of the cluster at the tip apex, but appears to be related to the overall shape of the tip. In addition, effects of patch fields whose origin is the change in work function of the planes surrounding the tip apex appear to have an influence on the divergence of point-source electron beams. Their measurements of the angular distribution of the emitted electrons were carried out over an extremely high dynamic range of total current ranging from a few electrons per second up to the highest emission rate of 10^{15} electrons/s.

The Heidelberg group has performed careful experiments on the energy distribution of the electrons emitted from point-source tips [19]. They have shown that single-atom tips emit electrons with a spread in energy of approximately 250 meV, which is in good agreement with standard field emission theory. A significant reduction of both the emission voltage and the spread of the energy distribution has been found in point source emitters that were covered with cesium atoms [20]. The cesium-covered point-source tip provides a highly monochromatic source of low-energy electrons with an energy spread of less than 100 meV. This reduction in energy spread by roughly a factor of three is attributed to the lower tunnel barrier. This is brought about by the chemisorption of cesium on tungsten, which causes strong dipoles at the emitter surface. These effects are also in accordance with established views of field emission. Nevertheless it is a remarkable and, for practical purposes, an important improvement of the monochromatism of the emitted electrons.

In the first low-voltage holography experiments [7], fringe separations as low as 0.3

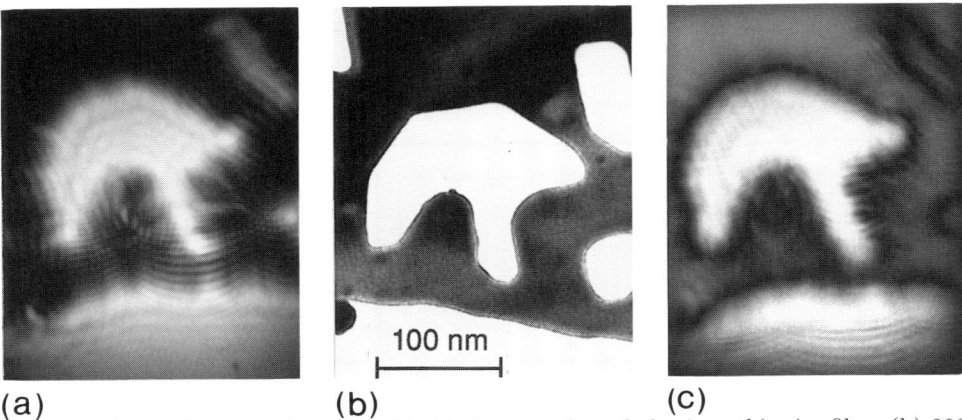

Figure 8. (a) Hologram taken with 50 eV electrons from holes in a thin Au film. (b) 200 keV TEM image of the same region. (c) Reconstruction of the object wave front, which compares well with TEM data.

to 0.4 nm were reported. These values have recently been confirmed by the Arizona group [21], who correlated this number to an upper value of the virtual source size. The first interferometric evaluation of the size of the electron point source has recently been reported by the Marseille group. They were able to incorporate a Moellenstedt biprism [22] into their lensless projection microscope and used this setup as an interferometer for a direct measurement of the coherence of the electron source [23]. Hence they showed directly that atomic sized point-source tips are associated with a value of 0.14 nm for the upper limit of the virtual source size.

Recently, the reconstruction of experimental holograms of hole structures in thin gold films obtained with low-energy electrons has been achieved (Figure 8). For comparison, TEM data for the same structures are also shown as well as holograms of similar but larger structures obtained with a coherent source of photons, a laser [24].

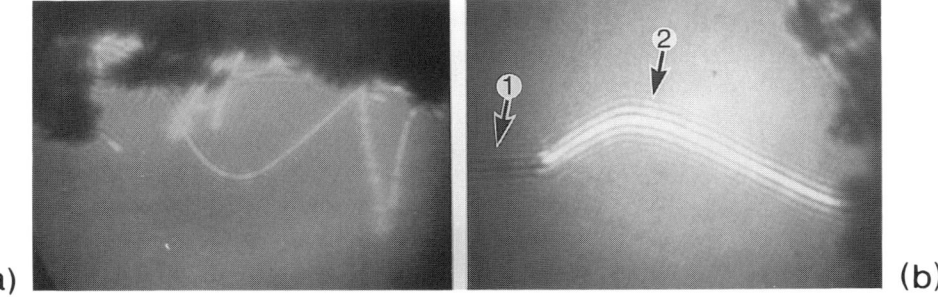

Figure 9. (a) Low magnification projection image of nanotubes; (b) "manipulation tip" (1) attached to an individual nanotube (2) for current measurements.

Some properties of nanotubes have been investigated using the LEEPS microscope together with the additional "manipulating tip" described above (see also Figure 9). It turns out that these modern carbon chemistry objects with diameters of the order of just one nanometer are able to transport currents of several microamperes, corresponding to current densities of 10^8 A/cm^2!

6. CONCLUSION

To summarize, we list what we believe are important features of the holographic LEEPS microscope:
- The low energy of the electrons used for holography provide high phase contrast for light atoms like carbon.
- The microscope produces highly magnified images without employing lenses because it utilizes a "directly" coherent electron source. There is no need to correct the holograms for spherical aberrations because they are absent. Numerical reconstruction of the holograms is therefore straightforward.
- Owing to the acceptance of electrons scattered in a large angular regime, there is realistic hope for a truly three-dimensional object representation (so far this has only been demonstrated in the LEEPS microscopy theory [11]).
- Radiation damage appears to be minor due to the low energy of the electrons.
- Image acquisition is fast due to the brightness of the source. A high-contrast hologram can be obtained within less than 10 μs.
- The holographic information can be combined with complementary data obtained *in situ* by ion projection microscopy using protons or helium ions. Ions can also be used *in situ* to machine objects in a controlled fashion.
- An additional "manipulating tip" allows the mechanical and electronic manipulation of nanometer-sized objects, for example to investigate such electron transport properties in one-dimensional conductors as nanotubes.

Acknowledgment. We thank Erica Williams for taking the TEM image of the gold film and are also grateful to Donald S. Bethune for providing us with the nanotube samples.

REFERENCES

1. D. Gabor, Nature (London) 161, 777 (1948).
2. M.E. Haine and T. Mulvey, J. Opt. Soc. Am. 42, 763 (1952).
3. For a recent summary, see: T. Mulvey, Euro. Microscopy and Analysis, 31 (Jan. 1994).
4. H.-W. Fink, IBM Journal of Research and Technology Vol. 30, No.5, 460 (1986).
5. H.-W. Fink, Physica Scripta 38, 260 (1988).
6. W. Stocker, H.-W. Fink, and R. Morin, Ultramicroscopy 31, 379 (1989).
7. H.-W. Fink, W. Stocker, and H. Schmid, Phys. Rev. Letters 65, 1204 (1990).
8. R. Morin, A. Gargani, F. Bel, Microsc. Microanal. Microstruct. 1, 289 (1990).
9. R. Morin, and A. Gargani, Phys. Rev. B 48 (9), 6643 (1993).
10. H.-W. Fink, H. Schmid, H. J. Kreuzer, and A. Wierzbicki, Phys. Rev. Letters 67, 1543 (1991).

11. H.J. Kreuzer, K. Nakamura, A. Wierzbicki, H.-W. Fink, and H. Schmid, Ultramicroscopy 45, 381 (1992).
12. H.J. Kreuzer, A. Wierzbicki, M.G.A. Crawford, and C.B. Roald in *Manipulation of Atoms Under High Fields and Temperatures: Applications*, Eds. V.T. Binh, N. Garcia, and K. Dransfeld, NATO ASI E Series Appl. Sci., Kluwer, Dordrecht, 1993.
13. H.J. Kreuzer, in *Atomic-Scale Imaging of Surfaces and Interfaces*, MRS Conference Proc. Vol. 295. Eds. D.K. Biegelsen, D.J. Smith, and S.Y. Tong Materials Research Society, Pittsburgh, pp. 235-242 (1993).
14. H.-W. Fink and H. Schmid, in *Proc. JRDC Int'l Symp. on Nanostructures and Quantum Effects*, Tskuba, Japan 1993, Springer, Heidelberg, 1994.
15. H.J. Kreuzer and H.-W. Fink, Science News 140, 23 (1992).
16. H. Schmid, and H.-W. Fink, Appl. Surf. Sci. 67, 436 (1993).
17. H. Schmid, and H.-W. Fink, Nanotechnology 5, 1, 26 (1994).
18. S. Horch, and R. Morin, J. Appl. Phys. 74 (6), 3652 (1993).
19. H.U. Mueller, B. Voelkel, M. Hofmann, Ch. Woell, and M. Grunze, Ultramicroscopy 50, 57 (1993).
20. R. Morin, and H.-W. Fink, J. Appl. Physics, in press.
21. J.C.H. Spence, W. Qian, and A. Melmed, Ultramicroscopy 52, 473 (1993).
22. G. Moellenstedt and H. Dueker, Z. Physik 145, 377 (1956).
23. A. Degiovanni, and R. Morin, Proceedings of ICEM 13, Electron Microscopy 1994, Les Editions de Physique, Vol. 1 (1994).
24. H.J. Kreuzer, H.-W. Fink, H. Schmid, S. Bonev, to be published in J. Microscopy.

On the reconstruction of low voltage point projection holograms

J. C. H. Spence, X. Zhang and W. Qian

Department of Physics and Astronomy, Arizona State University, Tempe, Arizona 85287-1504, USA

We consider the reconstruction of experimental in-line electron holograms obtained at 100 volts from carbon films containing holes. The films are treated as masks, whose transmission function is zero or unity. Thus, unlike weak-phase objects, no reference wave exists at small defocus deep in shadow regions, and non-linear terms may be large. The holograms are point-projection shadow images, obtained using a nanotip field-emitter placed a few hundred nanometers from the films, which act as the grounded anode. The holograms are shown to be equivalent to out-of-focus conventional images formed with collimated illumination. Using this equivalence reconstructions are made of thin fibers drawn across a hole, and compared with reconstructions from semi-infinite half-plane objects. The conditions for successful reconstruction are studied using simulated and experimental data as a function of Fresnel number N and object shape. We find that edge regions are sharply defined in the reconstruction and easily discerned against the background of the virtual image and (non-linear terms) for small and moderate values of N (Fraunhofer holography requires the more stringent condition N <<1). For semi-infinite half-planes a much poorer reconstruction is obtained. Finally, we derive expressions for the resolution-limiting effects of stray magnetic fields due to the Aharonov-Bohm phase shift.

1. Introduction

Under coherent illumination conditions, the Fresnel fringes seen at the edge of an opaque object in an out-of-focus image forms an in-line hologram, in which the unobstructed wave passing around the edge provides a reference wave. Fresnel fringes were first observed using electron beams by Boersch [1]. The interpretation of these fringes as a hologram (i.e. the realization that they could be removed by reconstruction or "focus restoration") came later [2]. The first experimental shadow images formed using a small electron source appear to be those of Boersch [3] (at high energy) and Morton and Ramberg [4] (at low energy, see also [5]).

The recent work of Shedd [6] suggests that most of the objects used for recent in-line low voltage transmission electron holography [7,8] (such as thin organic or gold films) may be treated as opaque masks containing holes. This is consistent with transmission measurements of the inelastic mean free path for self-supporting films of carbon which gives a value of 6-8Å for electrons of energy about 100 eV [9]. Any pin-holes in these films would make this an overestimate. These holograms may thus be useful for determining the shape of molecules drawn across holes in thin carbon films. Efforts continue to obtain images at lower voltage, where the inelastic mean free path increases [9] allowing greater penetration of the beam. (The lowest voltage so far is 7 eV [10]. Our lowest voltage is about 40 volts).

In this paper we describe reconstructions from in-line electron holograms of thin carbon films, recorded at about 100 volts. The holograms are point-projection shadow images, which may give sub-nanometer resolution. Details of our point projection microscope (PPM) are given elsewhere [8]. A tungsten field emitter at a potential of about -100 volts is placed a distance z_1 (a few thousand Angstroms) from a grounded sample, which acts as the anode. A shadow image is formed on a channel plate, distance $z_2 = 14$ cm away, with a magnification $M = z_2/z_1$.

Images of the fiber-optic face-plate were recorded using a liquid-nitrogen cooled CCD camera. Typical exposure conditions were 0.1 seconds with about 1 nA beam current at 100 volts.

2. Hologram formation

In earlier papers [11,8] we demonstrated that the point projection image is equivalent to an unaberrated conventional out-of-focus image, if the tip to sample distance is treated as the defocus setting. Thus the "transmission" and "projection" geometries used by Haine and Mulvey [12] in the first attempts at electron holography are equivalent, as ultimately recognized by Gabor [13]. Consider a mask illuminated by a spherical-wave originating from a source at distance z_1 from the mask. The transmission function for the sample is $q(x)$, but we cannot assume the weak phase object approximation for masks. Choose Cartesian coordinates with z along the beam path. Let the spherical wave incident on the sample be represented in the parabolic approximation by $t_{z1}(x) = \exp(-i \pi x^2/z_1 \lambda)$ (with Fourier transform $T_{z1}(u) = C \exp(i \pi z_1 \lambda u^2)$, where C is a complex constant) and let the electron wavefunction across the downstream side of the slab be $\psi_i(x)$. Then, with $u = \Theta/\lambda$, where Θ is the scattering angle,

$$\psi_i(x) = q(x) t_{z1}(x), \tag{1}$$

the detected wavefunction in the far-field will be the Fourier transform

$$\psi_h(u) = FT(\psi_i(x)) = Q(u)*T_{z1}(u), \tag{2}$$

where the asterisk denotes convolution. Evaluating the convolution in equation 2, and ignoring unimportant phase factors gives

$$\psi_h(u) = \int [Q(U)\exp(i\pi z_1 \lambda U^2)]\exp(-2\pi i z_1 \lambda u U)dU. \tag{3}$$

Now consider $z_1 \lambda u \ U = z_1 \lambda \ (\Theta/\lambda) \ U = z_1 \lambda \ (X/z_2\lambda) \ U = (X/M) \ U$, where X is the spatial coordinate on the detector and $M = z_2/z_1$ is the magnification of the shadow image. Hence

$$\psi_h(X = z_2 \lambda u) = q(X/M)*t_{z1}(X/M). \tag{4}$$

This important result establishes that for masks (or for any thin sample for which a transmission function can be defined), the shadow image consists of an "ideal" image which is out of focus by the tip-to-sample distance z_1. We have assumed that the emission angle from the tip is not limited. For a crystalline sample for which the transmission diffraction orders overlap, this analysis does not apply in the presence of multiple scattering, since then a transmission function cannot be used, as discussed elsewhere [11]. Equation 4 is identical to the expression for an out-of-focus high resolution transmission electron microscope (HREM) image in the absence of lens aberrations if this is formed using plane-wave illumination. However the point-projection image has been magnified by $M = z_2/z_1$ without the use of lenses. As may readily be confirmed using an optical laser and a slide transparency, increasing z_2 increases the overall magnification of a shadow image, but not the focus defect, which is fixed by z_1. No assumption of periodicity has been made for the sample. The effects of limiting the emission angle from the tip are equivalent to limiting the semiangle subtended by the objective aperture in HREM. The resolution of the shadow image is approximately equal to the effective source size if it is extended, and is limited by the angular range over which emission from the tip or probe is coherent. For Fresnel fringes at a shadow edge, the resolution and source size are approximately equal to the finest (outermost) Fresnel fringe. An in-focus image can only be obtained using equation 4 if $z_1 = 0$, in which case M is infinite and, if a physical emitter is used, field emission then becomes impossible. In addition, if $z_1 = 0$, zero contrast is predicted for the image of a transmission phase object.

It follows from equation 4 that existing algorithms for focus correction (under plane-wave illumination) can be used to reconstruct low voltage in-line holograms (formed with spherical-wave illumination). For masks, Fresnel edge fringes and Young's fringes between nearby holes are the main contrast features, and the aim of reconstruction is to remove these fringes, thus revealing the shape of the holes. The problem of focus correction (or hologram reconstruction) in the in-line geometry has been studied extensively both for optical and electron holography. The first solution to the unavoidable twin image problem which arises in this geometry may be that of B.J. Thompson and co-workers [15], who developed the optical method of in-line Fraunhofer holography for small particles. Early work using field-emission electron sources for in-line electron holography can be found in references [16,17]. Solutions to the twin image problem based on recording images at different defoci or lateral tip positions are described in [18]. Shadow images of masks have recently become important in the field of semiconductor lithography, where the inverse problem arises. For finer line-widths and sharper edges one requires the edge transmission function whose defocussed shadow image is most abrupt.

3. Reconstruction algorithm

For a mask object we define the transmission function

$$q(x) = 1 - p(x) = 1 \quad \text{within a hole, and}$$
$$= 0 \quad \text{within opaque regions.}$$

Then, assuming the equivalence between point-projection shadow images and conventional images, the wavefunction on a plane distance z_1 downstream from the sample is, for a magnification of unity [19],

$$\psi_h(x) = q(x) * t_{z1}(x). \tag{5}$$

The recorded hologram intensity is

$$I(x) = \psi_h(x)\psi_h^*(x) = 1 - p(x)*t_{z1}(x) - p^*(x)*t_{z1}^*(x) + |p(x)*t_{z1}(x)|^2. \tag{6}$$

To reconstruct the object from the hologram, first Fourier transform, then multiply by $T_{z1}^*(u)$, and finally perform an inverse transform:

$$FT^{-1}\{T_{z1}^* \cdot FT[I(x)]\} = 1 - p(x) - p^*(x)*t_{2z1}^*(x) + |p(x)*t_{z1}(x)|^2 * t_{z1}^*(x) \tag{7}$$
$$= 1 - p'(x).$$

We call $p'(x)$ the retrieved object function. The first two terms give a perfect reconstruction of the original transmission function $q(x)$, the real image. The third term gives the complement of the object, out-of focus by twice the tip-to-sample distance z_1. If the object (of width d) is small, and z_1 large enough, this virtual twin image will produce a broad and slowly varying background which allows the real image to be isolated (The requirement for this is $z_1 \gg d^2/\lambda$

(Fraunhofer holography)). The fourth term is a second order term (actually the reconstruction of the hologram of the complementary object) which is small for weak phase objects, but not necessarily small for masks. Physically, this term might be expected to be troublesome in regions well inside the shadow edge, where the "reference wave" is weak. The ratio of the wave scattered from the edge to the direct wave controls this term, and, for a finite object, this ratio depends on the Fresnel number $N(d) = \pi d^2 / (\lambda z_1)$. Thus reconstruction for masks introduces greater errors than for weakly scattering objects (wpo), however for the purpose of identifying simple discontinuities at edges, we shall see that the Fraunhofer condition can be greatly relaxed. We investigate these issues below using computational trials on simulated and experimental data.

In figure 1 we illustrate the use of the algorithm for simulated data. The simulations require periodic boundary conditions (with supercell period L). Interference from neighbouring cells can be avoided if the Fresnel number $N(L) \gg 1$. The reconstruction of two opaque square objects of width 1nm and 3nm is shown using 100 volt electrons. The hologram was simulated using equation 6 and reconstructed using equation 7. The edge of the objects can be readily located from the profile, despite the disturbance from the background.

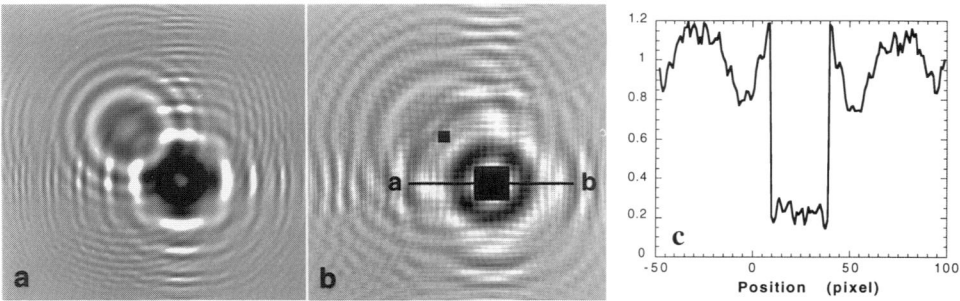

Figure 1. Simulated in-line holographic reconstruction of masks, showing sharp reconstruction. Defocus 60nm. Mask width 1nm and 3nm (Fresnel number 0.43 and 3.86). (a) Simulated hologram; (b) real part of the reconstructed wavefunction; (c) profile of intensity along a-b.

4. Results

Before attempting reconstruction, the magnification $M = z_2/z_1$ of the images must be known. Five methods may be used: 1. Use of a through-magnification series and an object (such as a grid bar) of known size. This is very inaccurate since erorrs between successive images compound. 2. Use of parallax-shift of images with lateral tip movement [7]. 3. Analysis of Fresnel edge fringes [14]. 4. Use of Young's fringes observed between two pin-holes of spacing D, using $d = z_2\lambda / D$, where d is the fringe spacing and z_2 the sample-to-screen distance. 5. Trial-and-error values of z_1 may be used in the reconstruction scheme until a sharp image is obtained. We have used methods 5 and 3.

Figure 2 shows two experimental point-projection holograms of a holey carbon film obtained under similar conditions at 90 volts. Figure 3 shows their reconstructions (based on equation 7) using different values of the defocus z_1. The holograms were reconstructed using 256×256 pixels. The program allows for the change of object pixel size with variation of defocus z_1, which changes magnification. An approximate value of z_1 was obtained using the above methods and the corresponding magnification M = 14 cm/ z_1 used. Since the computations are fairly rapid, the effect of varying the trial value of z_1 simulates changing experimental focus near Gaussian focus, and by comparing reconstructed amplitude and phase images, the in-focus image may readily be identified by eye, despite the background from the twin image and non-linear term. The focus correction needed to obtain the in-focus image was found to be $z_1 = 1850$ nm. This is therefore the experimental tip-to-sample distance. Note the reversal of fringe contrast with changes in defocus setting.

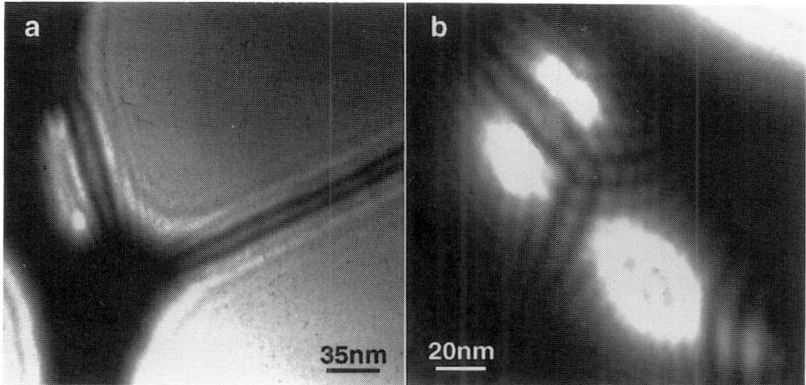

Figure 2. Experimental holograms of a holey carbon film obtained at 90V. (a) carbon fibers bridging holes; (b) three pinholes in the same foil.

It is of interest to compare reconstruction from small objects with that from a semi-infinte half-plane. Although still surrounded by Fresnel fringes, the in-focus reconstruction of the carbon fibers (small objects) in figure 3a has clearly revealed their edges; while in figure 3b the edges of the three holes (similar to semi-infinite half-planes) are not sharp, except in the region of the fibers separating them, where $N \approx 1.2$ and Fraunhofer conditions apply more nearly. The reconstruction shows that bright regions between holes in the original experimental image do not necessarily indicate transmission. Note the Young's fringes which appear between the pinholes in figure 3b, where three sets of fringes cross midway between the holes, a false impression of an atomic lattice image is given [6]. These fringes can be distinguished from atomic resolution lattice fringes by the fact that their spacing is independent of defocus.

Using the retrieved shape of the fibers from figure 3a, equation 6 was applied to simulate the experimental hologram. The result is shown in figure 4. The simulated hologram shows a close similarity to the experimental image.

Further simulations have been carried out to verify the experimental results and better understand the principles of reconstruction. Holograms of an isolated bar mask of width 14 nm and a mask with a circular hole of diameter 35 nm (these are about the same size as the objects in figure 2) were formed using equation 6, with the same defocus (1850nm) and electron energy (90V). The simulated holograms were subsequently reconstructed using equation 7, with the second order term and the twin image term manipulated separately (this can be done since the transmission function is known). The profiles are shown in figure 5. It is apparent that the sharp edge of the bar can be reconstructed, whereas the edge of the hole cannot, in agreement with experimental results. We also see that the role of the second order term is not (as for a wpo) only a disturbance. Although the object function plus its twin image provide a better indication of the edge position than does the retrieved object function, the second order term is neccessarily included (to make up the value of p'(x), which is the only function that can be reconstructed experimentally) to improve the fit to the original mask. The squared term thus improves the fit to the mask, but degrades resolution at edges in the reconstruction.

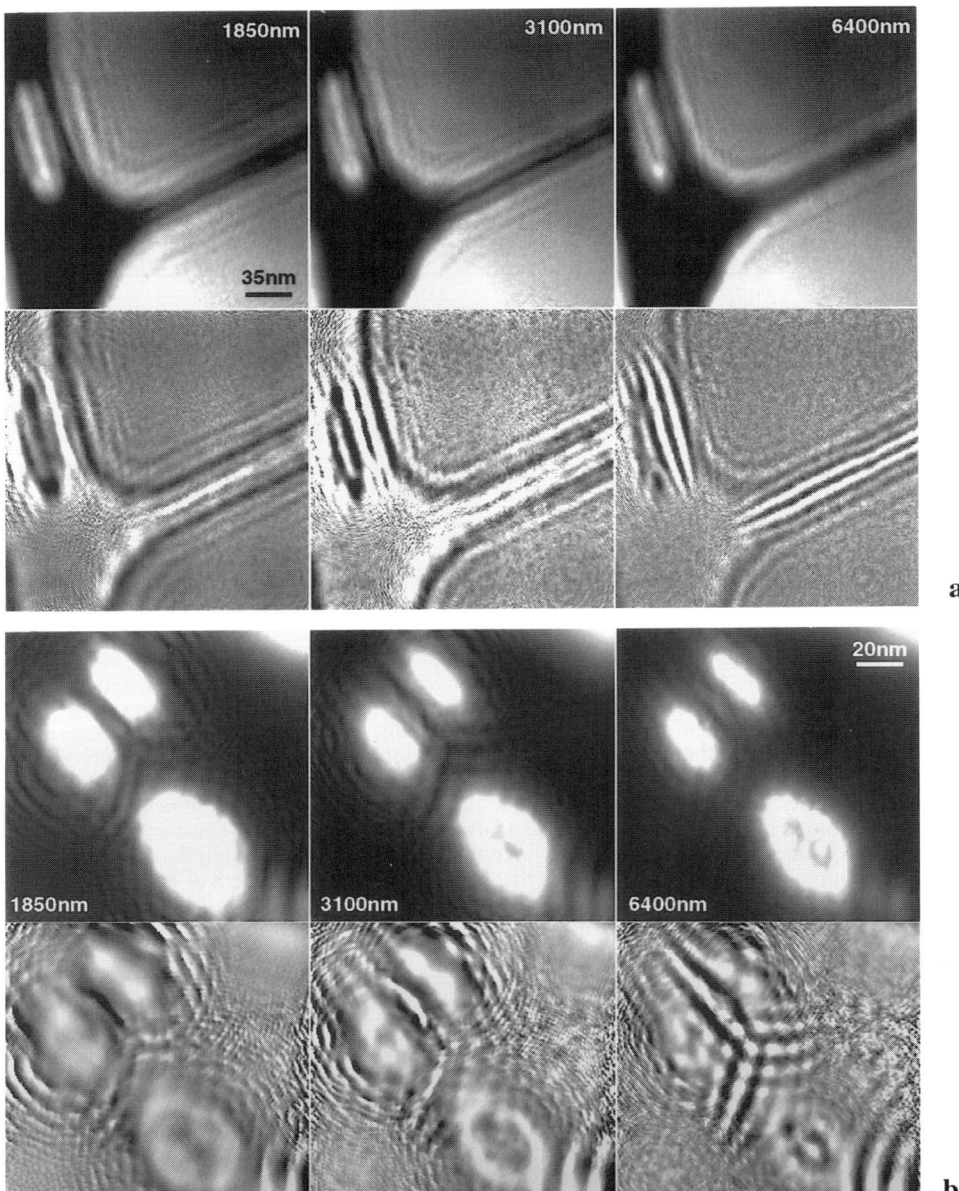

Figure 3. Reconstructed amplitudes (upper) and phases (lower) recovered from experimental holograms in fig. 2 for different defocus corrections. The in-focus image has defocus correction 1850 nm. It shows no contrast reversals within the image, sharper edges than the original and a broader band of Fresnel fringes outside (the twin image). (a) carbon fibers; (b) three pinholes.

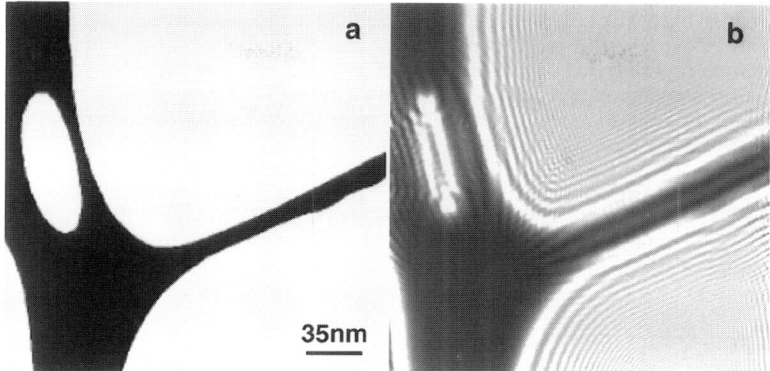

Figure 4. Object shapes retrieved from experimental images of figure 2a shown in (a) and, in figure (b), simulated hologram with defocus 1850nm. Image size 256×256 pixels.

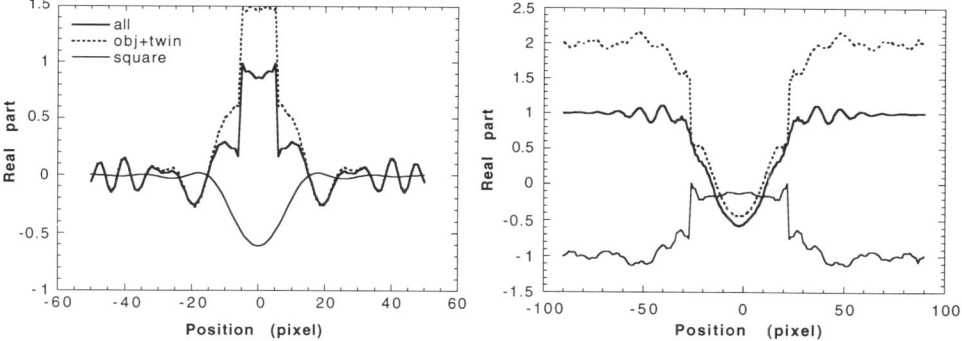

Figure 5. Profiles of simulated reconstruction, shown as real part of: retrieved object function p' ("all"), object function p plus its twin image term ("obj+twin"), and the second order term ("square") (refer to equation 7). (a) bar mask of width 12 pixels (1.2 nm/pixel), the profile line is drawn perpendicular to the bar; (b) mask with one hole of diameter 50 pixels (0.70 nm/pixel), the profile line is drawn across the hole passing the center. Image size 256×256 pixels.

The possibility of revealing sharp edges in the reconstruction depends on the Fresnel number $N = \pi d^2 / \lambda z_1$, which describes the effect of defocus z_1, wavelength λ, and object diameter d. This has been investigated by simulation of simple objects using different values of N. No general criterion has been found; but for a simple mask object the Fraunhofer condition $N \ll 1$ for elimination of the twin image is certainly an excessively stringent requirement. In figure 5 the bar has a $N = 2.6$, and it is not difficult to recover edges of an isolated object for N of tens or even larger under most reasonable combinations of z_1, d and λ. Fresnel number refers to an opaque region for a mask (but may be applied to holes in weak phase objects). For a small hole in a mask, however, Babinet's principle is not applicable since the holograms are not formed in the far-field of the whole illuminated area. In the simulation, holes of larger size

of the periodic continuation which results from the use of Fourier series, thus larger hole size means a smaller N for the opaque area. Smaller image size can also improve the result, for the same reason. In experiment, large transparent areas means better reconstruction, but only for isolated opaque regions. For an infinite half-plane mask such as the hole edges in figure 2b adjacent to the extended opaque regions, reconstruction may produce an image no better than the original image. In addition, any deviation from pure amplitude transmission at edges will also affect reconstruction.

In summary, we find that any sharp discontinuity (such as an edge) in a finite object will produce a corresponding sharp discontinuity in the reconstructed image at gaussian focus if N is not too large. As figure 3 shows, this focus setting (the tip-to-sample distance) can easily be identified from the a focal series of reconstructed images. This discontinuity, identifiable against the background of the twin image and the non-linear terms, easily allows the outline shape of opaque objects to be identified. Resolution in the reconstructed image will still be governed by the electron source size (which determines the highest order Fresnel fringe and hence the size of the hologram).

5. Resolution limits and stray fields

Finally we include a discussion of resolution limits of PPM. The effects on resolution of mechanical vibration of the tip and an enlarged emission area may be included by summing the image intensity distribution given by equation 4 over a range of source points using a suitable source intensity distribution function. This incorporates all partial spatial coherence effects.

The effects of a spread of energies ΔE in the beam has been discussed previously [14]. For Fresnel fringes, it was found that energy broadening of the source imposes a chromatic resolution limit such that the highest order fringe which may be observed has order $n_{max} = 2E/\Delta E$. The width of this fringe (referred to the object space) is

$$\Delta C_r = \sqrt{z_1 \lambda / (2n_{max})}, \qquad (8)$$

which, by substituting for n_{max}, gives the resolution limit due soley to the energy spread of the beam. For $z_1 = 1000$ nm and $\Delta E = 0.1$ eV we obtain $\Delta Cr = 0.17$ nm at $E = 100$ volts.

As shown in figure 6 at the left, a transverse time dependent magnetic field will introduce a time dependent phase shift between the optical paths SPD and SD. As for the Aharonov-Bohm effect, this phase shift is

$$\theta(t) = \frac{e}{\hbar} \oint \mathbf{B}(t) \bullet \mathbf{dS}, \qquad (9)$$

where the integral is carried out over the shaded area in the figure. This phase shift may be included in the Fresnel fringe analysis, and a time average of the intensity taken in order to compute the effects of stray fields. The result will depend sensitively on the order of the fringe. A rough estimate may be made by setting the maximum allowable phase shift equal to $\pi/4$ at the position of the n-th Fresnel fringe and solving for B. This fringe occurs at $X_n = M [2z_1 (n\lambda+\Delta)]^{1/2}$ on the detector, with $\Delta = 3/8\lambda$ the approximate phase shift on scattering at the edge [14]. This procedure gives

$$\langle B \rangle_{max} = (\pi/4)(\hbar/e)[2/(z_1 X_n)]. \qquad (10)$$

The fringe width δX_n at the detector is related to X_n by $\delta X_n = M^2 z_1 \lambda / X_n$. The smallest (highest order) fringe width is equal to the incoherent virtual source size. If we take this to be, say 1nm (referred to the specimen plane), then about six fringes will be seen. The maximum tolerable time-dependent field which allows observation of the sixth fringe is then $<B>_{max} = 6 \times 10^{-7}$ Tesla (a few milligauss) at 100 volts with $z_1 = 100$ nm and $z_2 = 14$ cm. Only the component of the field normal to the shaded area is important.

Figure 6 also suggests a construction for estimating the degrading effect of stray magnetic fields on the resolution of a general PPM image. If a homogeneous field $\mathbf{B}(t)$ fills the space between sample and detector in the direction shown (worst case), the electron's trajectory will be circular, and the rearward asymptotic extension of these deflected rays at the detector will define a displaced source point S'. The distance S-S' gives a measure of the source enlargement, and hence resolution loss, due to a time dependent field. We note that the SS' also contains a z-component, so that there is a defocussing effect. The constant earth's field (about 6×10^{-5} Tesla) can be ignored apart from a small aberration effect.

For steady fields, the angular deflection of a non-relativistic electron of energy eU_0 acted on over path-length L by a transverse electric field F is

$$\theta = FL/(2U_0), \tag{11}$$

while for a magnetic field with component B_t transverse to the beam the deflection is

$$\theta = (e/2m)^{1/2}[B_t L / U_0^{1/2}]. \tag{12}$$

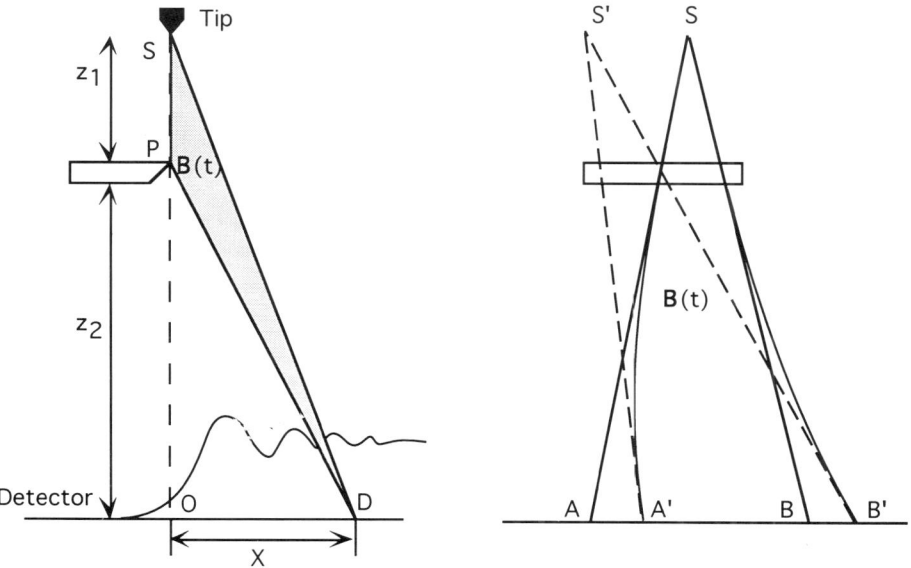

Figure 6. The effect of stray time-dependent magnetic fields at left on the intensity of a Fresnel fringe pattern, and at right on the resolution of the point projection microscope. $\mathbf{B}$ is normal to the page.

Acknowledgements

We are grateful to Dr. M. McCartney and W. Lo for assistance and to Profs. H. Lichte and A. Howie for useful conversations. Supported by NSF award DMR 91-16362.

References

1. H. Boersch, Phys. Zeit. 44 (1943) 202.
2. D. Gabor, Proc. Roy. Soc. A197 (1949) 454.
3. H. Boersch, Z. f. Techn. Physik. 12 (1939) 346.
4. G. A. Morton and E. G. Ramberg, Phys. Rev. 56 (1939) 705.
5. A. J. Melmed and J. Smit, J. Phys. E12 (1979) 355.
6. G. M. Shedd, J. Vac. Sci. Technol. A12 (1994) 2595.
7. H. W. Fink, H. Schmid, H. J. Kreuzer and A. Wierzbicki, Phys. Rev. Lett. 67 (1991) 1543.
8. J. C. H. Spence, W. Qian and A. J. Melmed, Ultramicroscopy 52 (1993) 473.
9. C. Martin, E. Arakawa, T. Callcott and R. Warmack, J. Electr. Spectr. Rel. Phenom. 42 (1987) 171.
10. R. Morin and A. Gargani, Phys. Rev. B48 (1993) 6643.
11. J. C. H. Spence, Optik 92 (1992) 57.
12. M. E. Haine and T. Mulvey, J. Opt. Soc. Am. 42, (1952) 763.
13. T. Mulvey, Personal communication (1994).
14. J. C. H. Spence, W. Qian and M. P. Silverman, J. Vac. Sci. Technol. A12 (1994) 542.
15. J. B. DeVelis, G. Parrent and B. J. Thompson, J. Opt. Soc. Am. 56 (1966) 423 and earlier papers.
16. A. Tonomura, A. Fukuhara, H. Watanabe and T. Komoda, Jap. J. Appl. Phys. 7 (1968) 295.
17. J. Munch, Optik 43 (1975) 79.
18. J. A. Lin and J. M. Cowley, Ultramic. 19 (1986) 31.
19. J. C. H. Spence, "Experimental High Resolution Electron Microscopy", Oxford Univ. Press, 1988.

COHERENT ELECTRON DIFFRACTION AND HOLOGRAPHY

J. W. Steeds, P. A. Midgley, P. Spellward and R. Vincent

Physics Department, University of Bristol, Bristol BS8 1TL, United Kingdom

1. INTRODUCTION

There has been a recent growth of interest in coherent electron diffraction (CED) now that commercial instruments may be used to obtain results of high quality more or less routinely [1-4]. Concurrent with these experimental developments there has been detailed attention to the interpretation of experimental results [5,6]. Rodenburg and his collaborators [5 and references therein] base their approach on the work of Hoppe [7] as well as that of Spence and Cowley [8]. Zuo and Spence [6] have revisited earlier work from the Arizona group and emphasised some of the approximations required for simple interpretation of the results of CED.

From an experimental point of view, the first CED results were demonstrated by Dowell and Goodman [9] and for many years afterwards the only subsequent experimental results came from Cowley in a long series of papers from Arizona [10]. These experiments were conducted on scanning transmission electron microscopes (STEMs) that were not optimised to perform diffraction experiments. It waited for the addition of cold field emission tips to conventional TEM columns, with versatile diffraction facilities, before results of high quality could be obtained from a wide range of materials. Recently, the ability to energy-filter diffraction patterns has further enhanced the quality of results [11].

It is evident from the exploratory work that has now been performed on CED that there are a number of fruitful avenues for more detailed investigation in the future. These will be discussed in this paper and include crystal structure determination, symmetry determination and the investigation of internal electric or magnetic fields. For effective exploitation of the technique it is necessary to demonstrate that methods exist for simple and direct interpretation of the results of CED experiments. It has also to be shown that important problems of materials science or crystallography can be tackled thereby; in some cases this has been achieved, in others it remains as a challenge for future investigation.

2. METHODS OF PERFORMING COHERENT ELECTRON DIFFRACTION

A number of different ways of performing CED have now been demonstrated. The most obvious approach is to illuminate the specimen with a fine focused probe created by a convergent beam of electrons that fills the angle limiting aperture coherently. If the convergence angle is 2α then for disc overlap from reflections that are separated by g (plane spacing d) we require $2\alpha k > g$ or $d > \frac{\lambda}{2\alpha}$. For an aberration-free lens the diameter of the Airy disc formed by a cone of angle 2α is $\Delta = 0.61\frac{\lambda}{\alpha}$. It follows that $d > \frac{\Delta}{1.22}$, that is the probe must be comparable with or smaller than the plane spacing.

The intensity then observed at a point in the overlap region between two adjoining reflections is the modulus squared of the sum of the two complex amplitudes arriving at that point and depends on their relative phases. It is important to note that this phase will depend on the position of the probe relative to the projected unit cell (of which d is a repeat distance in one particular direction) and as the probe is moved the intensity will change. In particular, constructive interference will change to destructive interference and vice versa if the probe is moved one half of a repeat distance of the projected cell. Probe instabilities will cause a modulation of the intensity in the overlap region. In general, the relative phase will also depend on the lens aberrations and the defocus from the ideal crossover condition. Spence & Cowley [6] pointed out that when the incident beam and the diffracted beam are symmetrically disposed either side of the axis for two-beam diffraction there exists a special orientation at the middle of the overlap where the relative phase is unaffected by electronic instabilities, defocus, spherical aberration or astigmatism.

A second method is to deliberately defocus the probe on the specimen (Fig.1) so as to observe fringes in the overlap region between the adjoining reflections. As the defocus is increased the number of fringes increases up to a limit that is related to the lateral coherence of the electron beam. As the figure illustrates, with defocus the different points within the overlap region no longer only coincide with different angles of incidence of the specimen but also correspond to electrons propagating through different positions on the specimen. The spatial dependence of intensity in the overlap region is mapped out directly [2]. This result was anticipated in the early work on STEM lattice imaging [8] and recently discussed directly by Zuo and Spence [6]. Beam instabilities will tend to reduce the visibility of the interference fringes. There are a number of advantages to this form of coherent diffraction for practical applications of the technique. These include the fact that the presence of the interference fringes is a good experimental check that coherent diffraction has been achieved. The visibility of the fringes is a good measure of the degree of coherence. Phase ambiguities are overcome [5] and symmetry properties of the specimen may be investigated [2]. Defects and interfaces can also be spatially resolved in the patterns [11, 12]. Astigmatism is easily recognized as a rotation of the fringes from their orientation perpendicular to g. Zuo and Spence [6] give a simple interpretation of the results, in the case of a low spherical aberration coefficient, as a point projection shadow image of the specimen. The visibility of the fringes is dependent on specimen thickness, decreasing as the thickness increases. This loss of visibility can be overcome by energy filtering the diffraction patterns. The zero-loss results then have fringes of high visibility once again. It is also interesting to note that fringes of low visibility may also be observed in the electrons that have suffered one or two plasmon losses [13].

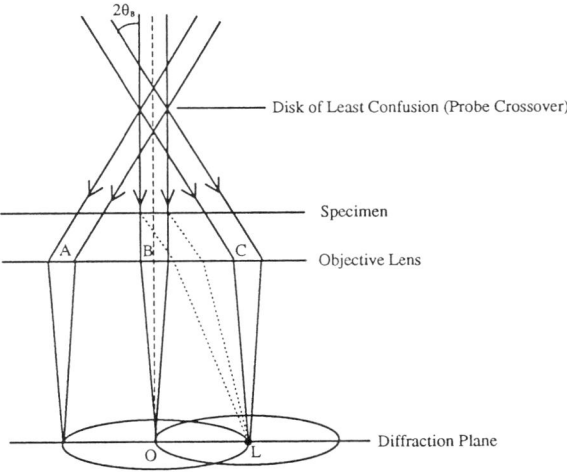

Figure 1. Schematic ray diagram showing coherent electron diffraction. By defocussing the probe, interference fringes fill the overlap region OL.

An example of a two dimensional diffraction pattern formed in this way from erbium pyrogermanate ($Er_2Ge_2O_7$) is shown in Fig. 2. Careful examination reveals that the vertically and horizontally oriented fringes (6.8Å spacing) suffer phase shifts of π on crossing the two perpendicular 2_1 screw axes that are orthogonal to the <001> direction of incidence. Fringes with spacings down to 2.8Å have been recorded in Bristol [11] and down to 1.9Å in Sendai [12].

An alternative version of the out-of-focus method that permits a much larger field of view may also be achieved by use of a very small aperture in the selected area diffraction plane [14]. This method is a direct adaptation of the method devised by Tanaka and co-workers [15] for large angle convergent beam electron diffraction (LACBED). By large defocus it is possible to separate the direct and diffracted beams in the area selecting plane and by encircling the two chosen beams with a small (~1μm) aperture, a pattern covered with more than 100 fringes may be observed. The size of the aperture required is dictated by the lateral coherence of the electron beam, estimated as 30 nm in our case (see [14] for further details). This method is particularly useful for defect or interface studies where a large area is required to identify the region of interest and also in space group determination where a specific orientation is required. It is restricted to particular diffraction vectors chosen by the positioning of the area selecting aperture. The large number of fringes obtained give rise to rather informative Fourier transforms on which a form of image processing similar to that of holography can be performed. Lens aberrations are clearly evident on account of the large angular range involved and may be corrected [14]. This method of interfering diffracted beams is quite distinct from that proposed by Pozzi [16] and achieved by Herring et al [17] and also from that of Rackham et al [18]. In neither of these latter cases is lateral coherence a requirement. A form of coherent diffraction that is more truly a form of holography may be performed when the defocused probe extends across the free edge of a specimen [11]. In this case, the part of the probe avoiding the specimen acts as a reference beam that is interfered with another part of the beam passing through the specimen but diffracted into the same direction as the reference beam (Fig. 3).

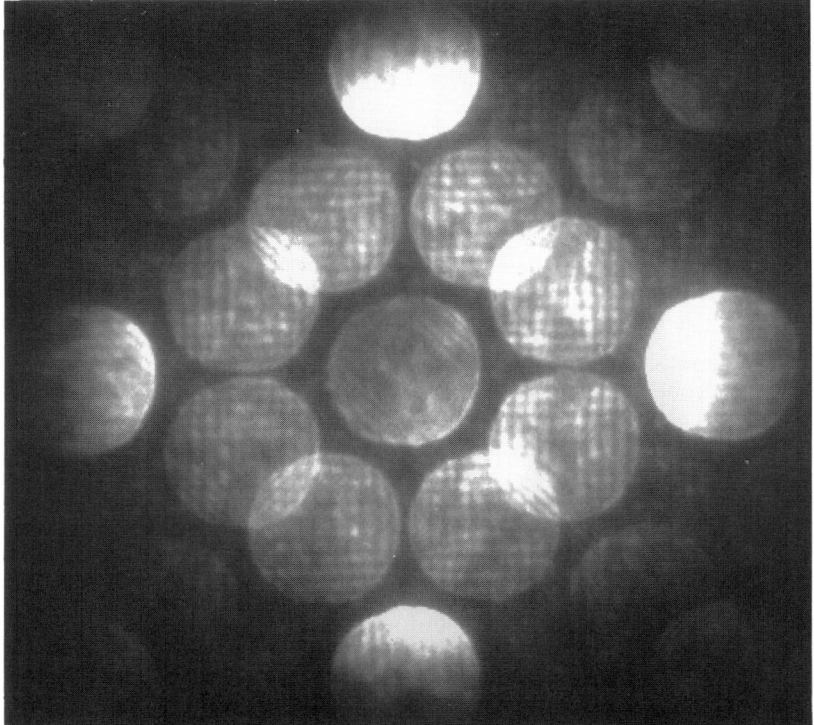

Figure 2. Coherent diffraction pattern taken at the [001] zone axis of $Er_2Ge_2O_7$. Fringes reverse contrast on crossing the <100> 2_1 screw axes.

One further form of coherent diffraction widely explored by Cowley and his co-workers [19] and useful in obtaining accurate values for the spherical aberration and defocus of the objective lens is the Ronchigram. Fig. 4 illustrates a Ronchigram from mica obtained with the Hitachi HF 2000 instrument in Bristol.

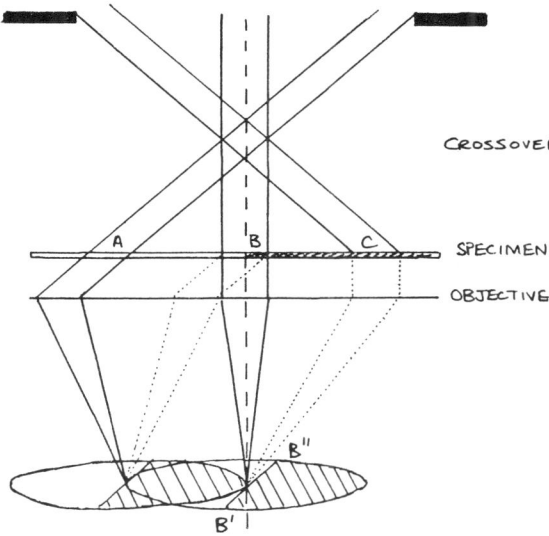

Figure 3. Schematic diagram showing coherent diffraction from the edge of the specimen. The hatched areas indicate the positions of the shadow images.

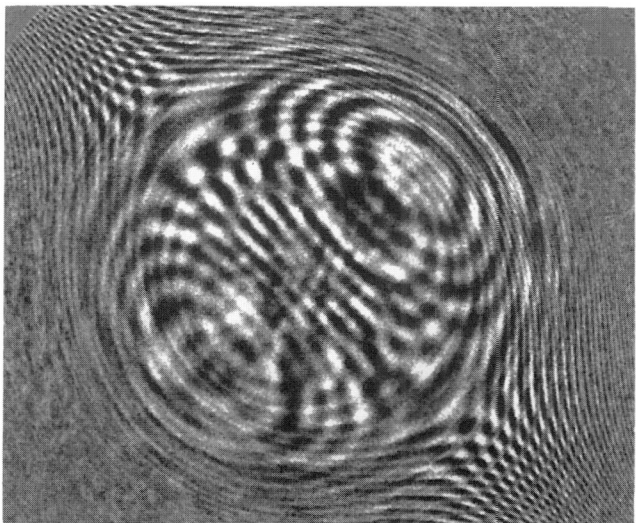

Figure 4. Ronchigram from a systematic row in mica (planar spacing 4.5Å). The pattern has been energy-filtered and the contrast enhanced by image processing.

3. DISCUSSION

We now discuss each of the three fields of application of CED that appear to have considerable potential for effective future application. The first of these is crystal structure determination. If the overlap regions between adjoining reflections can be analysed to give accurate values for the relative phases of the selected structure factors and the regions without overlap have intensities equal to the square of the amplitudes of the structure factors then Fourier inversion gives the projected structure. From two dimensional overlaps extending to large g values very high resolution information can be anticipated, far exceeding the present hopes of high resolution TEM. Reflections can be phased that correspond to small d spacings relative to the effective beam size. By including coherent diffraction from higher order Laue zones (as has already been demonstrated in [2]) three-dimensional structures could be evaluated, in principle, from a single zone axis. Unfortunately the excitement generated by such ideas has to be tempered by a number of practical difficulties discussed in detail by Rodenburg and co-workers [5] and Zuo and Spence [6].

In order to relate the results of CED experiments directly to the amplitudes and phases of the structure factors it is necessary to make the approximation of a flat Ewald sphere and a multiplicative object transmission function. The first of these approximations is not unreasonable for 200 kV electrons and a crystal with a relatively large projected unit cell in the zone axis direction. Only beams far from the centre of the pattern will be affected by the curvature of the Ewald sphere. In order to meet the condition of a multiplicative transmission function the object must be thin enough to satisfy the weak phase object approximation. There is no general condition that can easily be written down to define the maximum specimen thickness that is required for a particular zone axis and crystal structure. However, a first approximation to a condition might be derived from the zone axis string strength as defined by Steeds [20]. This may be written $\frac{\gamma \bar{Z} A}{d}$ where γ is the relativistic mass factor, $\bar{Z}$ is the mean atomic number arranged along the zone axis repeat distance d, for a string with a two dimensional Wigner Seitz cell of area A.

Even though this requirement for a weak phase object is rather limiting and hard to specify in detail there are a number of simple tests that one might apply to a particular CED pattern as a first measure of whether the specimen was thin enough. The first of these is that the intensities in all the overlap-free portions of the reflections should be uniform. Equally, within the overlap area the fringes should be of constant visibility and intensity. Any evidence of orientation dependent intensity in these regions invalidates the hope of a simple relationship between the phase of the fringes and the relative structure factors of the reflections. A second test is the parallelism of all the fringes in all the overlaps in the field of view once the astigmatism is corrected; all fringes should be perpendicular to the g vectors in the sense of overlap. If the fringes within a given overlap vary in orientation (for a perfect crystalline region) or if the orientation of the fringes varies from one overlap to another then this is a clear indication of the breakdown of the weak phase object approximation.

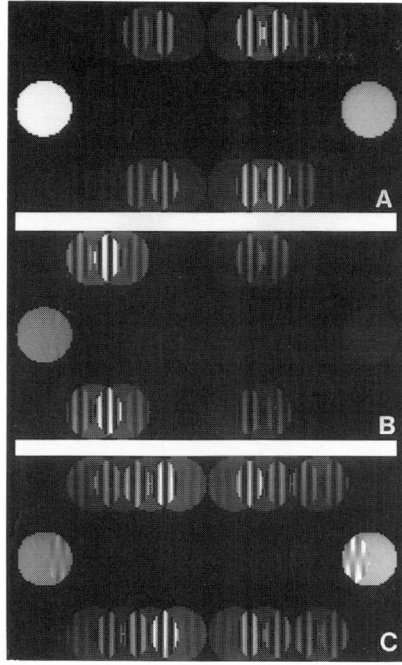
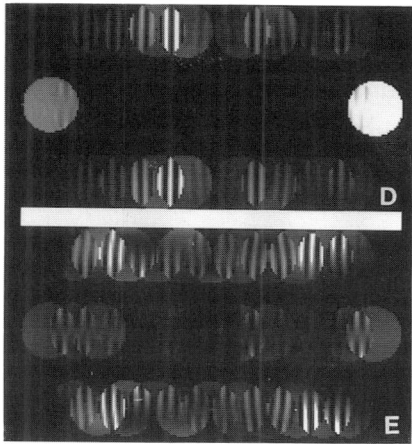

Figure 5. Simulated coherent diffraction patterns from the <11$\bar{2}$0> zone axis of 6H-SiC, thicknesses (a) 50Å, (b) 250Å, (c) 500Å, (d) 1000Å and (e) 2000Å. Note the fringe bending in (e).

We have made an investigation of the validity of the necessary approximations in the case of 6H SiC by simulating CED patterns (Fig. 5). We find, for example, that for specimens of 50Å thickness the approximations apply but by 500Å thickness they have broken down. Spence [21] found that for the 111 systematic row of ZnS at 1MeV major discrepancies between the dynamical and kinematical phases for certain overlaps by a specimen thickness of 80Å. We also investigated the importance of a change of orientation (and therefore curvature of the Ewald sphere) by calculating the phase differences introduced in the overlap regions by a fixed orientation change (Fig. 6). In this case no significant effect was detected.

Even if the approximations required for simple phase determination are satisfied we still require rapid, accurate methods for undertaking the task. One preliminary question concerns the degree of overlap required. The answer to this question seems to be the maximum possible without causing higher order overlaps. This criterion, although straightforward, is more demanding than may be obvious at first sight. Consider the simulation shown in comparison with an experimental result for [001] BaCuO$_2$ in Fig. 7. It is clear that as higher order overlaps occur the fringe patterns become rather complicated. Indeed this is what one would expect. For ever larger angles of coherent illumination the probe size becomes smaller both in absolute terms and also relative to the projected unit cell; the cell period becomes increasingly unimportant. However, the situation is clearly complicated in the case of zone axes with primitive translations g_1 and g_2. of very different magnitudes. In order to have first order overlap of the close reflections it is unlikely that the more widely spread reflections will overlap at all. For overlap of the more widely spaced reflections high order overlaps of the close reflections would occur. In these circumstances two separate results with different convergence angles are desirable.

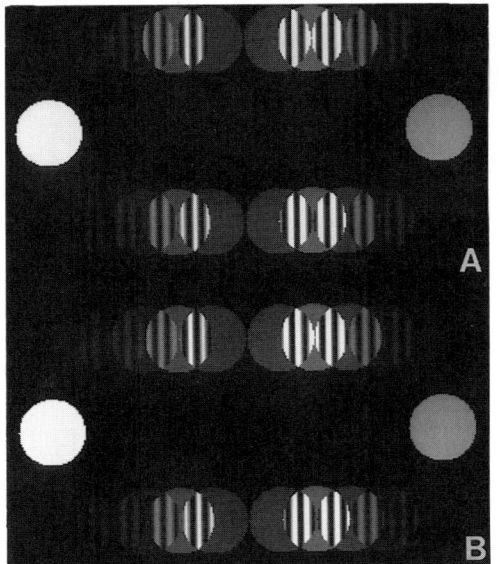

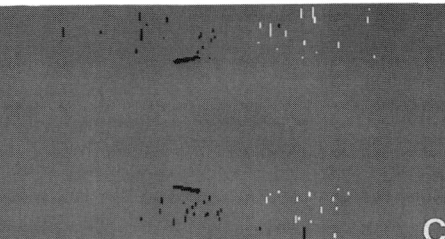

Figure 6. Simulated coherent diffraction patterns from 6H-SiC (thickness 50Å) showing the effect of a small mis-orientation; (a) exactly at the $<11\bar{2}0>$ zone axis and (b) 0.1° off axis. (c) is the difference between (a) and (b), where black is -1, white is +1 (out of 256).

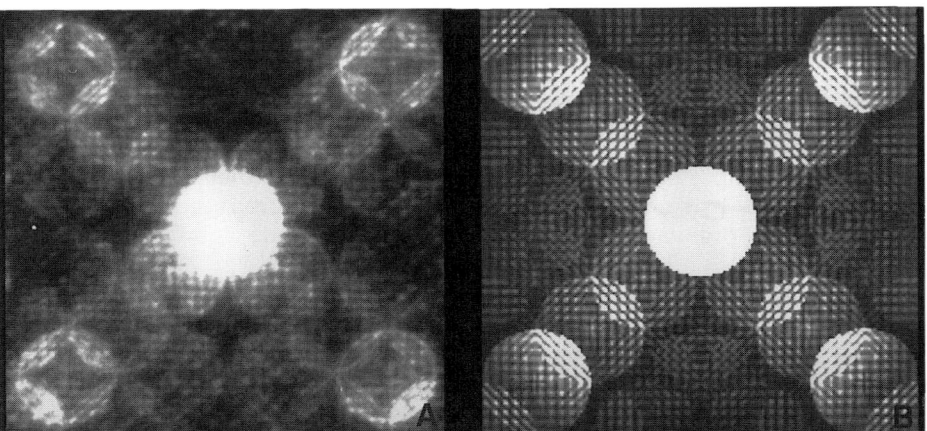

Figure 7. Coherent electron diffraction pattern from the [001] zone axis of $BaCuO_2$, (a) experiment, (b) simulation showing maximum overlap of nearest neighbours.

Having obtained a suitable set of results the next question that arises is how to measure the relative phases as accurately and quickly as possible. This issue was addressed in [5] and [11] and will not be pursued further here.

The second field of application is that of symmetry determination. Spence and Cowley [22, 23] have investigated the question of local symmetry dependent on the position of the focused beam within a projected unit cell. In the out-of-focus CED mode the issue is rather that of space group symmetry of crystalline materials. It has been demonstrated ([2] and Fig. 2) that glide planes and 2_1 screw axes may be determined by this technique. However, these operations may also be determined by conventional CBED, although perhaps with greater difficulty. It would be very exciting if some of the other screw axes, not determined by conventional CBED, could be determined in CED.

The third field of application where CED offers exciting prospects of new results is in the investigation of local electric or magnetic fields. Just as in the case of conventional electron holography [24] electric or magnetic fields with a horizontal component introduce additional phase changes into the electron beam that could be detected as additional phase shifts in the fringes of CED. There are a large number of practical problems where such measurements, performed at the atomic or near atomic scale could be of immense value. Unfortunately, there are a number of complicating factors. In the case of fields at the interface between dissimilar materials there is likely to be a change of mean potential across the interface and this itself will introduce a phase jump [24]. Unless the diffraction planes either side of the interface are in exactly the same orientation diffraction effects would introduce additional phase jumps. Even if the mean potentials and crystal plane orientation are identical there remains the possibility of a thickness change at the interface, produced during thinning for TEM examination, that would also cause a phase jump. Clearly the case of an extended defect with an associated electric field in an otherwise homogenous material, offers some simplification of the problems mentioned. Finally there is the possibility that stress relaxation, or other changes resulting from specimen preparation, could change the field that we aim to measure. In spite of all these difficulties, the unique potential offered, in principle, by CED is too attractive to be ignored.

One final topic deserves comment. Just how localised can the information be from CED? This question was discussed by Zuo and Spence [6]. Clearly the results are thickness dependent and also dependent to some extent on the diffraction angle of interest. For thick specimens and large Bragg angles the region explored by the electron beam increases. For zone axis orientations the diffraction is generally described by a few well-localized Bloch states. Then the column from which the information is obtained is approximately $\Delta + (\lambda t)^{\frac{1}{2}}$ where t is the specimen thickness.

REFERENCES

1. W. J. Vine, R. Vincent, P. Spellward and J. W. Steeds Ultramicroscopy 41 (1992) 42

2. J. W. Steeds, R. Vincent, W. J. Vine, P. Spellward and D. Cherns Hitachi News 24 (1992) 3

3. K. Tsuda, M. Terauchi, M. Tanaka, T. Kaneyama and T. Honda JEOL News 31E (1994) 12

4. N. Tanaka, M. Egi and K. Kimoto Proceedings of ICEM 13 - Paris Vol. 1 (1994) 869
5. B. C. McCallum and J. M. Rodenburg Ultramicroscopy 52 (1993) 85
6. J. M. Zuo and J. C. H. Spence Phil. Mag. A68 (1993) 1055
7. W. Hoppe Acta. Cryst. A25 (1969) 495
8. J. C. H. Spence and J. M. Cowley Optik 50 (1978) 129
9. W. C. T. Dowell and P. Goodman Phil. Mag. 28 (1973) 471
10. J. M. Cowley Electron Diffraction Techniques Vol. 1 (ed J. M. Cowley) Chapter on Coherent Convergent Beam Diffraction, Oxford Univ. Press 1992, p 439
11. J. W. Steeds, X. F. Duan, P. A. Midgley, P. Spellward and R. Vincent in MRS Vol 332, Determining Nanoscale Physical Properties of Materials by Microscopy and Spectroscopy, in the Press
12. K. Tsuda, M. Terauchi, M. Tanaka, T. Kaneyama and T. Honda Proceedings of ICEM 13 - Paris Vol. 1 (1994) 865
13. P. A. Midgley, C. G. Trevor and R. Vincent Inst.Phys.Conf.Series No. 138, 1993, 515.
14. R. Vincent, W. J. Vine, P. A. Midgley, P. Spellward and J. W. Steeds Ultramicroscopy 50 (1993) 365.
15. M. Tanaka, R. Saito, K. Ueno and Y. Harada J. Electron Microsc. 29 (1980) 408
16. G. Pozzi Optik 65 (1983) 77
17. R. A. Herring, G. Pozzi, T. Tanji and A. Tonomura Ultramicroscopy 50 (1993) 94
18. G. M. Rackham, J. E. Loveluck and J. W. Steeds in Electron Diffraction 1927-1977, Inst. Phys. Conf. Ser. No. 41, 1978, 435
19. J. A. Lin and J. M. Cowley Ultramicroscopy 19 (1986) 31
20. J. W. Steeds Ch.15 of 'Introduction to Analytical Electron Microscopy', Eds. J. J. Hren, J. I. Goldstein and D. C. Joy, Plenum Press 1979 p.387.
21. J. C. H. Spence Scanning Electron Microscopy 1978 Vol.1 (1978) 61
22. J. M. Cowley and J. C. H. Spence Ultramicroscopy 3 (1979) 433
23. H. J. Ou and J. M. Cowley Proc. 46th EMSA Meeting Ed. G. Bailey San Francisco Press (1988) p. 882
24. G. Matteucci, G. Missiroli, E. Nichelatti, A. Migliori, M. Vanzi and G. Pozzi J. Appl.Phys. 69 (1991) 1835

Modeling of convergent beam interferometry

R.A. Herring[a] and G. Pozzi[b]

[a]Tonomura Electron Wavefront Project, Exploratory Research for Advanced Technology, JRDC, P.O. Box 5, Hatoyama, Saitama 350-03 Japan

[b]Dept. of Physics, University of Bologna, Bologna 40126 Italy

A theory for modeling a method of interferometry which involves interfering convergent electron beams by means of an electron biprism, CBED+EBI, is presented. The CBED+EBI method requires an electron biprism being placed below the specimen and in between any two or more convergent beams. The electron biprism compensates the convergent beams deviation angle by means of an applied potential such that interference of the beams is made possible. The model uses a description of the CBED+EBI method being produced in the microscope's diffraction mode on the Fraunhofer plane. The theoretical description considers the defocus conditions, spherical aberration, beam tilt and electron biprism and, when compared to fringe thickness and phase distribution measurements, agrees well with the experimental data.

1. INTRODUCTION

Using electron optical theory Pozzi [1] proposed a method of interfering convergent electron beams where the coherency requirement would be overcome by placing an electron biprism below the beam splitter (the crystal specimen), at the level of the selected area aperture plane. This method was then achieved by Herring et. al. [2] and called, Convergent Beam Electron Diffraction Plus an Electron Biprism Interferometry, CBED+EBI, where it was shown that the good contrast of the interference fringes were possible for a wide range of experimental conditions. This paper expands on the original theoretical proposal by Pozzi [1] and shows how modeling of the CBED+EBI method using the theory agrees quite well with experimental results. For a more detailed description of the CBED+EBI method the reader is referred to [3] which will give a full description, with hints, of how to perform the method and considers the role of partial beam coherence for fringe contrast and its measurement.

2. THEORETICAL CONSIDERATIONS

In the CBED+EBI set-up (Figure 1a) the electron beam impinges on the crystal specimen after having traversed the prefield of the objective lens. The beam is diffracted by the single crystal and each diffracted beam travels first through the post-specimen field of the objective lens and then through the field-free space until it reaches the electron biprism, which is usually inserted in the selected area aperture holder which is fitted in the first intermediate image space. The electron biprism is inserted between the diffracted beams in such a way that it does not intercept any part of them but only deflects by opposite angles the beams traveling on opposite sides. The beams propagate through the microscope to the observation plane, where the image is recorded and where, under suitable electron optical conditions interference fringes (Figure 1b) can be observed in the overlapping regions.

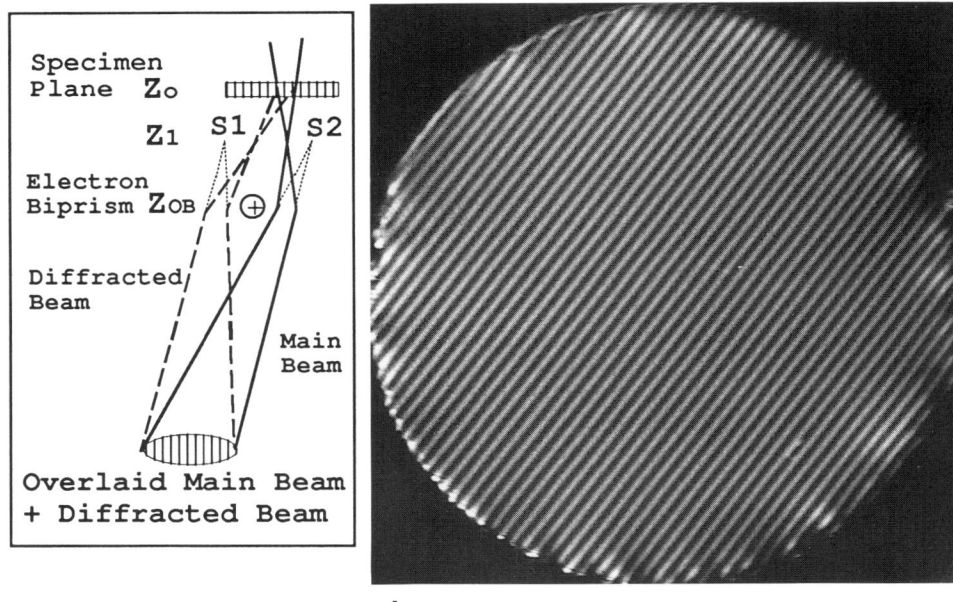

a b
Figure 1. Interference of the main beam with the diffracted beam by use of an electron biprism showing in a) a schematic of the interferometry method and in b) the main beam interfering with the 111 diffracted beam of GaAs.

If the optic axis is denoted by Z, then Z_O is the coordinate of the specimen plane, Z_1 and Z_{OB} are those of the planes conjugate to the electron source and electron biprism respectively, whereas the Fraunhofer plane is at an infinite distance from the above planes (Figure 1a). Special mention is given to the biprism defocus, Z_{OB}-Z_O, which can be varied by the objective lens current and is defined as the distance between the biprism and specimen plane. The biprism defocus in the diffraction mode has a different meaning than the defocus between the observed plane and specimen plane, as is used in the imaging mode of the TEM. S1 and S2 are the apparent focused probe positions. Note that the order in which these planes are located along the optic axis in the object space depends on the lens excitation and may not reflect the physical order with which the various elements are actually located in the microscope, i.e. source, specimen biprism etc. This physical order should be followed or remembered when the propagation of the electron beam from the source to the detector is investigated. The image wave function in the CBED+EBI mode is derived by considering the spatial frequency representation of the wave function, which is related to the real space representation by a Fourier transform. We now follow the propagation of the wave function from the source to the detector plane. If the electrons are emitted by a δ-like point source at $\mathbf{r_1}$ in the source plane (referred to the object space), then the complex amplitude of the $\mathbf{k}$ spatial frequency is given by

$$\exp[-2\pi i \mathbf{k r_1}] \tag{1}$$

Assuming ideal imaging in the condenser, and a centered circular aperture $B(\mathbf{k})$ which can be a top-hat function and equal to 1 for $|\mathbf{k}| < K_A$ and 0 for $|\mathbf{k}| > K_A$, cutting all spatial frequencies larger than K_A the amplitude entering the objective lens prefield is given by

$$S(\mathbf{k}) = B(\mathbf{k})\exp[-2\pi i \mathbf{k r_1}] \tag{2}$$

If we allow for the presence of a tilting stage inserted between the condenser and the objective lens, its net effect, provided $\mathbf{r_1}$ is unchanged, can be simply described by a rigid shift by $\mathbf{K_T}$ of the beam amplitude where the wave function in the spatial frequency representation is now given by,

$$S(\mathbf{k} - \mathbf{K_T}) = B(\mathbf{k})\exp[-2\pi i \mathbf{k r_1}]\exp[2\pi i \mathbf{K_T r_1}] \tag{3}$$

The effect of the aberrations of the objective prefield can be accounted for by C_{SA}, the spherical aberration constant of the objective lens prefield. The wave function impinging on the specimen should also include the defocusing factor, equal to $Z_O - Z_1$, which is the distance between the specimen plane and the focus probe position (Figure 1a), and is given by

$$S(\mathbf{k} - \mathbf{K_T})\exp[-i\gamma_A(\mathbf{k})] \tag{4}$$

where

$$\gamma_A(\mathbf{k}) = \frac{\pi}{2}C_{SA}\lambda^3 \mathbf{k}^4 + \pi\lambda(Z_O - Z_1)\mathbf{k}^2 \tag{5}$$

By considering the specimen to be an ideal single crystal, described by the transmission function

$$\sum_i A_i \exp[2\pi i \mathbf{g_i r}] \tag{6}$$

where A_i is the complex amplitude relative to the diffracted beam of reciprocal wave vector $\mathbf{g_i}$, it can be easily ascertained that its effect on the wave function is to generate the following output

$$\sum_i A_i S(\mathbf{k} - \mathbf{K_T} - \mathbf{g_i})\exp[-i\gamma_A(\mathbf{k} - \mathbf{g_i})] \tag{7}$$

i.e. an array of replicas of the incoming beam, each displaced by the reciprocal vector $\mathbf{g_i}$.

In the propagation from the specimen to the biprism, these beams first suffer the spherical aberration C_{SB} of the objective lens post-specimen field, then they propagate from the first intermediate plane to the biprism plane, so that in the corresponding defocusing factor should be substituted with Z_{OB}-Z_O, the biprism defocus. Both phase factors can be grouped in the term $\gamma_B(\mathbf{k})$.

The transmission function of the biprism is given in real space by

$$D_R(\mathbf{r})\exp\left[2\pi i \frac{\alpha_B}{\lambda}\mathbf{b}(\mathbf{r}-\mathbf{r_B})\right] + D_L(\mathbf{r})\exp\left[-2\pi i \frac{\alpha_B}{\lambda}\mathbf{b}(\mathbf{r}-\mathbf{r_B})\right] \tag{8}$$

where $\mathbf{r_B}$ is the coordinate of a generic point on the biprism axis, α_B is the angular deflection of the biprism, directly proportional to the applied voltage, and $\mathbf{b}$ is a unit vector perpendicular to the biprism axis. The two functions D_R and D_L represent the amplitude transmission of the two half-planes to the right (R) and left (L) of the biprism, of radius R, and are given by

$$D_R(\mathbf{r}) = Step\left[(\mathbf{r}-\mathbf{r_B})\mathbf{b} - R\right] \tag{9}$$

and

$$D_L(\mathbf{r}) = Step\left[-(\mathbf{r}-\mathbf{r_B})\mathbf{b} - R\right] \tag{10}$$

respectively, where $Step(x) = 1$ if $x > 0$; $Step(x) = 0$ if $x < 0$. Therefore, apart from the position dependent amplitude factors, we have still a transmission function of the form given in equation 6. It is important that electron optical conditions be realized so that the diffracted beam in the biprism plane do not strike the biprism wire, but pass either to the left or to the right. Then the net effect of the biprism is again that of a rigid displacement in the spatial frequency plane by $\pm \mathbf{K_B} = \pm \frac{\alpha_B}{\lambda}\mathbf{b}$, plus an additional phase factor $\exp\left[\pm 2\pi i \mathbf{K_B} \mathbf{r_B}\right]$.

More precisely, the image wave function is the sum of two contributions, one regarding the beams coming to the left of the wire, index i_L

$$\psi_L = \Sigma_{i_L} A_{i_L} \exp\left[2\pi i \mathbf{K_B} \mathbf{r_B}\right] S(\mathbf{k} + \mathbf{K_B} - \mathbf{K_T} - \mathbf{g}_{i_L})$$
$$\exp\left[-i\gamma_A(\mathbf{k} + \mathbf{K_B} - \mathbf{g}_{i_L})\right]\exp\left[-i\gamma_B(\mathbf{k} + \mathbf{K_B})\right] \tag{11}$$

and the other, those to the right, index i_R

$$\psi_R = \Sigma_{i_R} A_{i_R} \exp\left[-2\pi i \mathbf{K_B} \mathbf{r_B}\right] S(\mathbf{k} - \mathbf{K_B} - \mathbf{K_T} - \mathbf{g}_{i_R})$$
$$\exp\left[-i\gamma_A(\mathbf{k} - \mathbf{K_B} - \mathbf{g}_{i_R})\right]\exp\left[-i\gamma_B(\mathbf{k} - \mathbf{K_B})\right] \tag{12}$$

It should be noted that, as observations are carried out in the diffraction or Fraunhofer mode on the diffraction plane, the observed image intensity is simply given by

$$I = |\psi_R + \psi_L|^2 \tag{13}$$

3. ANALYSIS OF THE INTERFERENCE PHENOMENA

Let us investigate under which conditions two discs overlap and the main features of the interference fringes observed in the overlapping region. By considering two generic diffracted beams, one traveling to the left of the biprism and the other to the right, of reciprocal wave vectors $\mathbf{g_L}$ and $\mathbf{g_R}$ respectively, the image wave function across them is given by

$$A_L \exp\left[2\pi i \mathbf{K_B r_B}\right] S(\mathbf{k} + \mathbf{K_B} - \mathbf{K_T} - \mathbf{g_L}) \exp\left[-i\gamma_A(\mathbf{k} + \mathbf{K_B} - \mathbf{g_L})\right]$$

$$\exp\left[-i\gamma_B(\mathbf{k} + \mathbf{K_B})\right] + A_R \exp\left[-2\pi i \mathbf{K_B r_B}\right] S(\mathbf{k} - \mathbf{K_B} - \mathbf{K_T} - \mathbf{g_R}) \tag{14}$$

$$\exp\left[-i\gamma_A(\mathbf{k} - \mathbf{K_B} - \mathbf{g_R})\right] \exp\left[-i\gamma_B(\mathbf{k} - \mathbf{K_B})\right]$$

The analysis of the corresponding intensity, given by the square modulus of the above expression, looks very complicated, owing to the presence of so many factors. Nonetheless it can be ascertained at once that, owing to the form of the amplitude factors in S, perfect overlapping is achieved when

$$\mathbf{k} + \mathbf{K_B} - \mathbf{K_T} - \mathbf{g_L} = \mathbf{k} - \mathbf{K_B} - \mathbf{K_T} - \mathbf{g_R} \tag{15}$$

i.e. when

$$\mathbf{g_L} - \mathbf{g_R} = 2\mathbf{K_B} \tag{16}$$

This means that the biprism should be aligned perpendicular to the line joining the two centers of the diffracted beams, and its deflection α_B should be equal to the relative Bragg angle $\theta_B = \lambda|\mathbf{g_R} - \mathbf{g_L}|/2$. When this condition is realized, both the exponential factors containing the position $\mathbf{r_1}$ of the source and the aberrations γ_A of the prefield are identical in both left and right amplitudes, so that they drop out from the intensity and have no effect on it. This means, in particular, that in the case of perfect overlapping, the interference phenomena are not sensitive to the position of the source and can therefore be observed in principle also with extended sources of very poor lateral coherence.

If the simplifying assumption is made of neglecting the spherical aberration terms in the aberrations functions (i.e. restricting first to the case of ideal imaging), it is possible to calculate the image intensity without resorting to numerical simulation. By taking into account that non perfect alignment is the most common condition, by posing

$$\delta\mathbf{k} = \mathbf{g_L} - \mathbf{g_R} - 2\mathbf{K_B} \tag{17}$$

it turns out that the intensity in the overlapping region is given by

$$I = |A_L|^2 + |A_R|^2 + 2|A_L||A_R|\cos\begin{bmatrix} 4\pi K_B r_B + 2\pi r_1 \delta k - 4\pi\lambda(Z_{OB} - Z_O)kK_B + \\ 2\pi\lambda(Z_O - Z_1)k\delta k - \pi\lambda(Z_O - Z_1)(g_L + g_R)\delta k \\ +\beta_R - \beta_L \end{bmatrix} \quad (18)$$

where β_R and β_L are the intrinsic phases associated with the crystal's diffracted amplitudes A_R and A_L, respectively.

In the case of perfect overlapping, $\delta k = 0$, the spacing of the interference fringes is given by

$$\Delta = 1 \Big/ 2|K_B|\lambda(Z_{OB} - Z_O) \quad (19)$$

The spatial frequency plane is projected onto the final recording plane where the parameter Y is the spatial coordinate in the observation plane, which, as we are working in the diffraction mode, is related to the spatial frequency k by

$$Y = \lambda k L \quad (20)$$

where L is the microscope camera length. It turns out that the spacing, ΔY, on the plate is given by

$$\Delta Y = \frac{\lambda L}{2\theta_B(Z_{OB} - Z_O)} \quad (21)$$

This spacing remains almost the same even if $\delta k \neq 0$ as in the usual operating conditions the following inequality is satisfied

$$|(Z_O - Z_1)\delta k| \ll |(Z_{OB} - Z_O)K_B|. \quad (22)$$

In fact, both defocuses are of the same order of magnitude, but $|\delta k| \ll K_B$. Moreover, attention should be paid to the terms in the argument of the cosine which depends on the position r_1 of the source and on r_B of the biprism, since they become relevant in determining the contrast of the fringes if realistic factors are taken into account such as the finite dimension of the source, i.e. partial lateral coherence, and possible drift and/or vibration of the biprism wire.

4. FRINGE SPACING

If we consider the spacing of the fringes on the viewing screen, as given previously by equation 21, then for typical operating conditions of a microscope having a camera length of 3.7 m, $\lambda = 0.00251$ nm, $Z_{OB}-Z_O = 0.6$ mm, and $\theta_B = 4$ mrad, the fringe spacing ΔY is 1.9 mm on the viewing screen which is easily seen. Determining the plane Z_{OB} with respect to the specimen plane, Z_O, is performed by switching the modes of the microscope from the imaging zoom mode to the selected area imaging mode and measuring the objective defocus difference between an infocus specimen image and an infocus biprism image. Obtaining fine fringes with respect to the specimen are accomplished by either 1) increasing the defocus $Z_{OB}-Z_O$ by changing the objective lens currents or 2) interfering high spatial frequency beams, i.e., using high ordered diffracted beams, $2\theta_B$.

The spacing of the fringes are reduced, with respect to the probe size on the specimen, by defocusing, i.e., increasing $Z_{OB}-Z_O$. The fringe spacing as a function of biprism defocus follows equation 21 very closely (Figure 2). Therefore, because $Z_{OB}-Z_O$ is usually large, neglecting C_S is justified for giving the main features of the experimental data, as assumed in equation 21. It has been shown that highly defocused interferograms can have angular fringe widths of only ~0.5 mrad and ~1000 fringes on the observation screen [3]. Thus, very high carrier frequency fringes are possible, although their usefulness for carrying specimen information is limited by the diffusion of information away from the region of interest/object, given by the out-of-focus specimen condition [4]. However, the many angles offered by the convergent beam may be useful for looking three dimensionally at objects in the crystal. It should be noted that the CBED+EBI method has already been found capable of measuring good phase shifts at many different types of phase objects in crystals [5, 6, 7].

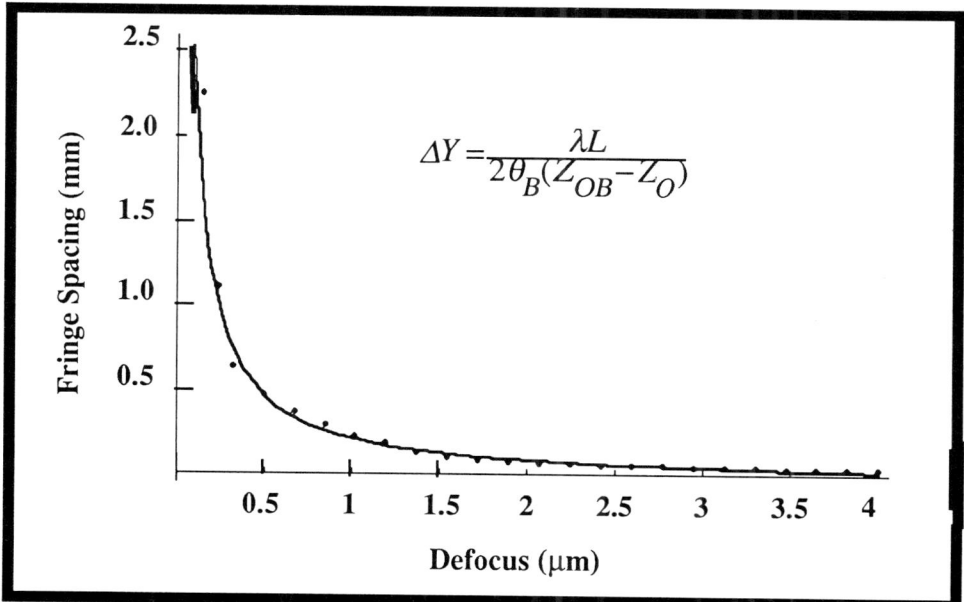

Figure 2. Fringe spacing of measured data, as viewed on the observation screen, and matching to the above equation as a function of biprism defocus.

Interfering higher spatial frequency beams results in finer fringes being produced in the interferogram. Thus far, the smallest spacing achievable, as a function of $2\theta_B$ and the main beam centered on the optic axis, for observation of fringes in an interferogram has been the 000 beam interfering with the 333 beam of Si, although the contrast was low, as expected from the information envelope of the HF 2000 FEG microscope using an objective pole piece having C_S of 0.69. In addition, because any two beams can be interfered by means of the biprism, interfering spatial frequencies coming from both sides of the optic axis are possible, e.g., interfering the symmetrical diffracting beams of 111 and $\bar{1}\bar{1}\bar{1}$. In these cases, because only two beams are used, their interference is clearly resolved, whereas HREM lattice imaging typically cannot resolve these higher order spatial frequencies from the strong low order spatial frequencies.

5. SPHERICAL ABERRATION CONSIDERATIONS

The fringe spacing in equations (19-21) has been calculated in the assumption of ideal imaging, however aberrations and most notably spherical aberration, are not negligible and their effect is displayed in a position dependent variation of the fringe spacing and orientation in the interference patterns shown before. However, especially for large defocuses, these variations are too small to be seen as observed in Figure 1b, but can be detected by either a careful measurement of the fringe spacing by a ruler or by a digital fringe analysis methods.

Another method for displaying and analyzing the effect of aberrations, especially for interferograms taken at relatively large defocus, is to consider the interferograms as holograms and to process them by holographic techniques. The processing of the interferogram can be presented in the form of contour maps of the experimental phase distribution. These results can be interpreted by including the spherical aberration in the intensity distribution (equation 18). As the defocus aberration is only responsible for the carrier fringe frequency, the resulting interferogram can be considered as a hologram of the fictitious phase object given by

$$\Delta\phi = 0.5\pi C_{SB}\lambda^3\{(\mathbf{k}-\mathbf{K_B})^4 - (\mathbf{k}+\mathbf{K_B})^4\} \qquad (23)$$

This is the phase distribution recovered in the processing of the interferograms. Figure 3a is a 16x amplified experimental image of the flattened phase for the interference of the 000/222 beams of GaAs. Figure 3b presents initial simulations using equation 23 where the contour map of the phase was also flattened at the center. This simulation was calculated for the following data set: $C_{SB} = 2.5$ mm and $\mathbf{K_B} = g_{111}$, for the interference of the $000/222$ beams of GaAs. Further comparison of simulations with experimental results and their details are given in [3]. It is shown that the main features of the experimental results are recovered, although the experimental images are perturbed by detrimental effects, like charging up of the aperture rim. The numerical simulations confirmed that these patterns do not depend on the defocus distance, i.e., the carrier fringe spacing. It is hoped that this method of fringe analysis may open interesting perspectives for aberrations analysis and measurement.

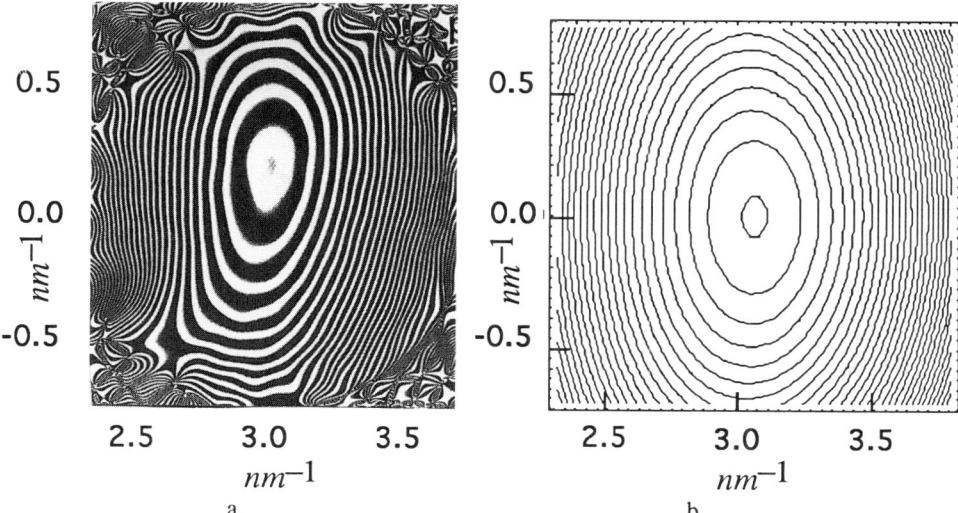

Figure 3. Phase distribution which has been 16x amplified ($\pi/8$ phase image) and which was recovered by reconstructing with a planar wave and flattened at the center where a) is the experimental image and b) is the simulation. See text.

6. SUMMARY

A theoretical model of CBED+EBI has been derived in detail for a defined set of electron optical conditions. The theory has considered the phase contributions of defocus, spherical aberration, beam tilt ($\mathbf{K_T}$), and the biprism. The experimental data of defocus and spherical aberration were modeled with good results using this theory.

The exponential factors containing the position $\mathbf{r_1}$ of the source are identical in both the left and right beams at the position of the biprism, so that they drop out from the intensity upon perfect overlapping of the beams and thus have no effect on the interference. This means, in particular, that in the case of perfect overlapping, the interference phenomena are not sensitive to the position of the source and can therefore be observed in principle also with extended sources of very poor lateral coherence. It is expected that the realization of the CBED+EBI method with standard thermionic sources will demand a more careful and critical adjustment of the overlapping beams.

Although the theoretical interpretation given here has concentrated on the out-of-focus specimen condition at the fraunhofer plane, future work will require modifications to this model since recent developments have enabled the focus plane to be varied from the Fraunhofer plane to the in-focus specimen plane and the convergence of the beam to vary from a highly convergent beam to a parallel beam [4], which all give promise for the CBED+EBI method, or a modified version of it, becoming a useful form of holography for material science research.

ACKNOWLEDGMENTS

The authors wish to thank Drs. A. Tonomura, T. Tanji and Q. Ru and Mr. T. Hirayama for their support during this ERATO project.

REFERENCES

1. G. Pozzi, Optik 65 (1983) 77.

2. R.A. Herring, G. Pozzi, T. Tanji and A. Tonomura, Ultramicroscopy 50 (1993) 94.

3. R.A. Herring, G. Pozzi, T. Tanji and A. Tonomura, Ultramicroscopy, submitted.

4. H. Lichte, Ultramicroscopy 38 (1991) 13.

5. R.A. Herring and T. Tanji, Proc. 51st Annual Meeting of Microscopy Society of America, G.W. Bailey and C. L. Rieder (eds.) (Cincinnati, OH, 1993) 1086.

6. R.A. Herring, J.E. Bonevich, T. Tanji and A. Tonomura, Mater. Res. Soc. Symp. U: Determining Nanoscale Physical Properties of Materials by Microscopy and Spectroscopy, (Boston, Mass. 1993), accepted.

7. R.A. Herring and T. Tanji, 13th Int. Congr. on Electron Microscopy (Paris, 1994), Vol. 1, J.P. Chevalier, F. Glas and P.W. Hawkes (eds.), p.325.

8. R.A. Herring et al., Ultramicroscopy, to be published.

INTERPRETING THE RECONSTRUCTED OBJECT WAVE

Dirk Van Dyck and Marc Op de Beeck

Department of Physics, University of Antwerp (RUCA)
Groenenborgerlaan 171, B-2020 Antwerpen, Belgium

ABSTRACT

It is shown that the wavefunction at the exit face of a crystal in a zone axis orientation can be interpreted in terms of the structure of the object by using a simple channelling theory in which the dynamical diffraction can simply be described in real space using the property that electrons are trapped in the electrostatic potential of the atomic columns. Due to this channelling effect, the electron diffraction can be highly dynamical inside each column, and at the same time the wavefunction retains a one-to-one relationship with the crystal structure. This description does not require the crystal to be periodic. Influence of adjacent columns can be treated using a perturbation theory.

If the crystal is sufficiently thin, i.e. of the order of 10 nm, and the accelerating voltage is not too high (e.g. 100-300 keV), the motion of the electron is almost perfectly periodic with depth. The theory shows how the depth periodicity is related to the mass/thickness of the column which allows the exit wavefunction to be parametrized in a simple analytical form. These results open perspectives to solve the inverse problem of how to derive the projected structure of the object from the exit wavefunction.

1. INTRODUCTION

The exit wave of an object can be reconstructed experimentally by using side-band holography [1,2] or focus variation methods [3,4,5]. The final goal is to obtain the structure of the object projected along the beam direction. However the relation between the wavefunction and the projected structure is not always straightforward. An example is shown in Figure 8 for $Ba_2NaNb_5O_{15}$ where the heavy atoms show up in the amplitude (Figure 8a) of the wavefunction and the light atoms in the phase (Figure 8b). In a sense, the dynamical scattering of the electrons in the object has thus to be reversed. In terms of the classical plane wave description for electron scattering, this is a tedious problem. Another problem is that, in the reconstruction procedure, the exact values of defocus, spherical aberration and astigmatism are not known with sufficient accuracy, so that the retrieved object function is still convoluted with a residual impulse response function. The fine correction of these effects requires prior knowledge of the object itself and can thus only be performed in the final structure retrieval step. Hence we need a simple and robust, albeit approximate analytical expression for the wavefunction in terms of the projected structure which holds for realistic crystal thicknesses and which can be used for the final fitting of the object structure. For this purpose we developed the single column channelling theory.

There is need for a simple intuitive theory that is valid for larger crystal thicknesses. In our view, a channelling theory fulfills this need. Indeed, it is well known that, when a crystal is viewed along a zone axis, i.e. parallel to the atom columns, the high resolution images often show a one-to-one correspondence with the configuration of columns provided the distance between the columns is large enough and the resolution of the instrument is sufficient. From this, it can be suggested that, for a crystal viewed along a zone axis with sufficient separation between the columns, the wave function at the exit face mainly depends on the projected structure, i.e. on the type of atom columns. Hence, the classical picture of electrons traversing the crystal as plane-like waves in the directions of the Bragg beams which stems from the X-ray diffraction picture and upon which most of the simulation programs are based, is in fact misleading. The physical reason for this "local" dynamical diffraction is the channelling of the electrons along the atom columns parallel to the beam direction. Due to the positive electrostatic potential of the atoms, a column acts as a guide or channel for the electron [6,7] within which the electron can scatter dynamically without leaving the column (Figure 1). It has been proposed [8] to exploit this so-called atom column approximation to speed up the dynamical diffraction calculations by assembling the wavefunction at the exit face using parts that have been calculated for each atom column separately.

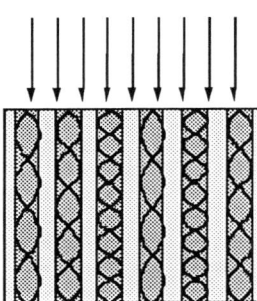

Figure 1. Schematical representation of electron channelling.

The importance of channelling for interpreting high resolution images has often been ignored or underestimated, probably because of the fact that for historical reasons, dynamical electron diffraction is often described in reciprocal space. However, since most of the high resolution images of crystals are taken in a zone axis orientation, in which the projected structure is the simplest, but in which the number of diffracted beams are the largest, we believe that a simple real-space channelling theory yields a much more useful and intuitive, albeit approximate, description of the dynamical diffraction, which allows to provide an intuitive interpretation of high resolution images, even for thicker objects.

2. THEORY

If we assume that the fast electron, in the direction of propagation (z-axis) behaves as a classical particle with velocity $v = \hbar k/m$ we can consider the z-axis as a time axis with

$$t = mz/\hbar k \tag{1}$$

Hence we can start from the time-dependent Schrödinger equation

$$-\frac{\hbar}{i}\frac{\partial}{\partial t}\Psi(\mathbf{R},t) = H\,\Psi(\mathbf{R},t) \tag{2}$$

with

$$H = -\frac{\hbar^2}{2m}\Delta_\mathbf{R} - eU(\mathbf{R},t) \tag{3}$$

with $U(\mathbf{R},t)$ the electrostatic crystal potential, m and k the relativistic electron mass and wavelength and $\Delta_\mathbf{R}$ the Laplacian operator acting in the plane ($\mathbf{R}$) perpendicular to z. Using (1) we then have

$$\frac{\partial}{\partial z}\Psi(\mathbf{R},z) = \frac{i}{4\pi k}\bigl(\Delta_\mathbf{R} + V(\mathbf{R},z)\bigr)\Psi(\mathbf{R},z) \tag{4}$$

with

$$V(\mathbf{R},z) = \frac{2me}{\hbar^2} U(\mathbf{R},z) \tag{5}$$

This is the well-known high energy equation in real space which can also be derived from the stationary Schrödinger equation in the forward scattering approximation [3]. The solution of (4) can be expanded in eigenfunctions of the Hamiltonian

$$\Psi(\mathbf{R},z) = \sum_n c_n \, \phi_n(\mathbf{R}) \exp[-i\pi k (E_n / E_o) z] \tag{6}$$

with H given by (3) and the incident electron energy :

$$E_o = \hbar^2 k^2 / 2m \tag{7}$$

$\mathbf{k}$ is the electron wavevector. For $E_n < 0$ the states are bound to the columns. We now rewrite (6) as

$$\Psi(\mathbf{R},z) = \sum_n c_n \, \phi_n(\mathbf{R}) \left\{ \begin{array}{l} (1 - i\pi k (E_n / E_o) z) \\ + \left(\exp[-i\pi k (E_n / E_o) z] - 1 + i\pi k (E_n / E_o) z \right) \end{array} \right\} \tag{8}$$

The coefficients c_n are determined from the boundary condition

$$\Psi(\mathbf{R},0) = \sum_n c_n \, \phi_n(\mathbf{R}) \tag{9}$$

from which

$$c_n = \int \phi_n^* \, \Psi(\mathbf{R},0) \, d\mathbf{R} \tag{10}$$

In case of plane wave incidence, one thus gets

$$\sum_n c_n \, \phi_n(\mathbf{R}) = 1 \tag{11}$$

and from (6) and (3)

$$\sum_n c_n \, \phi_n(\mathbf{R}) E_n = H\Psi(\mathbf{R},0) = H \cdot 1 = -eU(\mathbf{R}) \tag{12}$$

Now (8) becomes

$$\Psi(\mathbf{R},z) = 1 + i\pi k \{eU(\mathbf{R}) / E_o\} z \\ + \sum_n c_n \, \phi_n(\mathbf{R}) \{ \exp[-i\pi k (E_n / E_o) z] - 1 + i\pi k (E_n / E_o) z \} \tag{13}$$

The first two terms yield the well-known weak phase object approximation. In the third term only these states will appear in the summation for which

$$|E_n| \geq E_o / kz \tag{14}$$

In case the object is very thin, so that no state obeys (14), the weak phase object approximation is valid. For a thicker object, only bound states will appear with very deep energy levels, which are localised near the column cores. Furthermore, a two-dimensional projected column potential has only very few deep states, and when the overlap between adjacent columns is small only the radial symmetric states will be excited. In practice, for most types of atom columns, only one state appears, which can be compared with the 1S state of an atom.

In the case of an isolated column, taking the origin in the centre of the column, we then have

$$\Psi(\mathbf{R},z) = 1 + i\pi k\{eU(\mathbf{R})/E_o\}z \\ + c\phi(\mathbf{R})\{\exp[-i\pi k(E/E_o)z]-1+i\pi k(E/E_o)z\} \quad (15)$$

A very interesting consequence of this description is that, since the state φ is very localised at the atom cores, the wavefunction for the total crystal can be expressed as a superposition of the individual column functions

$$\Psi(\mathbf{R},z) = 1 + i\pi k\{e\hat{U}(\mathbf{R})/E_o\}z \\ + \Sigma_i c_i \phi_n(\mathbf{R}-\mathbf{R}_i)\{\exp[-i\pi k(E_i/E_o)z]-1+i\pi k(E_i/E_o)z\} \quad (16)$$

with

$$U(\mathbf{R}) = \Sigma_i U_i(\mathbf{R}-\mathbf{R}_i) \quad (17)$$

If all the states other than the 1S have very small energies i.e.

$$|E_n| \ll E_o/kz \quad (18)$$

then (8) can be simplified as

$$\Psi(\mathbf{R},z) = \Sigma_n c_n \phi_n(\mathbf{R}) + \Sigma_n c_n \phi_n(\mathbf{R})\{\exp[-i\pi k(E_n/E_o)z]-1\} \quad (19)$$

so that (15) becomes

$$\Psi(\mathbf{R},z) = 1 + c\phi(\mathbf{R})\{\exp[-i\pi k(E_i/E_o)z]-1\} \quad (20)$$

and (16) becomes

$$\Psi(\mathbf{R},z) = 1 + \Sigma_i c_i \phi_i(\mathbf{R}-\mathbf{R}_i)\{\exp[-i\pi k(E_i/E_o)z]-1\} \quad (21)$$

Expressions (16) and (21) are the basic result of the channelling theory.

The interpretation of (21) is simple. Each column i acts as a channel in which the wavefunction oscillates periodically with depth. The period is related to the "weight" of the column, i.e. proportional to the atomic number of the atoms in the column and inversely proportional to their distance along the column. The importance of these results lies in the fact that they describe the dynamical diffraction for larger thicknesses than the usual phase grating approximation and that they require only the knowledge of one function ϕ_i per column (which can be tabulated similar to atom scattering factors or potentials). Furthermore, even in the presence of dynamical scattering, the wavefunction at the exit face still retains a one-to-one relation with the configuration of columns. Hence this description is very useful for interpreting high resolution images and to provide a possible answer to the direct retrieval problem. Eq. (20) applies to light columns, such as Si(111) or Cu(100) with an accelerating voltage up to about 200 keV. When the atom columns are "heavier" and the accelerating voltage higher (which due to the relativistic correction also increases the effective strength of the potential), then (16) has to be used. This is for example the case for Au(100).

Figure 2 shows the electron density $|\Psi(\mathbf{R},z)|^2$ as a function of depth in a Au$_4$Mn alloy crystal for 200 keV incident electrons. The corners represent the projection of the Mn column. The square in the centre represents the four Au columns. The distance between adjacent columns is 0.2 nm. The periodicity along the direction of the column is 0.4 nm. From these results it is clear that the electron density in each column fluctuates nearly periodically with depth. For Au this periodicity is about 5 nm and for Mn 13 nm. These periodicities are nearly the same as for isolated columns so that the influence of neighbouring columns in this case is still small. The energies of the respective 1S states are respectively about 250 eV and 80 eV.

When the atoms are heavy and the accelerating voltage very high (0.5 to 1 MeV), a larger number of states come in and the result becomes more complicated. When the crystal is viewed along a higher index zone axis, the distance between adjacent columns decreases whereas the weight of the columns also decreases. Hence the bound states broaden and overlap between adjacent columns starts to occur. This can be incorporated in the theory using perturbation theory. When the overlap between columns is too large one has to consider them as a kind of molecule [8]. The localisation can also be improved by using higher voltages. It has to be stressed that the derived results are only valid in a perfect zone axis orientation. A slight tilt can destroy the symmetry and excite other, non-symmetric states, so that the results become much more complicated. It is interesting to note that channelling has usually been described in terms of Bloch waves [9,10]. However as follows from the foregoing, channelling is not a mere consequence of the periodicity of the crystal but occurs even in an isolated column parallel to the beam direction. In fact, even for an isolated column, the problem can be treated mathematically by making the column artificially periodical so as to generate a basis of functions (Bloch functions) to expand the wavefunction. In this view, the Bloch character is only of mathematical importance. Even in a crystal in which the distance between the adjacent columns is sufficient (e.g. 0.2 nm), this is the case. Bloch wave calculations then yield the same 1S states as found in our simplified treatment. Only when the overlap between columns increases or when the beam is inclined, the other Bloch states become physically important.

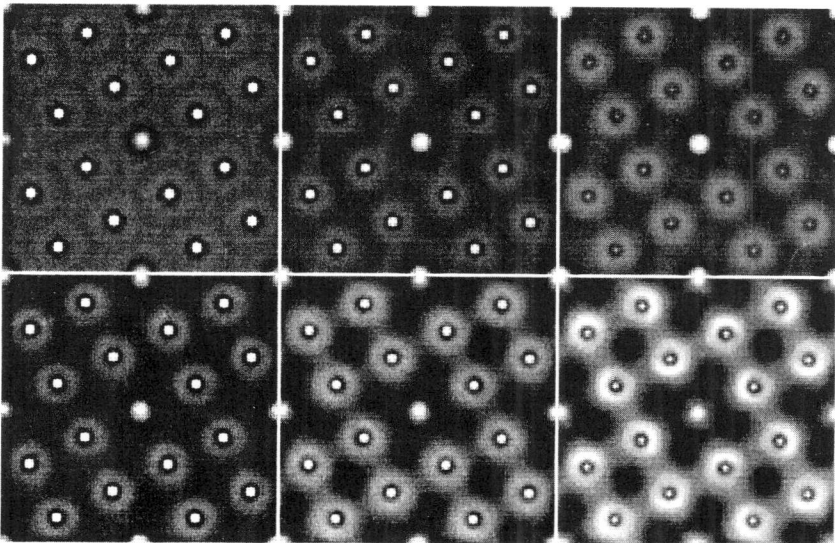

Figure 2. Electron density as a function of depth in Au$_4$Mn (thicknesses 2, 4, 6, 8, 10, 12 nm).

3. TEST OF THE VALIDITY

In order to test the validity of the theory, we compared the theoretical results with simulations using a real space slice program. In case of one single column of type i, centred at the origin, one expects, from (20), in case the column is not too heavy,

$$\Psi(\mathbf{R},z) - 1 = 2\phi_i(\mathbf{R})\sin(0.5\pi(E_i/E_o)kz)\exp(-0.5\pi i\{1+(E_i/E_o)kz\}) \qquad (22)$$

i.e.
- the amplitude decreases with increasing distance R from the centre of the columns with a slope given by $\phi_i(R)$, the 1S eigenfunction
- the amplitude oscillates with depth with a periodicity of $4E_o/kE_i$
- the phase increases linearly with depth, starting from $\pi/2$

Figure 3 shows the amplitude (A) and phase (B) of (Ψ-1) for one isolated Cu column, with periodicity 3.8 A along the ordered axis. These results are in complete agreement with (22). Figure 4 (A) and (B) show the amplitude and phase for an isolated heavier column of Au with repeat distance of 4.08 Å along the column. Here the isolation of the amplitude is superimposed onto a linearly increasing function. This is in agreement with the improved expression (15).

In order to demonstrate the validity of the theory when the columns are very close we performed a simulation of $|\Psi - 1|^2$ for an artificial unit cell (5 x 4 X 4 Å) consisting of three Cu (or Au) columns with an artificial distance of 1 Å (Figure 4). It is clear that the periodicity, as expected for an isolated column, holds for Cu (Au) up to a thickness of 30nm (10nm).

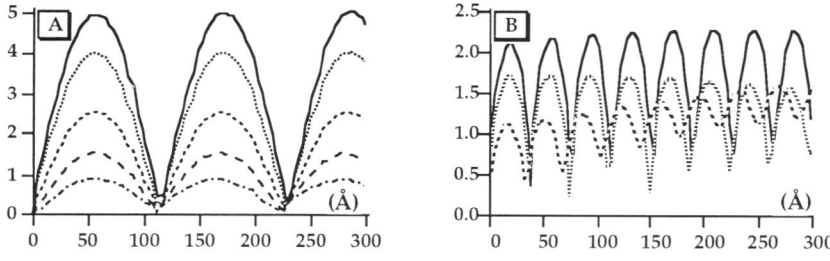

Figure 3. Amplitude of Ψ-1 for an isolated Cu (A) and Au (B) column (200 keV).

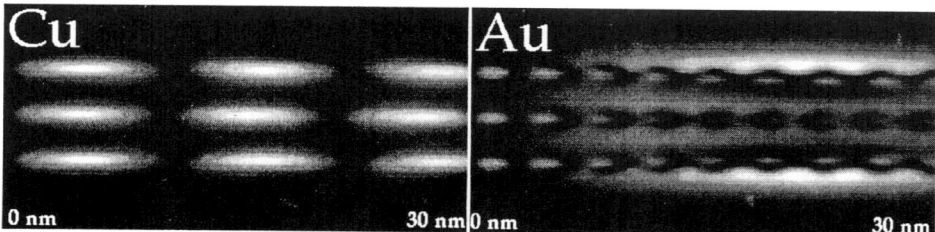

Figure 4. $|\Psi-1|^2$ for an artificial unit cell of (5 x 3 x 4 Å) consisting of three Cu (or Au) columns with inter-column distance of 1 Å.

4. UNIVERSAL SCALING

As is clear from Figures 5 and 6 the eigenfunctions scale with $\sqrt{|E|}$ and the eigenenergy E scales with $Z/d^{5/4}$ with Z the atomic mass and d the distance between the atoms in the column. This scaling behaviour allows to parametrize the wavefunction in the form:

$$\Psi(\mathbf{R},z) \approx 1 + \{\exp[-i\pi k(E/E_o)z] - 1\} E^{-1/2} \phi_o(\sqrt{|E|}\mathbf{R}) \qquad (23)$$

This expression can still be improved so as to hold for thicker objects and/or heavier columns:

$$\Psi(\mathbf{R},z) \approx 1 + |E|^{4/5} V_o(\mathbf{R})$$
$$+ \{\exp[-i\pi k(E/E_o)z] - 1 + i\pi k(E/E_o)z\} |E|^{-1/2} \phi_o(\sqrt{|E|}\mathbf{R}) \qquad (24)$$

where V_o and ϕ_o are universally scaled functions for the projected potential, respectively the 1S bound state. The advantage of this result is that the exit wavefunction is expressed analytically in terms of the parameters (Z, d, z, column position) that describe the crystal structure. This result enables to calculate image and diffraction patterns in an analytical (and thus very fast way) and also to solve the inverse problem of deriving the crystal structure by fitting with the experimentally reconstructed object wave.

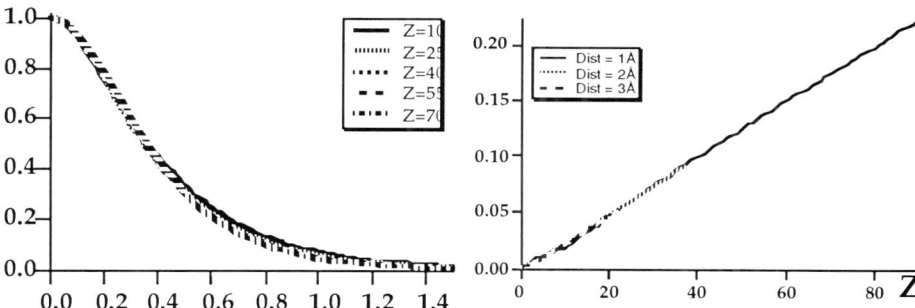

Figure 5. The scaled eigenstate $\phi(R)$ as a function of $|E|^{1/2}$ for various mass Z.

Figure 6. The linear scaling of the energy E as a function of $Z/d^{5/4}$.

5. DISCUSSION

5.1. Channelling and defects

The channelling effect still occurs in the presence of defects such as translation interfaces, twin interfaces, dislocations, provided the columns parallel to the incident beam are not disrupted. To demonstrate this we take again the Au$_4$Mn alloy of Figure 2 in which we now introduce translation interfaces (antiphase boundaries) (Figure 7). The electron density, calculated with the periodic continuation method is shown at the left (thicknesses 8, 10 nm). In the middle, we represent the channelling approximation, i.e. we perform the simulation for individual columns, which are then assembled according to the structure positions. If the channelling along the individual columns would not be affected by the interface, the electron density at both sides of the interface would be identical to that of the perfect crystal shifted over the displacement vector of the interface. In order to reveal the deviation from this ideal situation, we subtract this "ideal" electron density at both sides and display the difference with increased contrast (right). From this it is clear that the deviation from the ideal channelling condition is very small and is mainly due to very small asymmetric contributions to the image. These can be only be observed at thicknesses at which the gold columns are nearly extinct.

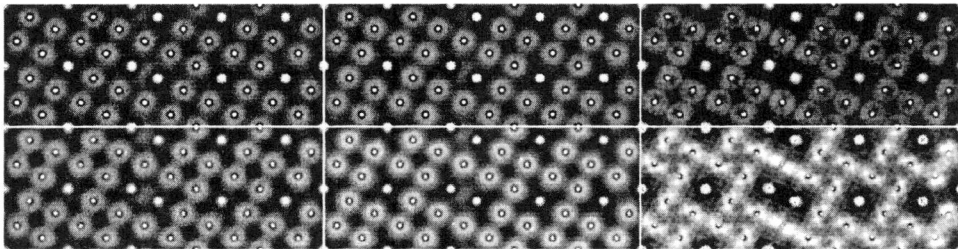

Figure 7. Deviation from the ideal channelling at a translation interface in Au_4Mn (8,10 nm).
Left : multi-slice calculation of the electron density at the exit face.
Centre : Channelling approximation. Right: difference image.

5.2. Diffraction pattern

Fourier transforming the wavefunction (20) at the exit face of the object yields the wavefunction in the diffraction plane, which can be written as

$$\Psi(\mathbf{g},z) = \delta(\mathbf{g}) + \sum_i \exp(-2\pi i \mathbf{g}\cdot\mathbf{R}_i) F_i(\mathbf{g},z) \tag{25}$$

(In the case of heavy columns we will have to use (16) instead). In a sense the simple kinematical expression for the diffraction amplitude holds, provided the scattering factor for the atoms is replaced by a dynamical scattering factor for the columns, in a sense as obtained in [8] and which is defined by

$$F_i(\mathbf{g},z) = \left\{ \exp(-i\pi k(E_i/E_o)z) - 1 \right\} c_i f_i(\mathbf{g}) \tag{26}$$

with $f_i(\mathbf{g})$ the Fourier transform $\phi_i(\mathbf{R})$. It is clear that the dynamical scattering factor varies periodically with depth. This periodicity may be different for different columns.

In case of a mono-atomic crystal, all Fi are identical. Hence $\Psi(\mathbf{g},z)$ varies perfectly periodically with depth. In a sense the electrons are periodically transferred from the central beam to the diffracted beams and back. The periodicity of this dynamical oscillation (which can be compared with the Pendelösung effect) is called the dynamical extinction distance. It has for instance been observed in Si(111). An important consequence of (21) is the fact that the diffraction pattern can still be described by a kinematical type of expression so that existing results and techniques that have been based on the kinematical theory remain valid to some extent for thicker crystals in zone orientation. Examples are

- diffraction at periodical stacking of translation interfaces, twin interfaces and mixed layer compounds
- diffuse scattering from substitutionally ordering alloys with a column structure
- diffraction contrast at defects. In particular the extinction rule based on the **g.R** criteria remains valid if the defect is parallel to the incident beam.

5.3. High resolution images

The wavefunction in the image plane can be written as the convolution product of the wavefunction at the exit face of the crystal with the impulse response function $t(\mathbf{R})$ of the electron microscope

$$\Psi(\mathbf{R},z) = 1 + \sum_i c_i \, \phi_n(\mathbf{R} - \mathbf{R}_i)\{\exp[-i\pi k(E_i/E_o)z] - 1\} \otimes t(\mathbf{R}) \quad (27)$$

If the microscope is operated close to optimum focus and in axial mode, the impulse response function is sharply peaked.

If the distance between the columns is larger than the width of the impulse response function $t(\mathbf{R})$, the overlap between convolution products $\phi_i * t(\mathbf{R})$ of adjacent sites can be assumed to be small so that each column is thus imaged separately. The contrast of a particular column varies periodically with thickness. The periodicity can be different for different types of columns. It is interesting to note that the functions ϕ_i as well as $t(\mathbf{R})$ are symmetrical around the origin, provided the objective aperture is centred around the optical axis. Hence, the image of a column is rotationally symmetric around the position $\mathbf{R}_i$ of the columns. The intensity at $\mathbf{R}_i$ is a maximum or a minimum. The positions of the columns can thus be determined from the positions of the intensity extreme.

In case the resolution of the microscope is insufficient to discriminate the individual columns, or the focus is not close to optimum, the overlap between the convolution products of adjacent columns cannot be avoided and the interpretation of the contrast is not straightforward.

5.4. Direct structure retrieval

Using holographic methods such as side-band holography [1,2] or focus variation [3-5], it is possible to reconstruct the exit wave of an object. In order to retrieve the crystal structure out of this object wave one can use this channelling theory. Indeed, expression (23) yields a simple analytical expression for the wavefunction at the exit of a column in terms of Z, d and z. Albeit approximately, this expression is sufficiently simple to be used in a final least squares fit from which these structural parameters can be obtained. If necessary, the residual imaging parameters such as C_s and defocus can be fitted simultaneously.

The focus variation and structure retrieval methods have been implemented on a CM20 ST FEG microscope in the framework of a Brite-Euram project. Figure 8a,b shows the reconstructed exit wave for $Ba_2NaNb_5O_{15}$ and Figure 8c shows the retrieved projected structure. The position of all atom columns (even oxygen) can be determined accurately and the mass approximately.

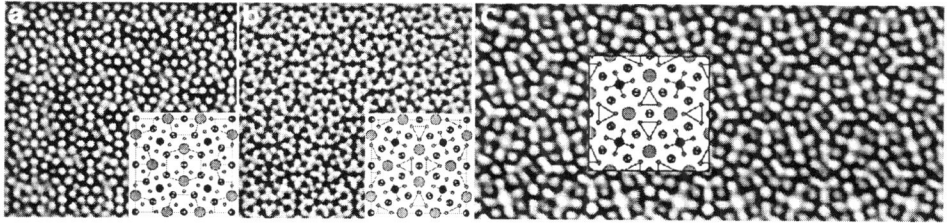

Figure 8. Experimentally retrieved amplitude (a), phase (b) and reconstructed structure + structure inset (c) for $Ba_2NaNb_5O_{15}$.

6. CONCLUSIONS

The real space channelling theory for isolated columns provides a simple and intuitive means to express the exact wavefunction of an object in terms of the projected column structure.

The theory is valid for most crystal thicknesses used in HREM situations. It provides a simple analytical tool to study dynamical scattering in a crystal and related effects and to solve the inverse problem of deducing the projected crystal structure out of the experimentally reconstructed exit wave.

7. ACKNOWLEDGEMENTS

This work has been performed within the framework of a Brite-Euram Project (nr 3322). One of us (Marc Op de Beeck) is indebted to the Belgian Fund for Scientific Research (N.F.W.O.) for financial support.

REFERENCES

1. H. Lichte, Ultramicroscopy, 20 (1986) 293.
2. T. Tanji and K. Ishizuka, Microscopy Society of America Bulletin, 24 (1994) 494.
3. D. Van Dyck, Proc. XIIth ICEM (Seattle), Francisco Press Inc., 1990, p.26.
4. W.M.J. Coene, A.J.E.M. Janssen, M. Op de Beeck and D. Van Dyck, Phys. Rev. Lett., 29 (1992) 3743.
5. D. Van Dyck, M. Op de Beeck and W.M.J. Coene, Optik, 93 (1993) 103.
6. D. Van Dyck, Proc. XIIthe ICEM (Seattle), Francisco Press Inc., 1990, p.64.
7. S. Amelinckz and D. Van Dyck, J.M. Cowley (ed.), Electron Diffraction Techniques, Vol.2, Oxford Science Publishers, 1993.
8. D. Van Dyck, J. Danckaert, W.M.J. Coene, E. Selderslaghs, D. Broddin, J. Van Landuyt and S. Amelinckx, W. Krakow and M. O'Keefe (eds.), Computer Simulation of Electron Microscope Diffraction and Images, 1989.
9. B.F. Buxton and P.T. Tremewan, Adv. Cryst., A36 (1980) 304.
10. K. Kambe, G. Lehmpfuhl and F. Fujimoto, Z. Natur., 29A (1974) 1034.

Electron Holography
A. Tonomura, L.F. Allard, G. Pozzi, D.C. Joy and Y.A. Ono (Editors)
© 1995 Elsevier Science B.V. All rights reserved.

Focal Series Wave Function Reconstruction in HRTEM

Marc Op de Beeck[a], Dirk Van Dyck[a] and Wim Coene[b]

[a] University of Antwerp (RUCA-EMAT), Groenenborgerlaan 171, B-2020 Antwerp
Belgium

[b] Philips Research Laboratories, Prof. Holstlaan 4, 5600 JA Eindhoven
The Netherlands

1. INTRODUCTION

The idea of improving the resolution in HRTEM from a focal series of HRTEM images of weakly scattering objects by computer-aided compensation of the lens aberrations for the reconstruction of the complex wave function information at the level of the specimen is already quite old [1,2,3]. A more advanced method for wave function restoration applicable to general and thicker objects, has been proposed by [4]. But only recently, the technical means became available to capture the images in a quantitative way by using CCD cameras and to process them on fast computer systems. Moreover, during the last few years the use of Field Emission Gun (FEG) microscopes has become widely spread. These FEG's have a lower energy spread ($\Delta E=0.4$eV) when compared to the thermo-ionic LaB_6 filaments ($\Delta E=0.8$eV). Also the brightness of these guns has been improved drastically. Therefore, the illuminating electron beam is much more spatially coherent and the effective information limit for the FEG-TEM will thus almost exclusively be determined by the focal spread (ρ_i = sqrt(.5 $\pi\lambda\Delta$)) and is far lower than for a LaB_6-TEM. This makes the idea of pushing the interpretable resolution down to the information limit using image processing techniques much more appealing.

Moreover, we believe that reconstruction techniques have not only become technically feasible, but even indispensable since most of the high resolution information beyond the point resolution of the instrument is highly delocalised within the image, so that direct interpretation cannot be done without a reliable reconstruction method. An other aspect which is sometimes overlooked is that a reconstruction method may lead towards a direct quantitative interpretation of the HRTEM information in terms of the crystal structure, which avoids cumbersome simulation techniques. In this paper, we will present the so-called parabola method for object wave function reconstruction, which is an elaboration of the method proposed in [5].

2. IMAGE FORMATION IN A TEM

The image formation process in HRTEM can be split in two distinct parts. In a first step, the incoming electron wave is modulated due to diffraction at the atoms in the specimen. In case of a zone-axis orientation, this yields a regular grating of atom columns. Especially for very thin crystals, this results mainly in a phase shift which is larger close to the atom cores, while the amplitude nearly remains unchanged. The specimen is then considered to be a "(pure) phase object". For thicker specimens, both the amplitude and the phase will be changed and the information about the crystal potential will be distributed over both. The resulting wave function at the exit plane of the specimen will be called "the object wave function". It can be shown that the detail in the object wave function, even for thicker specimens, is directly related to the projected column potentials.

In the second step of the image formation process, phase and amplitude of the object wave function are convoluted with a complex point spread function which accounts for the aberrations of the objective lens. Moreover, only the intensity of the resulting "image wave function" can be recorded, yet losing the phase information. Figure 1 represents the CTF for a typical 200 kV LaB_6-TEM and a FEG-TEM at Scherzer focus. For a FEG-TEM it clearly shows the large gap between the point resolution ρ_s (the first zero-crossing of the CTF) and the information resolution ρ_i. Beyond ρ_s the wave aberration function is rapidly oscillating. This strongly hampers the direct interpretation of the recorded HRTEM images, even for very thin specimens. Each high resolution image can therefore be considered as a highly blurred representation of the atomic structure of the specimen. The interference between the different diffracted beams can be varied by altering an electron-optical parameter which is reflected in a change of the applied microscope transfer function. The most common method is defocusing, often called 'through-focusing'. We will take advantage of the knowledge of the variation of the lens aberration function with the defocus settings so as to find the optimum object wave function which is compatible with the recorded images.

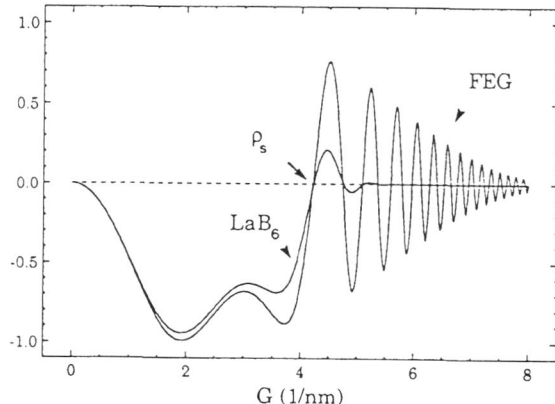

Figure 1: Contrast transfer function for a typical LaB_6 and a FEG TEM at 200 keV. Both exhibit the same point resolution (i.e. the first zero-crossing), but the FEG microscope allows transfer up to higher frequencies.

3. LEAST SQUARES WAVE FUNCTION RECONSTRUCTION

The principle of reconstructing the specimen exit face wave function $\Psi(g)$ out of a focal series of HRTEM images $I_{exp,n}(g)$ (n=1, ..., N) can be formulated within a maximum likelihood framework [3]. This involves an ordinary least squares functional under the assumption of gaussian noise characteristics:

$$S = \sum_{n=1}^{N} \int d\mathbf{g} |M_n(\mathbf{g})|^2 \qquad (1)$$

with the image difference $M_n(g)$ defined as:

$$M_n(\mathbf{g}) = I_{exp,n}(\mathbf{g}) - I_n(\Psi(\mathbf{g})) - N_n(\mathbf{g})\delta(\mathbf{g}) \qquad (2)$$

where $N_n(g) \delta(g)$ represents a background constant ('fog' level) which might be different for every recorded image. Minimising S with respect to $\Psi(g)$ and $\Psi^*(-g)$ leads to the following equation pair representing weighted correlations between the object wave function $\Psi(g)$ and the difference image $M_n(g)$:

$$\begin{cases} 0 = \sum_{n=1}^{N} \int d\mathbf{p}\, \Psi(\mathbf{p}) \quad T_n(\mathbf{p},\mathbf{g}) \quad M_n(\mathbf{g}-\mathbf{p}) \\ 0 = \sum_{n=1}^{N} \int d\mathbf{p}\, \Psi^*(-\mathbf{p}) \quad T_n(\mathbf{g},-\mathbf{p}) M_n(\mathbf{g}-\mathbf{p}) \end{cases} \qquad (3)$$

In this formulation, the function T_n is the so-called Transmission Cross Coefficient [6] which is used to estimate the image intensities $I_n(\Psi(g))$ at defocus z_n. Equations (3) are highly non-linear in the wave function $\Psi(g)$, therefore a solution in a closed form cannot be found.

This set of least squares requirements forms the basis for both the parabola and the MAximum Likelihood (MAL) methods. In the MAL method, this requirement is used in the form of a Picard operator: It starts from a proposal for the wave function Ψ, and checks whether this wave function fulfils the requirements (3) for all reciprocal g vectors. If not, the difference function is partially fed back in order to get an updated proposal for Ψ. This procedure is repeated until convergence is reached. In the parabola method on the other hand, the requirements (3) are approximated, and only the terms p=o are retained. We will further denote $\Psi(g) = \alpha\delta(g) + \Phi(g)$ with $\Phi(o)=0$. Physically, this is just a separation of the central beam α from the diffracted beams $\Phi(g)$. Neglecting the non-linear contribution to the image formation, i.e. the interference between the diffracted beams $\Phi(g)$ themselves, in a first instance, this procedure directly leads to the filter functions for $g \neq o$:

$$f_n(\mathbf{g}) = \frac{T_n(\mathbf{o},\mathbf{g}) - \left[P^*(\mathbf{g})/Q(-\mathbf{g})\right]T_n(-\mathbf{g},\mathbf{o})}{\left[Q(\mathbf{g})Q(-\mathbf{g}) - |P(\mathbf{g})|^2\right]/Q(-\mathbf{g}) + \eta(\mathbf{g})} \qquad (4)$$

where $\eta(g)$ is the Noise to Signal ratio and the P and Q functions are given by:

$$P(\mathbf{g}) = \sum_{n=1}^{N} T_n(\mathbf{g},\mathbf{o}) T_n(-\mathbf{g},\mathbf{o}) \qquad (5)$$

$$Q(\mathbf{g}) = \sum_{n=1}^{N} |T_n(\mathbf{g},\mathbf{o})|^2 \qquad (6)$$

Note that these filter functions are only valid for the determination of the interference between the central and the diffracted beams $\alpha^*\Phi(g)$. The central beam α can be calculated from the normalisation requirements. Different methods have to be applied for zero and non-zero average noise contributions [7].

It is clear that a minimum of two images (which are linearly independent in the mathematical sense) are required to reconstruct the exit wave function since each recorded image only shows the modulus squared of the image wave functions, yet omitting the phase information. Obviously, if we want to reconstruct the amplitude and phase of the object wavefunction, the number of data points should at least be equal to the number of unknowns. The use of more images will certainly improve the validity of the solution since the restrictions on the object wave function become more stringent.

3.1 Physical meaning of the Filter Functions

Performing an extra Fourier transform with respect to the reciprocal defocus axis, the filter function (4) can be shown to be a parabolic filter '$\zeta = .5\lambda g^2$' in the 3D reciprocal space (g, ζ) since in the limit of an infinite defocus range and zero beam divergence, $P(g)$ will vanish and $Q(g)$ will reduce to the square of the damping envelope due to chromatic aberration.

The first term in the numerator of (4) basically selects information on the parabola '$\zeta = .5\lambda g^2$', on which the linear wave function $\alpha^*\Phi(g)$ is located. The application of the transfer functions $T_n(o, g)$ will introduce phase shifts which can be interpreted as a 'back propagation' of the linear information of $\alpha^*\Phi(g)$ in the images at defoci z_n towards the exit plane of the specimen. Due to this first term, the conjugate wave function $\alpha\Phi^*(-g)$ on the parabola '$\zeta = -.5\lambda g^2$' equally exhibits the same phase shift, yielding a parabola at '$\zeta = \lambda g^2$'. In a practical application however, only a limited defocus range can be used. Therefore this parabola will be convoluted with a kind of spread function along the reciprocal ζ-axis, yielding a finite chance that the conjugate wave function will contribute to the information at $\zeta = 0$. This is corrected for by the second term.

The denominator of (4) basically represents how well the frequencies can be reconstructed for a specific defocus series. For zero beam convergence and in the case of an equidistant defocus series with step δz, the denominator can thus be rewritten as:

$$N e^{-(\pi\lambda\Delta g^2)^2} \left[1 - \left\{ \frac{1}{N} \frac{\sin(N \pi\lambda g^2 \delta z)}{\sin(\pi\lambda g^2 \delta z)} \right\}^2 \right] + N\eta(g) \tag{7}$$

This clearly shows that the low frequencies are difficult to retrieve. The lowest frequency which can easily be reconstructed is of the order of

$$g_{low} = 1/\sqrt{\lambda \Delta_z} \tag{8}$$

with Δ_z the total defocus range. In addition, if the defocus series contains a certain periodicity in the defocus step, the denominator rapidly goes to zero for frequencies which are resonant for this period. This will happen for the first time for the g_{max} value which satisfies the condition

$$g_{max} = 1/\sqrt{\lambda \delta_m} \qquad (9)$$

with δ_m the largest periodicity along the defocus series. Thus, in case one wants to reconstruct down to the information limit, the largest periodicity δ_m should be smaller than $.5\pi\Delta$.

The combination of both requirements puts some restrictions on the possible defocus values used in the reconstruction experiment. Tests on both simulations and experimental images have proven that a series of 20 images is more than sufficient for the proposed reconstruction procedure, provided an iterative procedure is applied to correct for the inaccuracies of the first linearised guess, using equation (4).

3.2 Correction for the Non-Linear Contribution

In case of (quasi-)linear imaging, i.e. in the Weak Phase Object approximation, the second order interference effects $I_n(\Phi(g))$, can be omitted as the non-linear terms $|\Phi(R)|^2$ (interference between scattered beams) can be neglected compared to $|\alpha^*\Phi(R)|$ (interference between central beam and one of the scattered beams). In the absence of noise and non-linear terms, equation (4) is equivalent to the result obtained by [2] for image reconstruction of a complex object in the regime of linear image formation.

In the general case of non-linear imaging, an error is made omitting $I_n(\Phi(g))$. This term is definitely correlated with the linear contribution and should therefore not be included in the noise. The linear parabola reconstruction formula, however, may be the ideal starting point for a further iteration procedure in order to account for non-linear effects. For the first iteration, the non-linear terms $I_n(\Phi(g))$ are neglected, yielding:

$$\alpha^*\Phi^o(g) = \alpha^*\Phi(g) + \sum_n F_n(g) I_n(\Phi(g)) \qquad (10)$$

where $\alpha^*\Phi(g)$ is the required final solution and the last term represents the erroneous contribution of the non-linear terms to the first guess. We will now try to find an optimised update for this approximate solution and therefore we write

$$\Psi_{j+1} = \Psi_j + \varphi \qquad (11)$$

where φ can be considered to be small in comparison with Φ_j. Again using the ordinary least squares technique we get up to first order in the deviation, which can, in a first order approximation be reduced to a 2x2 set of linear inhomogeneous equations, which can be solved as:

$$\varphi(\mathbf{g}) = \frac{c(\mathbf{g}) a^*(-\mathbf{g}) - c^*(-\mathbf{g}) b(\mathbf{g})}{a(\mathbf{g}) a^*(-\mathbf{g}) - b(\mathbf{g}) b^*(-\mathbf{g})} \quad (12)$$

where the coefficients are defined as:

$$a(\mathbf{g}) = \sum_n \int d\mathbf{p} |\Psi_j(\mathbf{p}) T_n(\mathbf{p},\mathbf{g})|^2$$
$$b(\mathbf{g}) = \sum_n \int d\mathbf{p}\, \Psi_j(\mathbf{p}) \Psi_j^*(-\mathbf{p}) T_n(\mathbf{p},\mathbf{g}) T_n(-\mathbf{p},-\mathbf{g})$$
$$c(\mathbf{g}) = \sum_n \int d\mathbf{p}\, \Psi_j(\mathbf{p}) T_n(\mathbf{p},\mathbf{g}) M_{n,j}(\mathbf{g}-\mathbf{p}) \quad (13)$$

This general approach has been suggested in [8] as the Self Consistent MAL procedure, and can be approximated to the parabola equivalent for $\mathbf{g} \neq 0$:

$$\alpha^* \Phi_{j+1}(\mathbf{g}) = \alpha^* \Phi_j(\mathbf{g}) + \sum_n F_n(\mathbf{g}) \left[I_{exp,n}(\mathbf{g}) - I_n(\Psi_j(\mathbf{g})) \right] \quad (14)$$

with $I_n(\Psi_j(\mathbf{g}))$ the theoretically estimated intensity for the reconstructed wave function in the j-th iteration step. It is clear that the reconstruction procedure should try to minimise the second term. This implicitly means that the parabola iterator is incomplete: the filter-function $F_n(\mathbf{g})$ will only select information on a parabola, which, in the limit of zero beam convergence and an infinite defocus interval, is a two dimensional section in the three dimensional reciprocal space. The iteration procedure, as described in (14), in fact only compensates for the non-linear contribution on this 2D section. Therefore, not all information in the defocus series is optimally used, which makes the parabola method less robust to noise than the MAL method in which all data in the 3D reciprocal space is used.

But exactly the fact that only the data on the parabolas is selected, allows us to speed up the calculations enormously since both linear and non-linear terms on this parabola feel exactly the same damping envelope due to chromatic aberration. Therefore, the time consuming convolutions in the Transmission Cross Coefficient can totally be avoided, certainly since the spatial damping envelopes $E_{\alpha,n}(\mathbf{g},0)$ can be described coherently for a FEG TEM. We thus first calculate the coherent image intensity including the spatial coherence with the modified transfer functions:

$$t_n(\mathbf{g},0) = e^{-i\chi(\mathbf{g})} E_{\alpha,n}(\mathbf{g},0) \quad (15)$$

The effect of the chromatic aberration on the parabolas can then be introduced by convoluting this intensity with the chromatic spread function. However, taking the specific form of the filter functions into account, this step can even be avoided. In the MAL approach, the chromatic aberration envelope has to be calculated correctly in the complete field of the 3D reciprocal space. However, this damping envelope can equally well be introduced by weighted averaging the coherent image intensities at different focus. It was proven that averaging over only a few (typically 10) equidistantly spaced images already yields a very good approximation. In this way, the algorithm could be speeded up with several orders of magnitude [8].

3.3 General Strategy in Practical Applications

Due to specimen drift and the finite recording time, the different experimental images of the defocus series may not be well aligned with respect to one another. This is reflected in a lateral image shift δr_n across the field of view. [3] has given a coupled solution for the correction of the defocus (δf_n) and lateral misalignment (δr_n), assuming a first order approximation and yet only small deviations from the provided starting values. In our experience, this procedure works well for sub-pixel lateral alignment, but requires a few iteration steps for the correction of an initial misalignment of more than one-to-two pixels. Using a Cross-Correlation-Function (CCF), the exact misalignment parameter can be found in only one iteration step.

Therefore, we proceed in the following way:

1) The images are aligned as well as possible using the CCF. As pointed out by Frank [9], the CCF becomes less and less accurate for increasing focus differences between the images due to contrast reversals etc. In our case, the defocus distance between the different images is of the order of the defocal spread (i.e. 4 nm), so that contrast reversals are unlikely to occur. In the case of a very thin specimen, more accurate results can be obtained with the improved CCF of Saxton [10], which then even is applicable for larger defocus distances. But in most general cases, where the object cannot be considered to be thin, only the mutual misalignment between two adjacent images can be measured accurately. Therefore, one has to be aware of the fact that there might still be a substantial alignment error between the first and the last images of the defocus series.

2) The (roughly) aligned focal series is used to calculate the first estimate of the wave function $a^*\Phi(g)$ using equation (4).

3) The contribution of the background 'fog' level $N_n(g)\,\delta(g)$ is estimated in combination with the contribution of the central beam $\alpha\,\delta(g)$, yielding the wave function $\Psi(g)$.

4) $\Psi(g)$ and $N_n(g)\,\delta(g)$ are used to calculate an estimation of all the different experimental images of the series. The estimated and experimental intensities are cross-correlated in order to refine the lateral alignment. Note that in this alignment step no contrast reversals occur since the experimental image is correlated with the estimated image for the same defocus.

5) The squared differences between the experimental and the estimated images are calculated to be able to judge the convergence.

6) We recalculate the estimate of the wave function $\alpha^*\Phi(g)$, including the non-linear corrections.

We then repeat steps 3) to 6) until convergence is obtained for both the position alignment and the Least Squares criterion. Experiments show perfect alignment and convergence in about four iterations.

4. EXPERIMENTAL VERIFICATION OF THE METHOD

The results we want to present here have been obtained with the first prototype of the Philips CM30 FEG-UltraTWIN which has been developed within the framework of the Brite-Euram project nr 3322. This microscope operates at 300 kV and has a point resolution of 1.7Å (C_s = 0.62 mm) and a theoretical information limit of 1Å. A series of 20 images was recorded digitally using a 1024^2 Slow Scan CCD camera. The equidistant defocus step of -30.41Å is of the order of the defocal spread of the microscope, and the start defocus was roughly 700Å under focus. The microscope and the CCD camera were driven by a TVIPS (Tietz Video and Image Processing Systems) computer running an extension of the TCL image processing software [11]. Special TCL procedures were developed which enable the full-automatic recording of the focal series through the remote-control of the CM-microscope. The total exposure time for one micrograph was of the order of 0.8-1.0 seconds. In order to reduce the read-out time by a factor of 4 to about 1.5 seconds, the CCD camera was used in binned mode, yielding 512^2 effective pixels per image so the recording of the whole focal series of 20 images required about one minute. As each pixel was represented by a short integer (2 bytes), the whole series required about 10 Mbyte of disk space. Flat field and dark field corrections were performed according to [12]. The electron-optical parameters like spherical aberration C_s and the defocus increment have been measured by the method of [13]. The parabola computations themselves were performed on an IBM-590, requiring a total reconstruction time of a few minutes.

The CoO sample was found to be crystalline up to the edge and the top grain is viewed along the [110] direction. The meta stable phase we visualise here has the NaCl structure. Therefore, in the [110] projection, the columns will be situated on a rectangular pattern with projected distances of 1.4 and 2 Angstrom (see figure 2). Although we are looking at the edge of the sample, we cannot guarantee that the top grain is thin, but since both Co and O are relatively light elements, we expect that the exit wave function will still directly reflect the column potential.

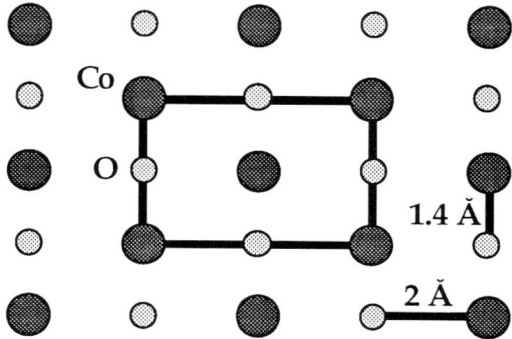

Figure 2 : Schematic representation of the metastable CoO structure along the [110] direction. The atoms are ordered in a line pattern. The inter-column distances are 1.4 and 2Å.

Figure 3 shows the reconstructed phase and amplitude. Both the phase and amplitude show a sequence of small and large spots on a line pattern. Unfortunately, the positions of these spots do not coincide. Therefore, a direct interpretation of this reconstructed wave function in terms of the crystal potential is not possible at all.

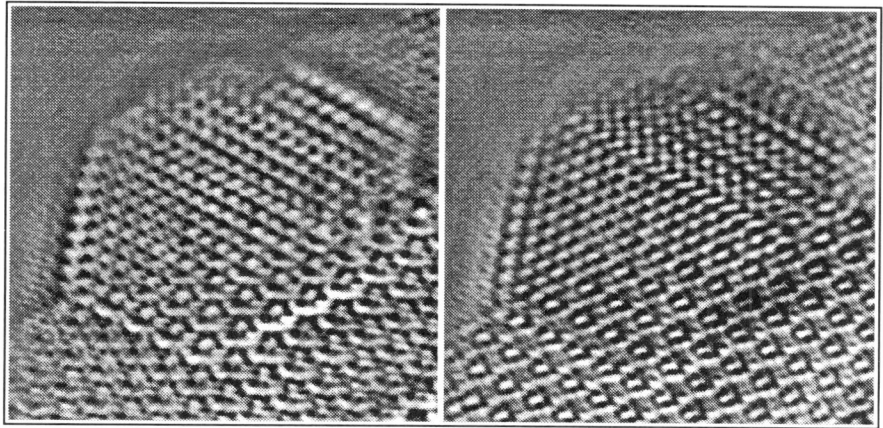

Figure 3: Reconstructed amplitude and phase of the CoO grain.

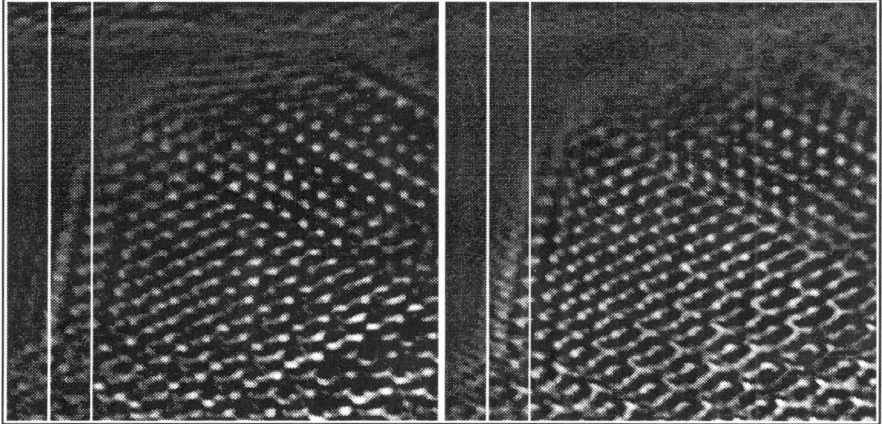

Figure 4: The effect of a wrongly chosen defocus offset on the localisation of the information in the crystalline grain. The image at the left is 20 nm out of focus, the on at the right is in focus.

In order to improve the interpretability of this reconstructed wave function, the diffraction inside the crystal has to be inverted in a sense. For a thicker object, where the weak phase approximation is not applicable, a new structure reconstruction technique has been developed, which will be presented in detail elsewhere [14]. This new method starts from the reconstructed exit wave function and optimises for a possible misjudgement of the defocus settings. This can only be achieved when sufficient knowledge about the crystal structure is present. In this case, we can exploit the fact that the transfer function blurs the exit wave function and will cause the distribution of apparently crystalline image detail into the vacuum, similar to fresnell fringes at lower resolution. From the Bloch wave diffraction theory, we know that the main diffracted intensities are situated in 1S eigenstates on the atomic columns in real space. We can therefore determine the absolute defocus very accurately by changing the applied transfer function so as to maximally localise the 1S intensity in the crystal grain (figure 4). The white dots in the right image show the position of the 1S eigenfunctions at the column sites. The difference between the Co and the O atoms can clearly be observed. We can therefore conclude that the combination of the wave function and structure reconstruction algorithms is very well capable of reconstructing the projected column map of the specimen. Here, these procedures allowed for a interpretable resolution improvement from 1.7 to better than 1.4 Angstrom

5. REFERENCES

1. P. Schiske, Proc. 4th EUREM, Rome, 1968, p.145.
2. W.O. Saxton, in Adv. in Electronics and Electron Physics, Academic Press, 1978.
3. E.J. Kirkland and B.M. Siegel, Ultramicroscopy, 16 (1984) 169.
4. D. Van Dyck and W.M.J. Coene, Optik 77 (1987) 125.
5. D. Van Dyck and M. Op de Beeck, Proc. 12the ICEM (Seattle), 1990, p.26.
6. K. Ishizuka, Ultramicroscopy, 5 (1980) 55.
7. M. Op de Beeck, D. Van Dyck and W.M.J. Coene, submitted to Ultramicroscopy.
8. W.M.J. Coene, A.J.E.M. Janssen, T.J.J. Denteneer, M. Op de Beeck and D. Van Dyck, Microscopy Society of America Bulletin, 24 (1994) 472.
9. J. Frank, in Computer Processing of Electron Microscope Images, 1980, p.187.
10. W.O. Saxton, J. Microscopy, in press (1994).
11. TCL, Technical Command Language, TNO-TUD, TPD, The Netherlands.
12. I. Daberkow, K.-H. Herrmann, L. Liu and W.D. Rau, Ultramicroscopy, 38 (1991) 215.
13. W.M.J. Coene and T.J.J. Denteneer, Ultramicroscopy 38 (1991) 225.
14. M. Op de Beeck and D. Van Dyck, submitted to Ultramicroscopy.

6. ACKNOWLEDGEMENTS.

This work has been performed in the framework of the European Brite-Euram project 3322. One of the authors (Marc Op de Beeck) is indebted to the Belgian I.W.O.N.L. and the Belgian National Fund for Scientific Research N.F.W.O. for financial support.

High-resolution tilted single-sideband holography

Kazuo Ishizuka

Tonomura Electron Wavefront Project, JRDC
c/o Toyo University, Kawagoe, Saitama, 350 Japan

Abstract
This theoretical investigation of tilted single-sideband (TSS) holography, where images are taken in the single-sideband mode by tilting an incident beam relative to the optic axis, shows that TSS does not require a highly coherent electron beam, and that the aberration is small with this method. It also shows that the allowable specimen thickness determined by Fresnel diffraction is larger than that for axial illumination imaging. Although this method can image only weakly scattering objects, this restriction may not be serious for ultrahigh-resolution microscopy for which the specimen should be very thin. Preliminary experiments using (011) silicon demonstrate the possibility that TSS holography is a very promising technique for improving resolution beyond 0.1 nm.

1. INTRODUCTION

Because the spherical aberration of an electron lens used for an electron microscope in contrast to that of an optical lens, cannot itself be corrected, Gabor [1] invented a technique (called holography) so that all the information of the object wave could be recorded on an micrograph and the spherical aberration could be corrected by processing the micrograph. This technique has not been widely used, however, because it requires a highly coherent electron beam. During the last few decades the resolution of an electron microscope has mainly been improved by decreasing the spherical aberration of the objective lens. But because this approach has nearly reached its limit, we will need a new approach to reach a resolution below 0.1 nm.

There are several ways to attain this resolution. One is by using high-voltage CTEM. Since there is a scaling relationship $C_s \lambda \approx const$ between the spherical aberration coefficient C_s and the wavelength λ, the resolution δ will increase with the square root of the wavelength: $\delta \propto \sqrt[4]{C_s \lambda^3} \propto \sqrt{\lambda}$. Another way is by using high-angle annular dark-field STEM, which is an incoherent imaging by using electrons inelastically scattered at large angles [2]. Since a resolution for incoherent imaging is higher than that for coherent imaging, a higher resolution may be achieved even when using the same objective lens. The resolution can also be improved through image processing by correcting the aberration as proposed by Gabor. Since off-axis image holography requires only one image to reconstruct the object wave, it has great potential in this field [3]. The object wave may also be reconstructed from a through-focus-series[4,5]. However, increasing resolution by correcting the aberration requires an accurate estimate of the wave aberration. This is discussed later in this paper.

Single-sideband holography can also restore an object wave. Although this method has been severely restricted by charge-up of the aperture due to a poor vacuum [6], satisfactory results have recently been obtained [7]. In this method the sideband mode is obtained with axial illumination, and thus an aberration problem similar to that in off-axis holography should be solved. One readily imagines that the resolution can be improved by an aperture synthesis using bright-field images taken at tilted illumination on a standard CTEM.

This report will show that a technique called tilted single-sideband (TSS) holography, combining the single-sideband method and the tilted illumination technique, is very promising for improving resolution.

2. DIFFICULTY WITH AXIAL-ILLUMINATION IMAGING

To correct aberration by image processing, we need to be able to estimate a wave aberration function with sufficient accuracy. Since this function may be approximated by using a few parameters (aberration coefficients), it is the accurate estimations of these parameters that is essential. When the spherical aberration and defocusing are included, the aberration function can be written as

$$\chi(K) = \frac{1}{4}K^4 - \frac{1}{2}ZK^2,$$

using the reduced coordinates $K = k\sqrt[4]{C_s\lambda^3}$ (in Scherzer) and $Z = z/\sqrt{C_s\lambda}$ (in Glaser) [8]. Here, the aberration is expressed in terms of numbers of waves.

The spherical aberration is balanced out by defocusing under the Scherzer condition at the Scherzer resolution limit, $K_{Sch} = \sqrt[4]{6}$, as shown in Fig. 1, but the spherical aberration itself is 1.5. When the Rayleigh's quarter-wavelength rule [9] is used as an upper limit of aberration, the spherical aberration coefficient should be estimated to within one part in six. Since the coefficient is easily estimated within this accuracy, and does not vary so much with a small excitation change of the objective lens during an experiment, the spherical aberration can be easily corrected within the Scherzer limit by using a predetermined coefficient.

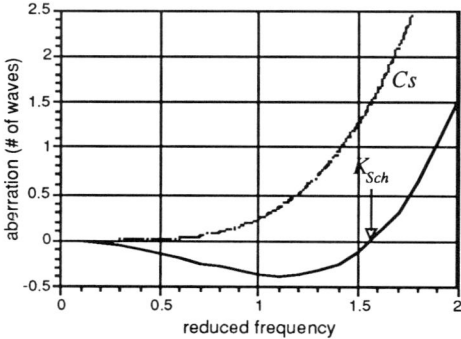

Figure 1. Aberration function under the Scherzer condition for axial illumination. Spherical aberration is balanced out by defocusing at the Scherzer resolution limit as shown by the solid line. However, the spherical aberration itself (dotted line) is 1.5 waves.

At twice the Scherzer limit, however, this spherical aberration increases to 24, because it grows as the fourth power of resolution. The spherical aberration coefficient should therefore be estimated to within one part in 96, that is about 1% (Table 1). The accuracy required for the defocus estimation for each anticipated resolution is also listed in Table 1. Note that the denominator corresponds to one Glaser, which is usually a few tens of nanometers. Meeting the requirements for twice the Scherzer resolution may therefore be difficult in practice.

Table 1
Required accuracy of aberration coefficient estimation for axial illumination.

Resolution	Spherical aberration		Defocusing aberration	
R	Maximum	Accuracy	Maximum	Accuracy
K_{Sch}	3/2	1/6	$\sqrt{3/2}Z$	$1/2\sqrt{6}$
$\sqrt{2}K_{Sch}$	6	1/24	$\sqrt{6}Z$	$1/4\sqrt{6}$
$2K_{Sch}$	24	1/96	$2\sqrt{6}Z$	$1/8\sqrt{6}$

where $K_{Sch} = \sqrt[4]{6}$ and $Z_{Sch} = \sqrt{3/2}$.

The conditions in the middle row of this table, corresponding to a resolution of $\sqrt{2}$ times the Scherzer limit, may be realized using axial illumination. It should be noted that these required accuracies are applied for through-focus-series reconstruction and off-axis image holography, because these techniques use images obtained with axial illumination.

3. TILTED SINGLE-SIDEBAND HOLOGRAPHY

3.1. Single-sideband method

In normal electron microscopy both scattered waves A and B, which are symmetrically located relative to the transmitted wave, contribute to the image formation as shown in Fig. 2a. In the single-sideband mode, on the other hand, only A or B passes through a semicircular aperture and each contributes to a different image.

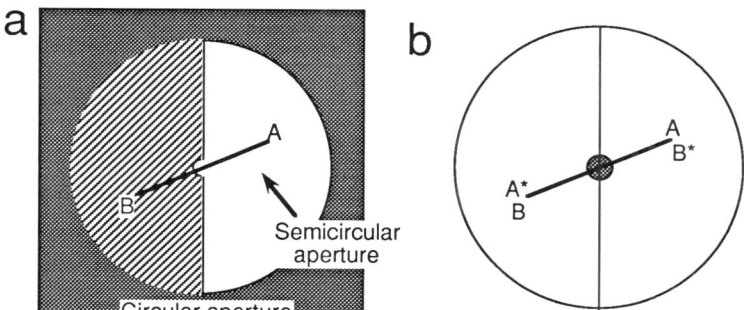

Figure 2. (a) Circular aperture for normal electron microscopy and semicircular aperture for single-sideband holography. (b) Fourier transform of an image intensity.

Here, the scattering is assumed to be weak, and image amplitude is written as $1 + \phi(r)$. Then, the image intensity is approximated by neglecting a second-order term in ϕ:

$$i(r) = |1 + \phi(r)|^2 \approx 1 + \phi(r) + \phi^*(r).$$

It is evident that ϕ cannot be uniquely determined in the image space because of the presence of its complex conjugate ϕ^*. Since a Fourier transform of a complex conjugate function is a complex conjugate of an inverted Fourier transform [10], the image intensity gives the following Fourier transform:

$I(h) \approx \delta(h) + \Phi(h) + \Phi^*(-h).$

If we include B for image formation, then its complex conjugate B* appears at the same position of A (Fig. 2b). Thus, Fourier component A cannot be uniquely determined even in the Fourier space.

But because there is no B* in the single-sideband mode, the Fourier component A can be uniquely determined. The Fourier component B is determined from another image obtained using a conjugate aperture, and the whole information for Φ is determined by combining information obtained from these two images. An inverse Fourier transform of Φ gives a complex object wave φ; that is, the amplitude and phase information of the object wave. This method is therefore a type of holography.

It is noteworthy that if the specimen is a weak-phase object (WPO) or an amplitude object, then the object function is a pure imaginary or real function, and thus its Fourier transform is symmetric or antisymmetric. In this case, information over one half of a Fourier plane needs to be collected. If the specimen is represented by a complex function, then information over a whole Fourier plane should be collected. Thus, a general phase object, even one for which the amplitude change is small, can be treated in single-sideband holography. This sideband mode, however, suffers from the same aberration problem discussed in Section 2, because it uses axial illumination.

3.2. Tilted illumination imaging

When an incident beam is tilted relative to the optic axis, as shown in Fig. 3, the scattering distribution on the back focal plane of the objective lens is shifted according to the tilt. Each TSS image is formed using different scattered beams selected by a fixed aperture, and for each of which the wave aberration is the same. If the incident beam is tilted close to the aperture edge, the imaging mode becomes single-sideband. The aperture is located at the optic axis, since the aberration is optimized for a symmetric aperture around the optic axis.

A Fourier transform of one TSS image is shown schematically in Fig. 4. A light gray region corresponds to a single-sideband area. An opposite sideband region is simply the complex conjugate to this sideband and does not give any new information. A double-sideband region corresponds to the areas where these two sidebands overlap. As in normal electron microscopy, information over the double-sideband cannot be uniquely determined. Note that since the origin of the aberration function is close to the center of the sideband, the wave aberration over the sideband is small. The accuracy required for aberration coefficient

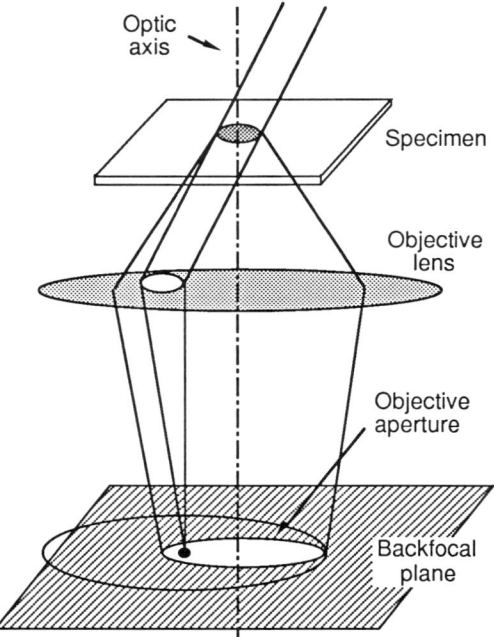

Figure 3. Tilted single-sideband holography. All images are recorded with tilted illumination using the aperture centered at the optic axis. If the incident beam is tilted close to the aperture edge, the imaging mode becomes single-sideband.

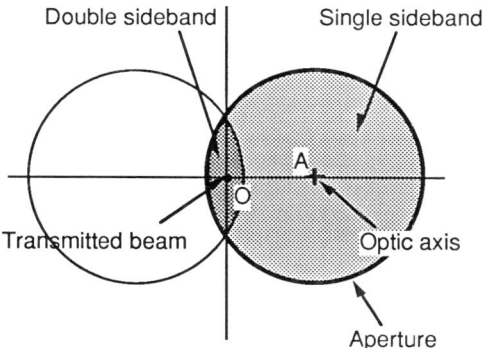

Figure 4. Fourier transform of a TSS image. It consists with two single-sidebands and one double-sideband.

estimation is therefore low as explained in the following subsection, and the aberration will be corrected easily.

3.3. Required accuracy of aberration coefficient estimation

This technique requires that the incident beam should be tilted accurately in order to minimize an aberration correction error. The required tilting accuracy is minimized, when the incident beam direction K_t is tilted to the minimum of an under-focused aberration function. Therefore, the tilting angle should satisfy $K_t = \sqrt{Z}$ for the aberration function including the spherical aberration and defocusing. The aperture edge should be close to K_t. The attainable resolution is therefore $2K_t$ for the aperture of radius K_t placed symmetrically around the optic axis. Thus, a higher resolution is attained by using a larger under-focus and a larger tilt angle.

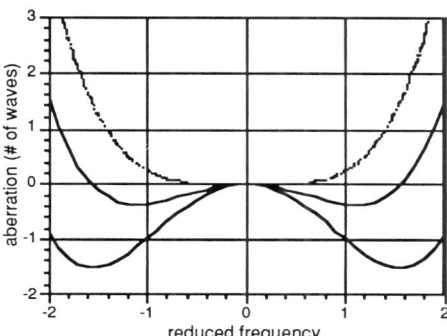

Figure 5. Aberration optimized for tilting illumination. The incident beam is tilted to the point of the aberration function minimum. Higher resolution will be obtained by using a larger defocus. The spherical aberration is shown by a dotted line.

For example, let us consider the case corresponding to a resolution twice that of the Scherzer limit (i.e., the middle row of Table 2) realized by using an under-focus of twice the Scherzer defocus. Since the aperture size for this case is the same as that for the Scherzer limit using an axial illumination, the accuracy required for the aberration coefficient estimation is the same.

Since this accuracy is easily attained (as discussed above), a resolution twice that of the Scherzer limit will be readily obtained. The accuracy required for the tilt angle ΔK_t is 0.169, which is evaluated from the maximum derivative of the wave aberration appearing at $\sqrt{Z/3}$. This accuracy is also easily attainable in practice.

Table 2
Required accuracy of aberration coefficient estimation for tilted illumination.

Resolution R	Defocus Z	Spherical aberration		Defocusing aberration		Tilt angle
		Maximum	Accuracy	Maximum	Accuracy	Accuracy
$\sqrt{2}K_{Sch}$	Z_{Sch}	3/8	3/2	$\frac{1}{2}\sqrt{3/2}Z$	$1/\sqrt{6}$	$1/8(1/6)^{\frac{3}{4}}$
$2K_{Sch}$	$2Z_{Sch}$	3/2	1/6	$\sqrt{3/2}Z$	$1/2\sqrt{6}$	$1/8(2/3)^{\frac{3}{4}}$
$2\sqrt{2}K_{Sch}$	$4Z_{Sch}$	6	1/24	$\sqrt{6}Z$	$1/4\sqrt{6}$	$1/8(8/3)^{\frac{3}{4}}$

where $K_{Sch} = \sqrt[4]{6}$ and $Z_{Sch} = \sqrt{3/2}$.

It may be noted that the entries in the bottom row of Table 2 correspond to the middle-row entries in Table 1 for the axial illumination. Since this condition may be attained practically, resolution can be improved to nearly triple the Scherzer limit by using the tilted illumination. TSS holography is therefore a very promising technique for ultrahigh resolution.

3.4. Required coherency

The spatial coherency required for TSS holography is considerably lower than that required for off-axis holography. This is because the former is amplitude division holography, while the latter is wavefront division holography. The coherency requirement is quantitatively determined by envelope functions that describe contrast transfer damping resulted from partial coherency. The envelope functions for tilted illumination imaging have been investigated by McFarlane and Cochran [11] and Wade and Jenkins [12] to explain selective information transfer observed on images taken from amorphous thin films. The same envelope functions can be used for TSS holography, since the weak scattering is assumed similarly. These envelope functions are special cases of the transmission cross coefficient [13].

The temporal-coherency envelope function has an achromatic circle centered on the optic axis where the function takes a minimum. The spatial-coherency envelope function over each single-sideband becomes identical to that for axial illumination, when the incident beam direction is tilted to the minimum of an under-focused aberration function. This is because the gradient of the aberration function applied for the incident beam becomes zero as in the case for axial illumination. Then the envelope function gives contrast transfer of one at the optic axis and on the achromatic circle of the temporal-coherency envelope function. The total envelope function with this tilt is thus symmetric around the optic axis. The spatial and temporal coherency required for TSS holography corresponding to the resolution twice that of the Scherzer limit (the middle row of Table 2) are therefore as low as those for axial illumination microscopy under the Scherzer condition.

3.5. Aperture synthesis

In the tilted single-sideband mode, the resolution is improved by an aperture synthesis in Fourier space. Figure 6a shows information obtained from four images taken with the same tilt angle and four azimuths separated by 90 degrees. The light gray region indicates information obtained within the resolution defined in Table 2. Information within the circle bounded by the thick solid line is isotropically determined from the four images. The dark region is a double sideband area. Although information within the double sideband can not be uniquely determined, it corresponds to low frequencies and is often not important.

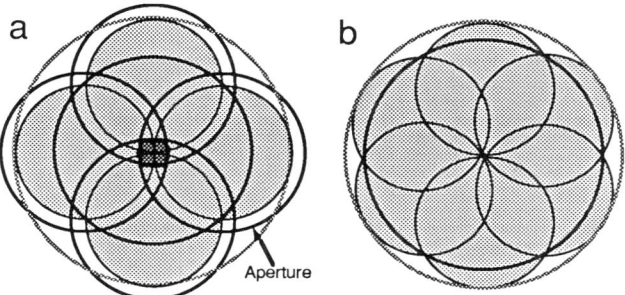

Figure 6. Aperture syntheses in Fourier space using (a) four and (b) six images. Information over the gray region is collected, but the circle shown by the thick solid line indicates the information limit obtained isotropically.

There are two possibilities for collecting information isotropically up to the desired resolution R. If a larger aperture with a radius of $\sqrt{2}R$ is used by increasing the defocus, information up to a resolution R can be collected using only four TSS images. Since the condition for $R=2\sqrt{2}K_{sch}$ for the tilted illumination may be attained in practice, information up to $2K_{sch}$ will actually be collected when this scheme is used. The other possibility is to use many images taken with equally spaced azimuths. As shown in Fig. 6b, information can be collected almost isotropically up to the desired resolution by using six images. Although the latter scheme needs more images, it has an advantage when the images are aligned by using a cross-correlation analysis between adjacent images. This is because its alignment accuracy is higher, since the adjacent images have a larger common area in Fourier space than they would have with the former scheme.

An aperture synthesis with tilted illumination has a further advantage. Fresnel diffraction causes the scattering from upper atoms to spread at the specimen bottom surface, thus degrading the resolution for thick specimens [11]. This can be accounted for by the scattering diagram for an axial illumination (Fig. 7a). An Ewald sphere, on which energy conservation holds during diffraction, has a finite curvature for a wave with a finite wavelength, and a shape transform of a finite-thickness specimen has a limited extent perpendicular to specimen surface. Therefore, the Ewald sphere will not cut higher-order reflections, and thus these reflections will

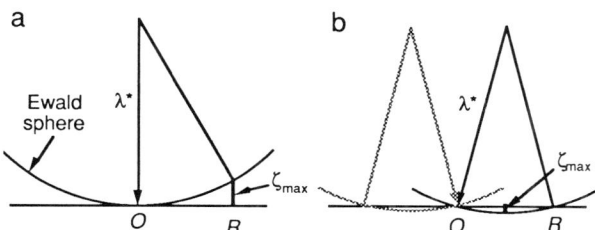

Figure 7. Scattering diagrams for (a) an axial illumination and (b) a tilted illumination. The maximum excitation error for the tilted illumination is one fourth of that for the axial illumination.

not be excited. An excitation error (a distance from the reflection to the Ewald sphere) becomes maximum for the outermost reflection R and is given by $\zeta = (1/2)\lambda R^2$. Since the shape transform extends approximately to $1/t$, where t is a specimen thickness, for axial illumination the specimen thickness should satisfy the following relation:

$$t \ll 1/\lambda R^2.$$

for an expected resolution R. A similar relation has been derived by Cowley [14].

Figure 7b shows a scattering diagram for a tilted illumination corresponding to the attainable resolution is R. The maximum excitation error, which appears at a distance $R/2$ from the origin, is 1/4 of that for the axial illumination. Therefore, the specimen thickness allowable for this tilted illumination is four times that for the axial illumination. Note that the excitation errors for lower-order diffractions are higher than those for the axial illumination and those for higher-order diffractions are lower. Therefore, the lower-order diffractions become weaker than those for the axial illumination and the higher-order diffractions become stronger, resulting in an image for which high-resolution information is accentuated.

4. PRELIMINARY RESULTS

4.1. Data acquisition

Preliminary TSS holography experiments were performed using a Topcon EM-002B electron microscope operated at 200 kV. The spherical aberration coefficient was 0.4 mm, and thus the Scherzer limit was 5.6 nm^{-1}. The tilting angle and the defocus value were 7.7 mrad and 29 nm. Figure 8 shows a corresponding aberration function. The size of the aperture used in this experiment was 3.4 nm^{-1} (8.5 mrad).

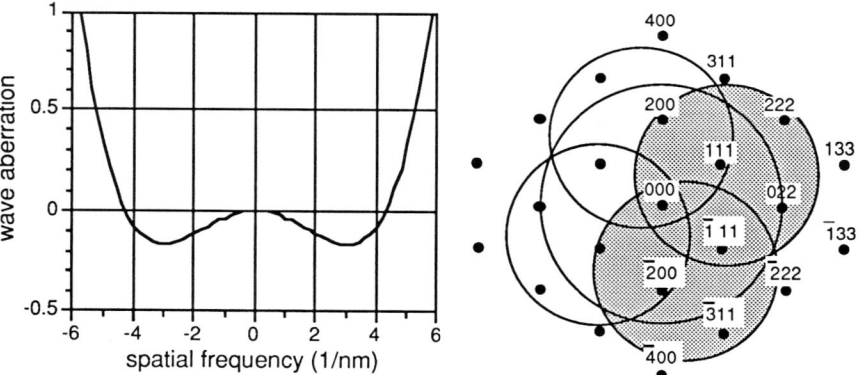

Figure 8. The aberration function corresponding to the experiment using a Topcon EM-002B: $C_s = 0.4$ mm and $z = 29$ nm. The radius of the aperture used in this experiment was 3.39 nm^{-1} (8.5 mrad), and the tilting angle was 7.7 mrad.

Figure 9. The aperture positions relative to the reflections of (011) silicon. The large circle corresponds to the Scherzer limit (5.55 nm^{-1}).

The TSS images of thin (011)-oriented silicon crystals were recorded using a Gatan slow-scan CCD camera (1024^2 pixels) [15] at a magnification 480 kx on the detector. Aperture positions, relative to the reflections of (110) silicon, are shown in Fig. 9. Beam tilt was controlled through DigitalMicrograph [16]. The center circle shows the Scherzer limit of this microscope. Since the used aperture size was not optimized in this preliminary experiment, the maximum resolution improvement relative to the Scherzer limit was only 35%. However, some reflections from {022}, {311}, and {222} may be able to contribute to the image formation.

4.2. Object wave reconstruction

Figure 10 reproduces the same parts (256 x 256 pixels) extracted from the two TSS images tilted to the +x and +y directions and the Fourier transforms of them. The relative aperture positions for (a) and (b) respectively correspond to upper right and bottom right apertures shown in Fig. 9. Due to a small specimen misalignment, reflection 022 in (a) contributes with sufficient strength, but reflection $\bar{3}11$ in (b) contributes only weakly. It should be noted that these spatial frequencies are actually contributing the image formation, since they cannot result

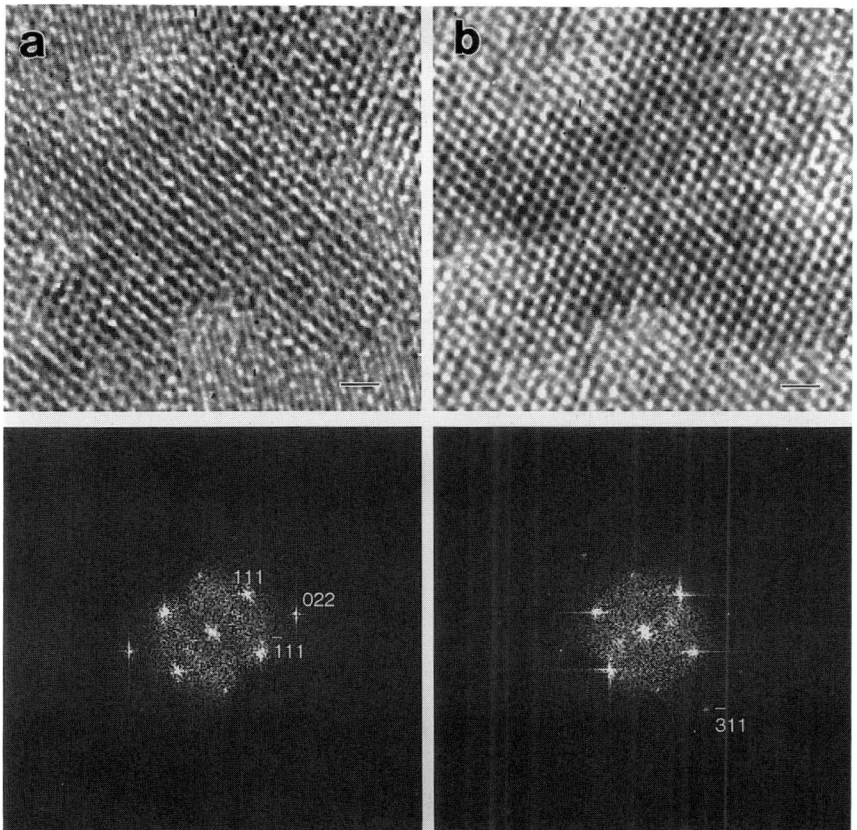

Figure 10. TSS images (top) and their Fourier transforms (bottom) of (011) silicon. (a) and (b) correspond to the TSS images tilted to +x and +y directions, respectively. The same parts (256 x 256 pixels) extracted for subsequent processing are reproduced. The scale bar indicates 1 nm. Note that reflections 022 in (a) and $\bar{3}11$ in (b) are actually contributing to the image formation.

from interference between any combination of the reflections contributing to the image formation [13]. The faint $0\bar{2}2$ reflection appearing in (b), however, must result from interference between the strong two reflections (e.g. $\bar{1}\bar{1}\bar{1}$ and $\bar{1}11$), because it was not included within the aperture.

The object wave was reconstructed, under WPO approximation, using the two TSS images shown in Fig. 10. Each TSS images was first Fourier transformed, and the wave aberration was corrected over the sideband by assuming the given optical parameters. Then, an intermediate complex wave was calculated for each TSS image using an inverse Fourier transform of the aberration corrected sideband, and the imaginary part was extracted from the complex wave. The final reconstructed image was synthesized in Fourier space from the two obtained imaginary parts by taking into account the multiplicity of the reflections.

The axial and reconstructed images are illustrated in Fig. 11. Each black spot in these images corresponds to a pair of silicon atom columns. The black spots in the reconstructed image are thinner than those in the axial image, which results mainly from the strong contribution of reflection 022.

The resolution improvement relative to the Scherzer limit was only 35%, because the aperture size used was not optimized in this preliminary experiment. If we use a proper aperture and align the specimen orientation, however, the silicon atom pairs could be resolved by using an electron microscope in the 200-kV class.

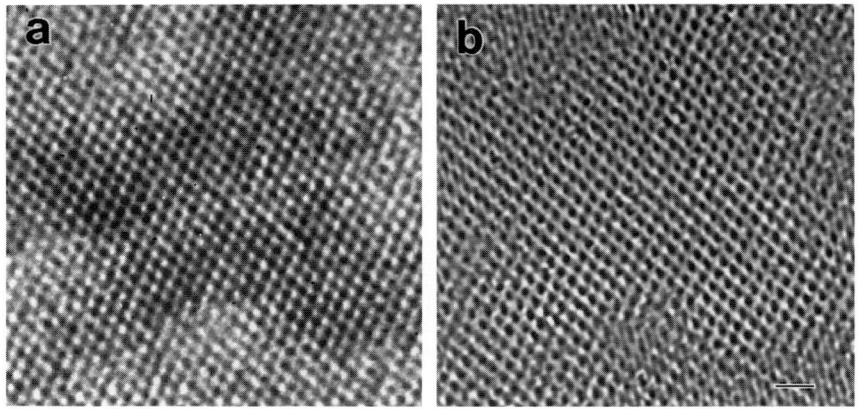

Figure 11. The axial image (a) and the reconstructed image (b). The object wave was reconstructed from the two TSS images shown in Fig. 10. Each black spot in (b), corresponding to a pair of silicon atom columns, are thinner than those in (a).

5. CONCLUSIONS

This theoretical investigation of TSS holography has shown that this method does not require the highly coherent electron beam needed for off-axis image holography. Moreover, aberration correction is easy, because there is little aberration with this method. The allowable specimen thickness determined by Fresnel diffraction is larger than that for an axial illumination imaging. Although TSS holography can be used only to image a weakly scattering object, this may not be a serious restriction in ultrahigh-resolution work for which the specimen should be very thin. TSS holography is therefore extremely suitable for ultrahigh-resolution microscopy, and offers a practical way to improve resolution beyond 0.1 nm. Preliminary experiments have demonstrated this potential.

ACKNOWLEDGMENTS

The author would like to thank to Mr. T. Ohno for assistance with computation, and several members of Topcon for their help in the experiments using the EM-002B.

REFERENCES

[1] D. Gabor, Proc. Royal Soc. London, Ser. A197 (1949) 454.
[2] S.J. Pennycook and D.E. Jesson, Ultramicroscopy 37 (1991) 14.
[3] H. Lichte, in Advances in Optical and Electron Microscopy, Vol. 12, Academic Press, London (1991) 25.
[4] D. van Dyck and M. Op de Beeck, Proc. 12th Int'l Congr. Electron Microsc., Seattle, 1 (1990) 26.
[5] T. Ikuta, J. Electron Microsc. 38 (1989) 415.
[6] K.H. Downing and B.M. Siegel, Optik 38 (1973) 21.
[7] M. Hohenstein, Ultramicroscopy 35 (1991) 119.
[8] P.W. Hawkes, Ultramicroscopy 5 (1980) 67.
[9] M. Born and E. Wolf, Principles of Optics, Pergamon Press, Oxford (1970) p.468.
[10] e.g. E. O. Brigham, The Fast Fourier Transform, Prentice-Hall, London (1974).
[11] S.C. McFarlane and W. Cochran, J. Phys. C8 (1975) 1311.
[12] R.H. Wade and W.K. Jenkins, Optik 50 (1978) 1.
[13] K.Ishizuka, Ultramicroscopy 5 (1980) 55.
[14] J.M. Cowley, Diffraction Physics, 2nd ed., North-Holland, Amsterdam (1981).
[15] O.L. Krivanek and P.E. Mooney, Ultramicroscopy 49 (1993) 95.
[16] Gatan R&D, 6678 Owens Dr., Pleasanton, CA 94588, USA.

STEM Holography of Magnetic Materials

Marian Mankos[a], P. de Haan[b], V. Kambersky[c], G. Matteucci[d], M. R. McCartney[e], Z. Yang[a], M. R. Scheinfein[a] and J. M. Cowley[a]

[a]Department of Physics and Astronomy, Arizona State University, Box 871504, Tempe, AZ 85287-1504, U.S.A.

[b]MESA Research Institute, University of Twente, P.O. Box 217, 7500 AE Enschede, The Netherlands

[c]Institute of Physics, Academy of Science, Cukrovarnicka 10, Praha 6, Czech Republic

[d]Department of Physics, University of Bologna, via Irnerio 46, 40126 Bologna, Italy

[e]Center for Solid State Science, Arizona State University, Tempe, AZ 85287-1704, U.S.A.

1. ABSTRACT

We have been using electron holography in a HB 5 scanning transmission electron microscope (STEM) to perform quantitative investigations of magnetic microstructure in thin magnetic specimens. Holograms are acquired with the scanning switched off. The objective lens is operated at large defocus, such that a relatively large area of the specimen is illuminated. Our system allows for flexible operation of off-axis holography modes, Fresnel and Differential Phase Contrast (DPC) modes of Lorentz microscopy. This combination of micromagnetic analysis techniques in one instrument provides a valuable tool for the investigation of magnetic microstructure at nm spatial resolution. STEM holography accompanied by conventional Lorentz microscopy techniques has been used to characterize thin magnetic films, magnetic multilayer structures and small magnetic particles.

2. INTRODUCTION

Recent developments in magnetic information storage technology including magnetic recording media, magnetic recording heads and magnetic sensors can be characterized by the trend to control the geometry, chemical composition and magnetic microstructure at the near-atomic level. New sample preparation techniques, such as Molecular Beam Epitaxy (MBE) and Electron Beam Lithography (EBL), permit the fabrication of magnetic devices with properties such as coercivity, micromagnetic features, local magnetic moment and anisotropy controlled at the nanometer level [1]. For example, multilayered structures, composed of alternating magnetic and non-magnetic thin films of monolayer thickness, exhibit long range oscillatory coupling [2] and giant magnetoresistance related to antiferromagnetic coupling [3], which depend critically on composition and structure. Similarly, magnetic properties of epitaxial films of magnetic elements grown on crystalline metallic, semiconducting and insulating substrates strongly depend on the microstructure : the easy axis of orientation of magnetization in a material can change from in-plane to perpendicular with subtle changes in structure and vary the shape of the hysteresis loop [4]. Surface steps, defects, thin film stresses and alloy segregation influence the magnetic microstrucure of these predominantly two-dimensional systems. Magnetic properties of small particles, which are used in magnetic recording technology, are strongly dependent on the size, morphology and magnetic microstructure. The need for a high spatial resolution technique, capable of determining qualitative and *quantitative* micromagnetic structure has become eminent. High spatial resolution magnetic contrast imaging techniques, available in the STEM, provide a valuable tool for exploring the structure-properties relationships in these materials.

3. TECHNIQUE

A detailed description of the holography instrumentation and theory is given elsewhere [5,6]; here we present a brief review only. In the STEM implementation of electron holography, we split the partially coherent electron source into two virtual electron sources through the use of a Möllenstedt electron biprism [7]. The two electron wave packets are transferred by the condenser lens and focused by the objective lens into two fine electron probes coherently illuminating the specimen. The separation of the two virtual sources can be varied by changing the voltage applied to the biprism or by simply changing the excitation of the condenser and/or objective lenses. This system has expanded flexibility when compared to standard TEM based holography methods. With the beam held stationary and the objective lens operated at a relatively large defocus, a shadow image of a relatively large area can be observed. The two wave packets interact with the specimen and form an interference pattern (hologram), which appears as a fringe modulated (twin-)image in the detector plane. From the hologram, recorded on a slow-scan CCD camera, the amplitude and relative phase of the two electron waves can be extracted using methods standard in TEM off-axis holography. A fast Fourier transform yields a diffractogram with two characteristic sidebands, whose separation depends upon the fringe spacing. One of the sidebands is isolated and its inverse Fourier transform reveals the amplitude and phase.

The amplitude image contains information about the thickness and bulk inelastic mean-free-path of electrons in the solid [8]. The relative phase difference $\Delta\varphi$, due to electromagnetic potentials present in the specimen region, can be expressed as [9]

$$\Delta\varphi = \frac{1}{\hbar}\oint m\mathbf{v}d\mathbf{l} - \frac{e}{\hbar}\iint \mathbf{B}d\mathbf{S}, \qquad (1)$$

where **v** is the electron's velocity, **B** the magnetic induction and **S** the oriented area enclosed by the two electron paths. From the first term in equation (1), the mean inner potential can be determined for a specimen of known geometry. The second term of equation (1), involves an area integral that contains information about the magnetization distribution in a magnetic specimen. In thin magnetic films of constant thickness, the first term contributes a constant phase shift, while a magnetic phase shift results from the second term. The phase difference $\Delta\varphi$ is then proportional to the magnetic flux enclosed by the two beam paths as they traverse the specimen. In a magnetic specimen of varying thickness, such as a small particle, the electrostatic contribution, which cannot be neglected, can be eliminated from the phase image with the help of the amplitude image. This can be done when the magnetization does not vary along the beam direction, since both terms of the phase difference in equation (1) are linearly dependent on the projected thickness. The magnetic phase image can be retrieved by dividing the phase image by the thickness distribution (determined from the amplitude image).

We have developed two distinct holography modes : the absolute and differential [5]. In the absolute mode (Fig. 1a), one of the two electron probes travels through vacuum, while the

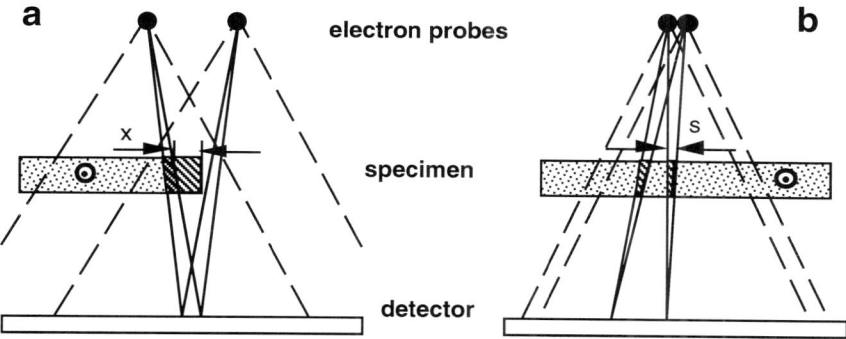

Figure 1. Absolute (a) and differential mode (b) of STEM holography.

other passes through the specimen. Assuming zero phase in vacuum we can absolutely determine the phase shift caused by the electromagnetic fields present in the specimen. For a magnetic specimen the phase shift $\Delta\varphi$ is proportional to the magnetic flux enclosed by the two beam paths. In a uniformly magnetized domain located near the edge of a specimen of constant thickness, the phase difference $\Delta\varphi$ changes linearly with increasing distance from the edge and $\Delta\varphi \propto \iint \mathbf{B}\,\mathrm{d}\mathbf{S} \cong B_n x t$, where B_n is the component of the magnetic field normal to the plane determined by the wave vectors of the two split electron waves, x is the distance from the edge and t is the (constant) thickness. The gradient of the phase determines the magnitude of B_n (averaged over the film thickness) in the domain. For a film of nearly constant thickness we can neglect the contribution of the constant phase of the electrostatic field present in the specimen, since quantitative information is derived from the gradient of the phase difference and the phase variation due to the electrostatic potential is small (approximately 0.1 rad/nm of film thickness). This straightforward interpretation of the phase image can only be done for magnetic fields which are confined to the plane of the magnetic specimen.

In the differential mode (Fig. 1.b), both electron probes pass through the specimen. The separation of the beam paths, which is adjustable by the biprism voltage as well as the excitation of the condensor and/or objective lenses [6], can be made as small as ten nanometers. The area defining the enclosed magnetic flux is approximately constant for every point in the hologram. In this mode the phase of an uniformly magnetized domain in a specimen of constant thickness is constant, in contrast to the absolute mode, where the same domain has a linearly varying phase. The phase difference $\Delta\varphi \propto \iint \mathbf{B}\,\mathrm{d}\mathbf{S} \cong B_n s t$, where B_n is the component of the magnetic field normal to the plane determined by the wave vectors of the two split electron probes, s is the separation of the beam paths projected into the specimen plane and t is the (constant) thickness. The differential mode is advantageous for the investigation of magnetic domain wall profiles and allows straightforward interpretation of the magnetic microstructure [5]. Since this mode does not require a vacuum reference wave, no assumption has to be made about the flatness of the phase in vacuum. This becomes advantageous in the case of strong leakage fields and for the observation of features far away from any holes in the specimen.

In far-out-of-focus holography artefactual contrast is contained in the reconstructed phase and amplitude images that is a direct consequence of the defocus Δf. Fresnel fringes appear near edges or holes and may obscure the magnetic contrast of reconstructed holograms. The

Figure 2. Defocus correction.
(a - right) Through-focal series of amplitude and phase generated from a sideband of the hologram. Top row - overfocus, center - in focus, lower row - underfocus.
(b - below) Line scans of five phase images corrected by different defoci near the optimum defocus.

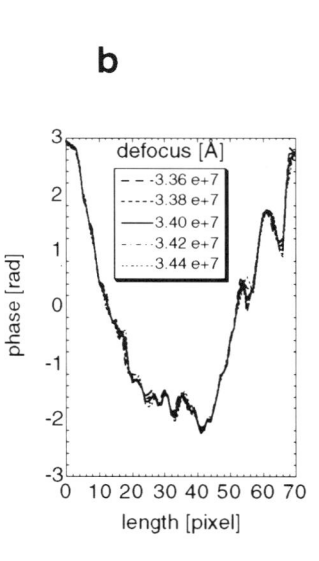
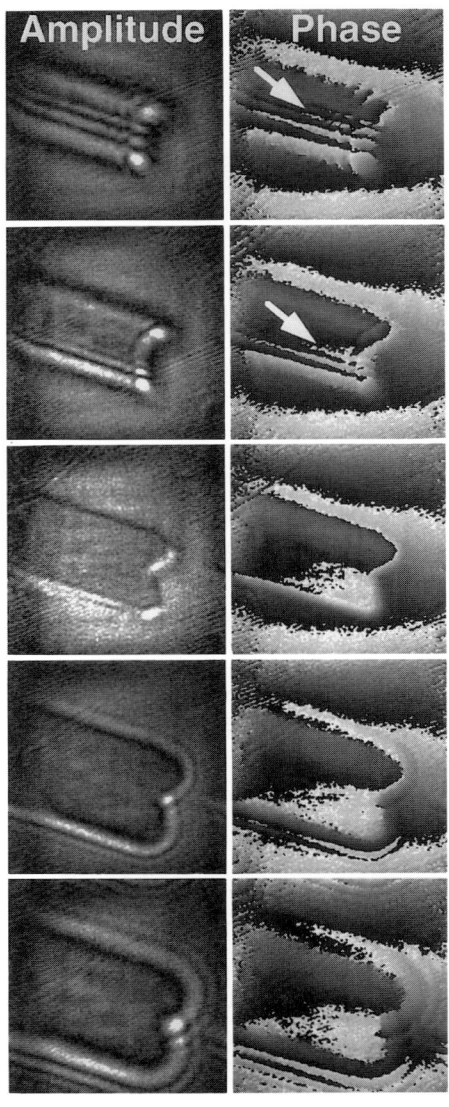

defocus aberration can be eliminated by multiplying the isolated sideband by a function equivalent to the inverse contrast transfer function of the objective lens, $e^{-2\pi i \chi(q)}$ with a defocus phase shift given by $\chi(q) = \lambda^2 \Delta f q^2 / 2$, where q is the spatial frequency and λ the electron's wavelength [10]. The spherical aberration contribution to the phase shift can be neglected for the electron-optical conditions employed here (long focal length and resolution limit ~ 1nm). A through focal series, shown in Fig. 2a, has been generated from the hologram of a small elongated CrO_2 particle by applying successive defocus phase corrections. Note the false phase line contrast present in both under- and overfocus images (arrows). The minimization of the standard deviation of the amplitude in vacuum has been found to be the most reliable criterion for the determination of optimum defocus, which can be loosely described as the condition for the vanishing of Fresnel fringes. A series of five line-scans of phase images corrected with defoci near the optimum value has been used for a quantitative evaluation of the phase error as a function of defocus. A plot of five line scans (Fig. 2b) reveals a maximum phase variation of 0.3 rad within a ±2% interval of the apparent correct defocus. This level of phase uncertainty is of the same order of magnitude as that due to noise in local phase measurements in vacuum.

4. APPLICATIONS

Both STEM holography modes, accompanied by the Fresnel and Differential Phase Contrast modes of Lorentz microscopy have been used to characterize thin magnetic films, magnetic multilayer structures and small magnetic particles. Thin magnetic films have been described in detail earlier [5]; here we concentrate on multilayer structures and small particles.

4.1. Multilayer Structures

4.1.1. Co/Pd

A Co/Pd superlattice, Pd(20nm)/[Co(1nm)Pd(1.1nm)]$_{10}$, which was grown on an amorphous carbon film was imaged using STEM electron holography. Hysteresis loop measurements indicated a dominant in-plane magnetization with saturation magnetization M_S =

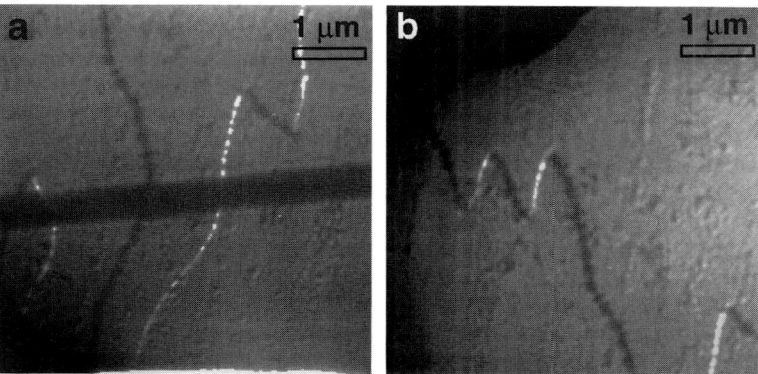

Figure 3 (a) and (b). Fresnel mode images of Co/Pd multilayer structures.

1600 emu/cm^3 and coercitive field $H_{C\parallel}$ = 88 Oe.

The Fresnel mode images (Fig. 3a, b) display a typical distribution of magnetic domain walls appearing as white and dark lines (note in Fig. 3a the broad dark biprism shadow and the edge which is parallel to the biprism at the bottom of Fig. 3a). Near an edge or hole, the domain walls become nearly parallel to each other, running approximately perpendicularly to the specimen's edge. Further away from the edge, the magnetization begins to curl forming typical 'w' shaped domain walls (Fig. 3b). A phase image, reconstructed and unwrapped from a hologram acquired in the absolute mode of STEM holography (Fig. 4a) and a three-dimensional map of the marked area (Fig. 4b) shows that the magnetization is oriented perpendicular to the edge of the sample and rotates by 180° when crossing the domain wall, a result consistantly observed in different specimen regions. The overlapping rectangles in Fig. 4a are remnant of the unwrapping process. The rectangles are pasted sections from phase images with a successively added or subtracted constant phase value. A line scan of the phase, taken along the edge and averaged over a uniform region 150nm across (Fig.4c) shows the linear dependence of the phase inside the domains I and II and the location of the domain wall. The slope of the phase absolutely determines the magnitude of magnetization inside the domains for uniform thickness films. In this case the phase gradient is 28.1 mrad/nm in domain I and 11.7 mrad/nm in domain II. The value in domain I agrees well with the theoretically predicted value for all Co layers ferromagnetically aligned throughout the superlattice stack. Assuming a total Co thickness of 10 x 1.0 nm = 10 nm with uniform bulk saturation magnetization of 1440 emu/cm^3, i.e. (x 4π 10^{-4} =) B = 1.8096 T, the phase gradient in units of rad/nm equals

$$\frac{\Delta \varphi}{\Delta x} = \frac{e}{\hbar} \iint \mathbf{B} \cdot d\mathbf{S} / \Delta x = \frac{e}{\hbar} Bt = 27.49 \text{ mrad/nm}, \qquad (2)$$

i.e. within 2% of the measured value. The magnetization in domain II is ~ 42% of the expected ferromagnetically aligned bulk value. This suggests that not all magnetic layers in domain II are

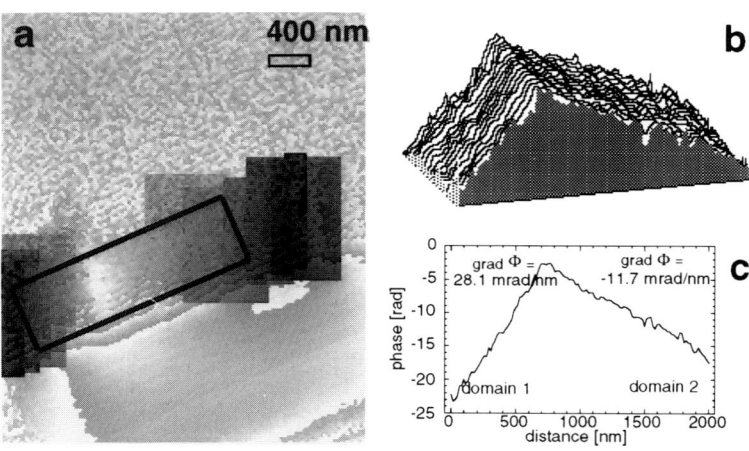

Figure 4. Partially unwrapped phase image (a) of domain structure near the edge of a Co/Pd multilayer, (b) three-dimensional plot of the phase in the region marked in a, (c) line-scan of phase along specimen edge, averaged across 150nm.

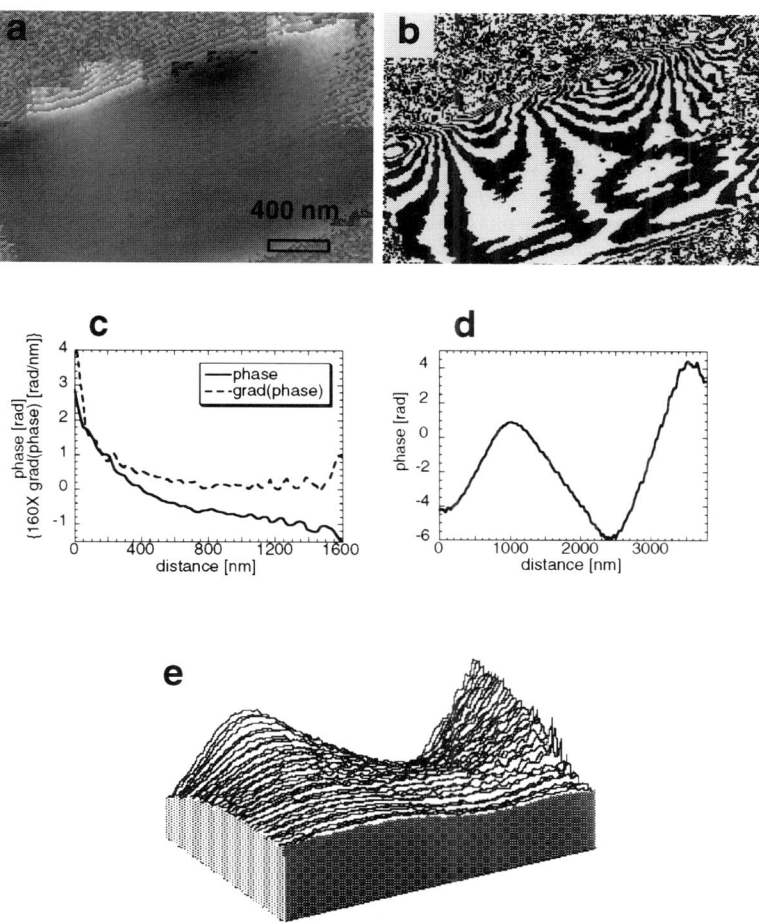

Figure 5. Leakage fields in Co/Pd multilayer films : (a) unwrapped phase image, (b) contour image of same area as in (a) where 1 contour corresponds to $\pi/9$ rad, (c) line-scan of phase perpendicular to edge and its gradient, (d) line-scan of phase parallel to the edge, (e) three-dimensional plot of phase outside the specimen.

magnetized in the same direction (assuming only in-plane magnetization). The measured value indicates that the magnetization vectors in the layers must be oriented with 7 layers in one direction and three layers in the opposite direction producing a net integrated magnetization of 40% of the saturated value.

While observing the magnetic structure near the specimen edge, a strong magnetic flux leak was observed in the surrounding vacuum (Fig. 5). The reconstructed, unwrapped phase (Fig. 5a) and contour image (Fig. 5b) of the same area displays the periodically changing phase; the contours are equimagnetic-induction lines and make the magnetization flow more visible. The line-scan in Fig. 5c, taken in a direction perpendicular to the film edge, shows the decay of the leakage field. The gradient of the phase, which is proportional to the projected component of the magnetic induction parallel to the edge, reveals that this field falls to 1/e of its maximum value at a distance approximately 150nm from the edge. The ripple in the right part of the profile is due to the Fresnel fringes of the biprism, which in principle can be removed by subtracting a phase image of free space. A comparison of the line-scan parallel to the film's edge (Fig. 5d) and a Lorentz image of the same area shows that the domain walls terminate at inflexion points of the phase curve and near the center of a domain the phase is at maximum or minimum. A three-dimensional plot of the phase in the space near the edge is shown in Fig. 5e.

Investigations of the magnetic microstructure in regions far away from a hole are carried out in the differential mode of STEM holography. In this mode, the phase represents a direct

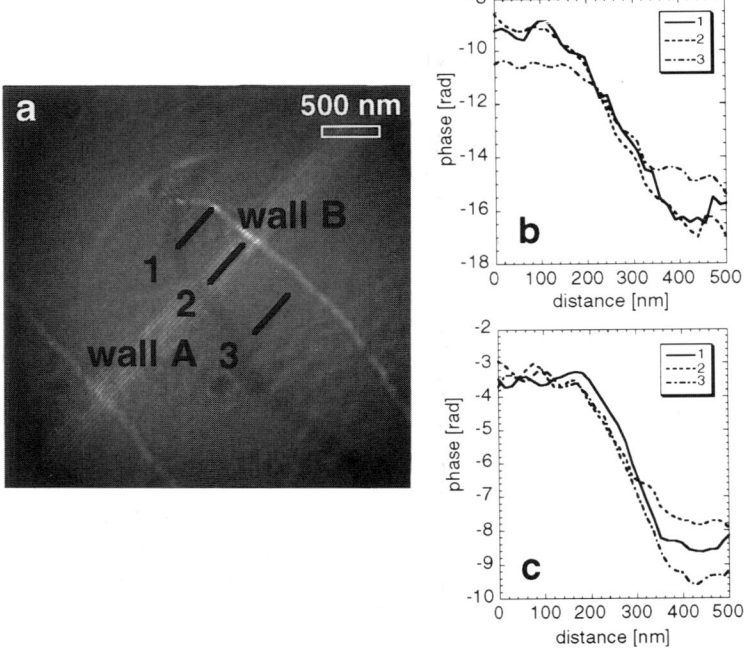

Figure 6. Domain walls in Co/Pd multilayer films : (a) hologram with marked walls and line profile positions, (c) profiles of broader wall A, (c) profiles of wall B.

measure of the magnetic field in the specimen and displays a constant phase value in regions of constant magnetization. This is advantageous for the investigation of domain wall profiles. A series of holograms (positions 1 through 3 in Fig. 6a) yields a set of 3 domain wall profiles for each of the two marked domain walls A and B, acquired in a direction parallel to the interference fringes. The average domain wall width is 245 nm (wall A) and 200 nm (wall B). The difference in mean domain wall width is likely related to the presence of partial antiferromagnetic coupling within the superlattice stack.

4.1.2. Co/Cu

Magnetic coupling between adjacent ferromagnetic layers in a superlattice composed of alternating ferromagnetic and nonmagnetic layers is present when giant magnetoresistance is observed. A series of multilayer structures with varying seed layer thickness, number of bilayers and bilayer geometry have been grown under UHV conditions. Samples grown on thin amorphous holey carbon films are observed in the Fresnel mode with the beam

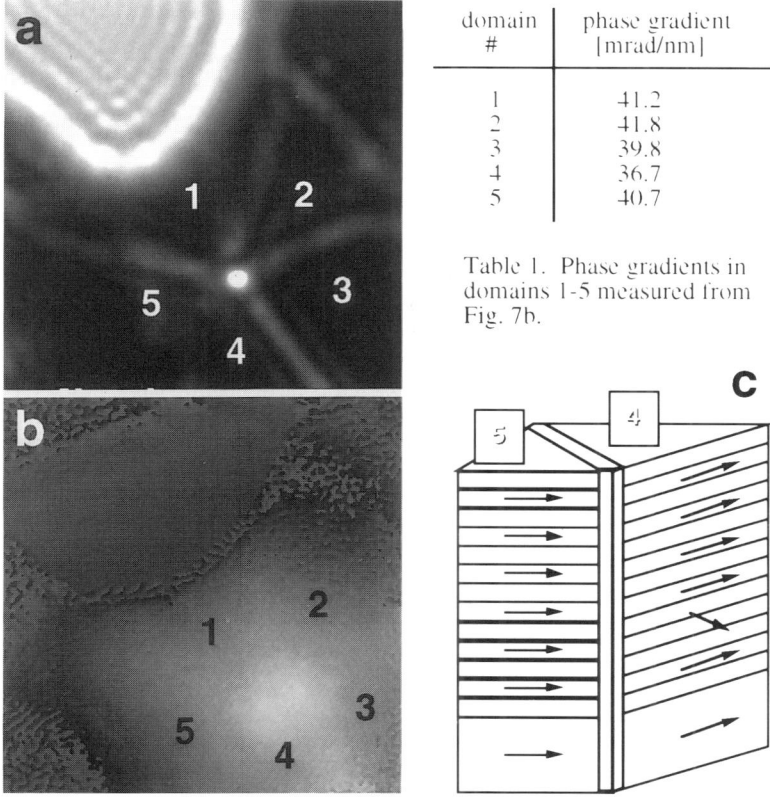

domain #	phase gradient [mrad/nm]
1	41.2
2	41.8
3	39.8
4	36.7
5	40.7

Table 1. Phase gradients in domains 1-5 measured from Fig. 7b.

Figure 7. Underfocus Fresnel image (a, 1.4μm²) and partially unwrapped absolute phase image (b) of a Co/Cu multilayer. The proposed micromagnetic structure is shown in (c).

perpendicular to the layers of the superlattice (not cross-sectional view). Images reveal the position of domain walls as bright or dark bands and holograms of identical regions yield quantitative information on the domain. The variation of the magnitude of magnetization can be used to determine the interlayer coupling [10] assuming in-plane magnetization. For example, the Fresnel image of a Co(6nm)/[Cu(3nm)Co(1.5nm)]$_6$/Cu(3nm) superlattice shows five domains aligned in a flux vortex (Fig. 7a). From the hologram, acquired in the absolute mode (Fig. 7b), the maximum phase gradients are determined in each of the five domains (table 1). The average maximum phase gradient in domains 1, 2, 3 and 5 is 41.0 ± 0.6 mrad, which differs from the predicted bulk value for a cobalt film of 15 nm total thickness (41.4 mrad/nm) by less than 1%. This indicates that the domains penetrate the sandwich and are uniformly (ferromagnetically) aligned. The phase gradient in domain 4 is 36.7 mrad/nm, which is approximately 90% of the expected uniformly magnetized value (37.3 mrad/nm). A proposed explanation of this magnetization amplitude loss is outlined in Fig. 7c. The magnetization in one of the layers (10% of the active thickness) is rotated by 90° with respect to the magnetization in the other layers. The amplitude must then be calculated as a vector sum, i.e. the magnetization amplitude equals to $\sqrt{0.1^2 + 0.9^2} \cong 0.906$ and therefore approximately 10% lower than the aligned value. If a single layer were antiferromangetically aligned in the superlattice stack within this domain, the phase gradient would have to be 80% of the maximum value. The existence of 90° coupling between layers has been confirmed by hysteresis loop measurements performed on the same sample [11]. This confirms that we are able to determine the orientation of domains in a superlattice and thereby are capable of correlating macroscopic giant magnetoresistance measurements with micromagnetic structure.

4.2. Small Particles

Magnetic properties of small magnetic particles, which are used in magnetic recording technology, are strongly influenced by their magnetic microstructure, size and morphology. Direct quantitative investigations of the magnetic microstructure in small magnetic particles has been limited by the relatively low spatial resolution of commonly used micromagnetic analysis techniques (magneto-optical methods, magnetic force microscopy, Bitter-pattern method, etc.). The Fresnel and Differential Phase Contrast (DPC) modes of Lorentz microscopy reveal the in-plane component of magnetization as well as the local microstructure at high spatial resolution, but do not allow accurate quantitative measurements. The deflection angle in the DPC mode is too small ($\sim 10^{-5}$ rad) and in the Fresnel mode, only strong magnetization changes (domain walls, ripple) are observed; an image deconvolution with an exact value of defocus and known wall profile would be required to explicitly extract the micromagnetic structure. Electron holography, carried out in a STEM, provides quantitative information about the magnetization in the specimen at nanometer resolution and therefore represents a valuable tool for the determination of the magnetic microstructure in small particles. Reconstructed amplitude and phase images of particles smaller than $\sim 1\mu m$ are strongly obscured by Fresnel fringes. These fringes must be removed with a defocus correction during the holography reconstruction process.

4.2.1. CrO$_2$

CrO$_2$ particles are elongated 'stretch shaped' particles (approx. 50-100 nm wide and 300-400 nm long) with a magnetic moment of 93 emu/g ($\sim$400 emu/cm^3) and a coercive field of 405 Oe. Holograms of single particles acquired in the absolute mode confirm the prediction that the particles are nearly uniformly magnetized. Micromagnetic calculations of the three-dimensional magnetic fields and phase differences have been carried out for comparison with experimental phase images since phase images of three-dimensional fields from small particles are difficult to interpret directly. Fig. 8 shows the calculated magnetic phase image of a 640nmx80nmx40nm particle, magnetized uniformly (saturation magnetization 400 emu/cm^3) along the x axis. The phase is obtained by integrating the B$_x$ component along y and z (Fig. 8a),

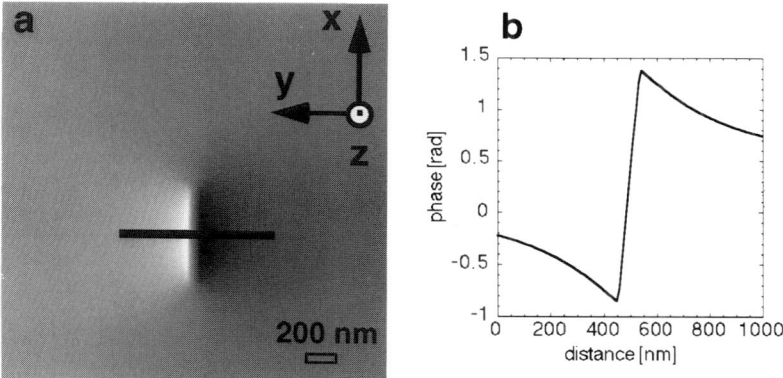

Figure 8. Calculated phase image (a) of a 640nmx80nmx40nm magnetic particle uniformly magnetized in the x direction. The line-scan (b) is taken along the line in a.

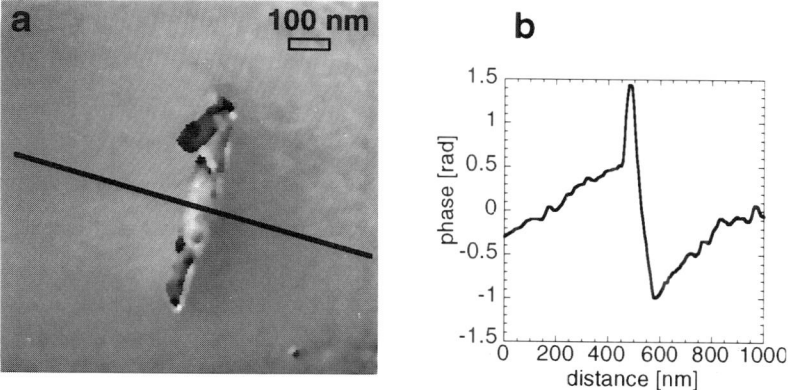

Figure 9. Absolute phase image of a CrO_2 particle (a) and line-scan (b) along line in a.

where the beams propagate along the z axis and the reference wave is assumed to have zero phase. The constant electrostatic contribution is neglected. Typical bright and dark contrast observed outside the particles is in good qualitative agreement with the experimental image (Fig. 9a). Absolute phase shifts caused by the magnetic dipole can be determined. The experimental line-scan across the particle (Fig. 9b) shows that the phase gradient inside the particle is approximately 26 mrad/nm, which compares favorably with the calculated image (25 mrad/nm). We note that an accurate value of the thickness is difficult to obtain from the experiment

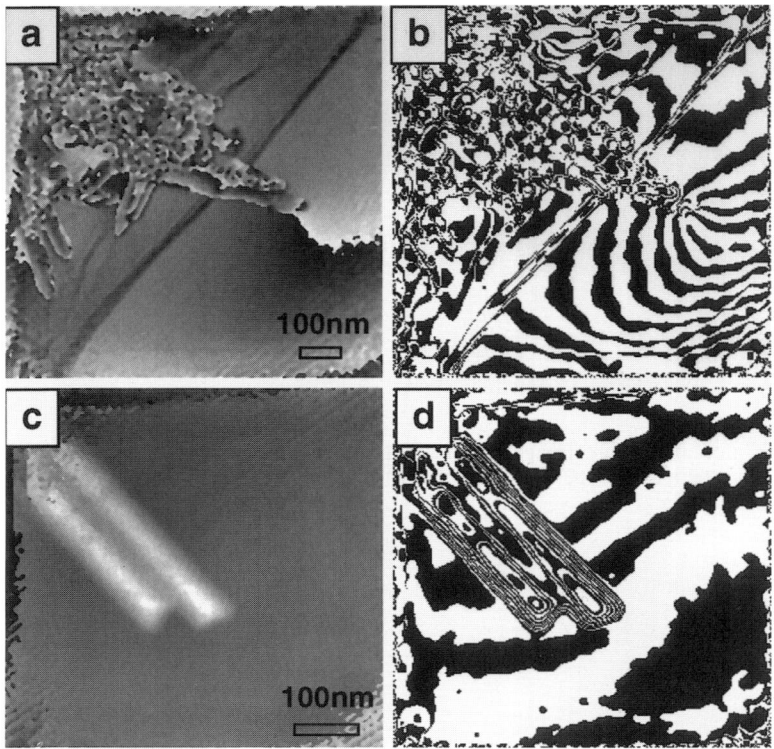

Figure 10. Interaction of CrO_2 particles : (a) and (c) - phase images, (b) and (d) - contour images, 1contour corresponds to $\pi/10$ rad.

(large tilt or knowledge of the mean free path is required). Differences in the line-scans may be due to small deviations from the ideal uniform magnetization of the particles due to surface effects, thickness changes or scattering effects at the particle surface (edges).

The interaction of CrO_2 particles is illustrated in Fig. 10. The magnetizations of the two CrO_2 particles near the edge of the supporting carbon film (Fig. 10 (a) and (b) - phase and contour image) are aligned parallel, which manifests itself as a strong leakage field. By comparison, the two CrO_2 particles in Fig. 10 (c) and (d) (phase and contour image) are aligned antiparallel, the magnetic flux is closed, and no flux leak is observed.

4.2.2. FeB

FeB particles (H_C=605 Oe, magnetic moment 52.8 emu/g) have a hexagonal platelet form, typically 100 nm in diameter and ~20nm thick. In Fig. 11, the STEM (a), amplitude (b) and phase image (c) of an FeB particle are compared. The phase images display a nearly constant phase difference across and around the whole region of the particle confirming that the magnetization is perpendicular to the platelet.

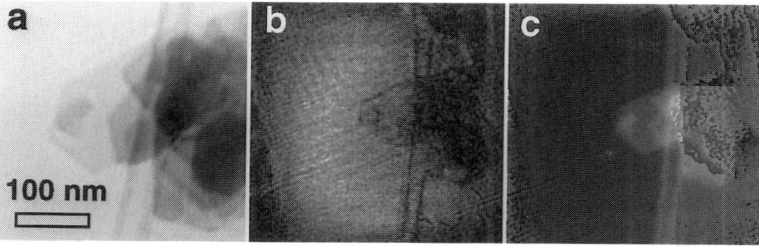

Figure 11. A STEM (a), amplitude (b) and phase (c) image of FeB platelets.

5. ACKNOWLEGMENTS

This work was supported by the NSF under grant DMR-9110386, and the Office of Naval research under grant #N00014-93-1-0099. The electron microscopy was performed in the Center for High-Resolution Electron Microscopy at ASU supported by the NSF grant DMR91-15680. We are grateful to A. A. Higgs for the flawless operation of the HB 5 microscope, P. Perkes for help with the image processing, J. C. Wheatley and C. Weiss for inspiration and help during specimen preparation and Dr. J. K. Weiss of Emispec Systems for creative image acquisition software.

REFERENCES

1. L.M. Falicov et al., J.Mater.Res. **5** (1990) 1299.
2. S.S.P. Parkin, R. Bhadra and K.P. Roche, Phys. Rev. Lett. **66** (1991) 2152.
3. M.N. Baibich, J.M. Broto, A. Fert, F. Nguyen Van Dau, F. Petroff, P. Etienne, G. Creuzet, A. Friederich and J. Chazelas, Phys. Rev. Lett. **61** (1988) 2472.
4. S.D. Bader, J. Magn. Magn. Mater. **100** (1991) 440.
5. M. Mankos, M. R. Scheinfein and J. M. Cowley, J. Appl. Phys. **75** (1994) 7418-24.
6. M. Mankos, A. A. Higgs, M. R. Scheinfein and J. M. Cowley, Ultramicroscopy (in press).
7. G. Möllenstedt and H. Düker, Z. Phys. **145** (1956) 377.
8. M. R. McCartney and M. Gajdardziska-Josifovska, Ultramicroscopy **53** (1994) 283.
9. Y. Aharonov and D. Bohm, Phys. Rev. **115** (1959) 485.
10 M. Mankos, Z. Yang, M. R. Scheinfein and J. M. Cowley, IEEE Trans. Magn. (in press).
11. Z. Yang and M. R. Scheinfein, Appl. Phys. Lett. (submitted).

Electron Holography
A. Tonomura, L.F. Allard, G. Pozzi, D.C. Joy and Y.A. Ono (Editors)
© 1995 Elsevier Science B.V. All rights reserved.

Amplitude-division electron holography

Q. Ru

Tonomura Electron Wavefront Project, ERATO
Research Development Corporation of Japan (JRDC)
P.O. Box 5, Hatoyama, Saitama 350-03, Japan

An amplitude-division electron holography is achieved in a non-biprism transmission electron microscope using a single crystal thin film to split the incident electron beam. Techniques and procedures for hologram formation and reconstruction are described and demonstrated with experiments. The relation between the spatial coherence of the illumination and the resolution of the reconstructed image is theoretically derived and compared with that of the conventional wavefront-division holography achieved with an electron biprism. Experimental results obtained using a gold crystal thin film as a beam splitter show that the amplitude-division electron holography permits any coherent or incoherent illumination when high resolution is not strongly required.

1. INTRODUCTION

The electron biprism originally developed by Mollenstedt and Duker has made it possible to achieve off-axis electron holography in a transmission electron microscope (TEM). [1] Most of the subsequent works on electron holography reported so far are based on use of this biprism. Since the wavefront, not the amplitude, of the incident electron beam is divided by the biprism, high spatial coherence is required to make interference fringes between the divided beams. Therefore, a field-emission gun (FEG) that can emit a high-coherence electron beam is indispensable for holographic experiments, and the majority of the results reported so far were obtained from the TEMs equipped with both an FEG and a biprism. [2]

Although the FEG has greatly facilitated the formation of off-axis electron holograms, the high-coherence requirement coming from the wavefront division interference system still remains. In fact, most of the coherence of the incident beam is required to form carrier fringes in an off-axis hologram and only a little part of the coherence contributes to recording the phase shift due to the object itself.

It has been shown in optical interferometry that the coherence requirement for forming carrier fringes can be removed by dividing the amplitude, not the wavefront, of the incident beam. Pozzi and Matteucci et al. [3,4] have shown an interesting method for achieving amplitude-division electron holography using a crystal thin film, not a biprism, to split the electron beam. Since the lattice fringes of a crystal film are used as the carrier fringes of an off-axis hologram, the coherence requirement can be greatly reduced. Their experimental results have essentially shown that even a TEM with a pointed filament electron source having low coherence can also form off-axis holograms.

Although the method has great potential to be widely used, no subsequent research has been reported by other groups, perhaps because a specimen needs to be placed in the selector-aperture plane of a microscope so that the column vacuum is broken every time a specimen is inserted, and the objective lens cannot be used for imaging the object. In order to achieve high resolution, the possibility of inserting specimen into the normal object plane before the objective lens has also been discussed by Pozzi. [3] The conclusion predicts that the high resolution scheme is practically achievable.

Recently, we have initially and preliminarily reported some experimental results obtained with the high resolution scheme. [5] This paper describes the method with both theoretical expressions and experimental results. Although a strict and complicated theoretical analysis has been made by Pozzi, a simple expression based on using an ideal objective lens is given to achieve a simple and fundamental relation between the coherence of the illuminating electron beam and the resolution of the reconstructed image. The result has also been discussed by comparing it with the wavefront-division type. Both theoretical and experimental results show that incoherent electron sources are also permitted to do electron holographic experiments.

2. HOLOGRAM FORMATION PROCESS

The electron optics illustrated in Figure 1 show how a Fresnel hologram of a particle is formed. Consider a case of observing small particles supported by a thin carbon film and mounted on a commercially sold 3-mm diameter grid. Prepare a single crystal thin film, also normally mounted on a 3-mm diameter grid, and place the crystal grid on the specimen grid in the specimen holder. Insert the holder in its normal position and then image and record the lattice fringes of the crystal. The incident beam is tilted so that the Bragg reflection is strongly excited by the single crystal. Because a gap between the crystal plane and the specimen plane is spontaneously introduced by the supporting grids of the specimen and crystal films, twin defocused images carried by the direct beam and the Bragg-reflected beam are formed in a laterally shifted position for each other in the hologram plane. If the distance between the twin images is greater than the object size, the two images separate perfectly and interfere with adjacent lateral plane waves to form an off-axis hologram.

The distance of the twin images is given by

$$l = \lambda z/d, \tag{1}$$

where λ s the wavelength, z the gap depth between the crystal and specimen films, and d the lattice spacing of the single crystal.

Because the single crystal is in focus and the object to be investigated is out of focus, a Fresnel hologram of the object is obtained. Although defocusing will be corrected in the hologram reconstruction process, an unnecessarily wide gap between the specimen and the crystal should be avoided because an unnecessarily large defocusing of the object limits the resolution of the reconstructed object image. The gap can be easily but roughly adjusted during specimen setting. If the film side of the crystal grid is attached to the object side of the specimen grid, a narrow gap introduced by the bend in the crystal film, or object film, can be obtained and used for measuring small objects. On the other hand, if the grid sides of the crystal and specimen grids face each other, a relatively wide gap introduced by the supporting grids can be obtained and used for measuring large objects.

In the following experiments, a single gold crystal thin film of 0.204-nm lattice spacing was used to split the incident electron beam, and a Hitachi HF-2000 electron microscope having a field-emission gun was used to take holograms. The electron biprism equipped with the microscope is used to check the coherence of illuminations. Figure 2 shows a hologram of a polystyrene latex particle of about 120 nm in diameter. The right image having higher contrast is formed by the direct beam; the left image is by the Bragg-reflected beam. The distance between the gold crystal plane and the latex particle was about 10 nm . There are as many as 1,500 fringes recorded in the hologram. An enlarged area of the hologram is shown in bottom of Figure 2 so that the fringes may be seen clearly.

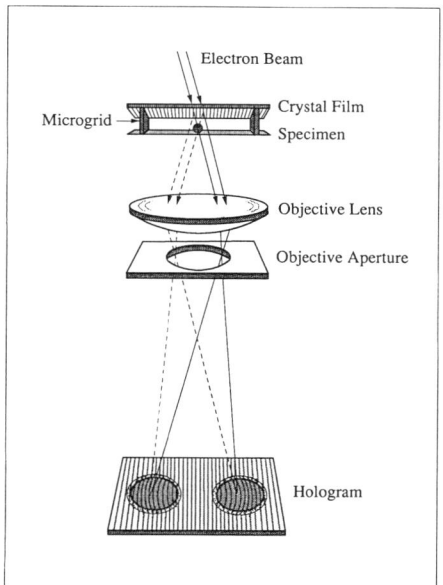

Figure 1. Hologram formation setup.

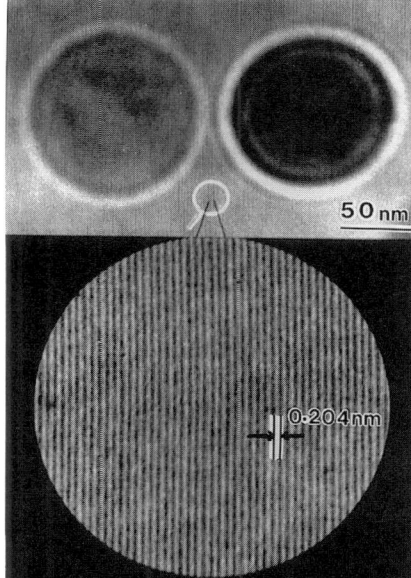

Figure 2. Hologram of a latex sphere.

Because the crystal plane is focused in the photographic emulsion plane, coherence requirement is greatly reduced for taking holograms. This means that any conventional electron microscope without a field-emission gun can be used to take holograms. Coherence requirement arises only when the defocus is to be corrected to reconstruct a sharp in-focus image from the Fresnel hologram. Therefore, the necessary coherence width depends on the resolution of the image to be reconstructed. If the defocus is not to be corrected, no coherence will be required, while just a defocused image will be reconstructed. On the other hand, if the defocus is to be corrected during reconstruction process, corresponding coherence for producing Fresnel fringes of specimen is required so that the in-focus image can be reconstructed from the Fresnel fringes.

Although our microscope is equipped with a field-emission gun, we chose the largest condenser aperture and used convergent illumination in the experiments to reduce the coherency to close to the conditions of a conventional microscope. Therefore, bright holograms including 1,500 fringes were obtained with short exposure times of 3-5 seconds. The coherence ot the illumination was so low that we could not form any fringes with a biprism under the same illumination conditions.

3. HOLOGRAM RECONSTRUCTION PROCESS

Illuminating the Fresnel hologram by a laser beam, we can see the in-focus object image locating before or after the hologram plate. In order to measure the phase distribution of the object wave, we need to arrange an interferometer. Since the phase shift due to an electron microscopic object is usually less than a wavelength, highly accurate measurement in phase measurement is required. In practical experiments, the surfaces of photographic films used to record electron holograms are usually not optically flat. The uneven surfaces can shift the phase of the illuminated laser beam for several wavelengths or more, and will obstruct the exact measurement of the phase distribution of the object wave.

An optical method has been reported to eliminate the phase error in the reconstruction of image holograms that are generally obtained with a biprism system. [6] This reconstruction method is based on illuminating the image hologram with two laser beams, which are tilted in symmetrical directions about the optical axis to reconstruct two conjugate object waves. When the hologram plane, i.e., the image plane of the reconstructed object waves, is focused, the phase shift due to the uneven surfaces of the hologram plate is canceled, but the phase shift due to the object is twice amplified by the interference of the two conjugate object waves. The method, however, cannot be used to reconstruct the Fresnel holograms. If a Fresnel hologram is illuminated with the two laser beams, two images of the object will be reconstructed in two different planes, one located before the hologram plane and the other after the hologram plane. When one of the two images is focused, the other object wave always forms a twice-defocused image in the observation plane to form a complicated interference fringe pattern.

Here is described a way to avoid the above-mentioned problems. The lattice fringes in a different area of the single crystal film where no objects lie on the lower specimen film are recorded as a reference hologram before or after the object hologram is recorded. The object and reference holograms are then superposed on a different photographic plate by double-exposure photography. When the superposed hologram is illuminated with a laser beam, as shown in Figure 3, two waves are reconstructed. One is a reference wave reconstructed from the reference hologram and the other is the object wave from the object hologram. The image plane of the reconstructed object wave is focused on the observation plane. The phase distribution of the in-focus image of the reconstructed object wave can therefore be measured from the interference fringes between the object and reference waves. Since the uneven surfaces of the hologram plate cause the same phase error in both of the waves, the phase error has been canceled by the interference.

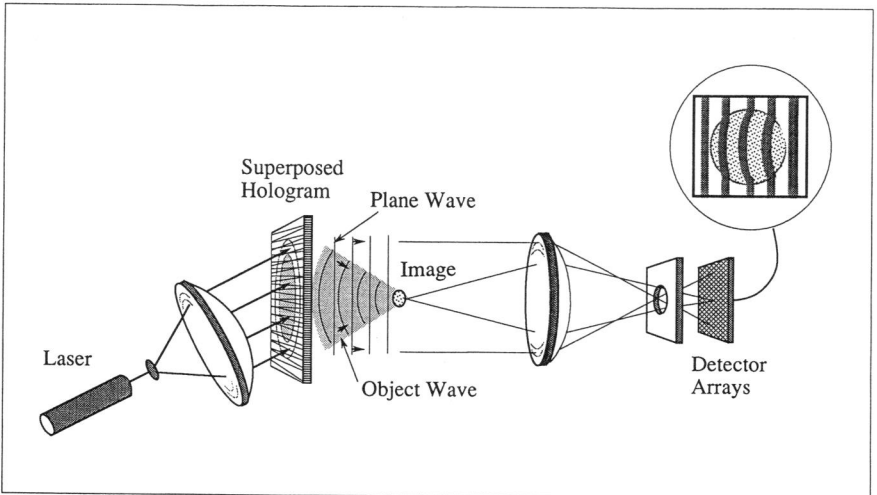

Figure 3. Optical setup for reconstructing Fresnel holograms.

The systematic phase errors caused by the aberrations of electron lenses in hologram formation process and the aberrations of optical lenses in hologram reconstruction process have also been removed by the technique. Even a perfect single crystal is used to form holograms, no perfectly straight lattice fringes can be obtained because of the distortions of the electron lenses. Since the reference hologram is recorded under the same electron optical conditions as the object hologram, the distortion has been canceled by the superposed hologram.

If the reference and the object holograms are superposed so that the lattice fringes in both of the holograms are parallel to each other, contour fringes corresponding to the phase distribution of the object wave will appear in the reconstructed image plane. Although the contour fringes directly show the phase distribution, it is difficult to observe when the phase shift of the object wave is less than a wavelength. In order to observe this sub-wavelength phase shift, the reference hologram is slightly rotated with respect to the object hologram in the double-exposure process. Therefore, proper carrier fringes can be introduced into the reconstructed image to clearly show sub-wavelength phase shift. The fringe pattern can be considered as an off-axis image hologram, and from that the amplitude and the phase distributions of the object wave can be separated and determined quantitatively using the conventional digital reconstruction methods. [7,8]

The above-mentioned reconstruction processes including defocus correction can also be done numerically by using a computer. In this case, Fresnel transform, in place of a Fourier transform, of the Fresnel hologram should be computed. Of course, a Fresnel transform can be easily obtained by using the FFT algorithm.

4. RELATION OF COHERENCE AND RESOLUTION

Since the single crystal film is focused in the hologram plane, the lattice fringes, i.e., the carrier fringes of the off-axis hologram, can always be formed, independent of the spatial coherence of the illuminating beam. The quality of the defocused object image, however, is dependent on illumination conditions. Here the relationship is described between the coherence of the illuminating beam and the resolution of the reconstructed image.

Although many elements, such as condenser lenses, condenser apertures and beam tilts, are practically used in a TEM to control the illuminating beam, arbitrary illumination conditions can be described using the simple model as illustrated in Figure 4. A practical electron source, or its virtual image formed using the condenser lenses, can be considered as an incoherent superposition of many point sources that produce many collimated incident beams with various intensities and incident angles in the specimen plane. The intensity variation of the collimated beams can be characterized using the function $f(a_x, a_y)$, where a_x and a_y denote the incident angles with respect to the x and y directions. Each of the beams does not interfere with any other, but a part of each beam interferes with another part of the same beam.

The single crystal film split each beam into two coherent beams, one passing through the object to be observed and the other passing through the neighboring area where no objects exist. Since the crystal film plane is focused on the hologram recording plane, the object wave is defocused. If the object represented by $u_1(x,y)$ is illuminated using a collimated beam passing along the optical axis, the defocused object wave can be derived from the Fresnel approximation as a convolution [9]

$$u_2(x,y) = f(0,0)u_1(x,y) * g(x,y) \qquad (2)$$

where

$$g(x,y) = \frac{\exp(ikz)}{i\lambda z}\exp\left[i\frac{k}{2z}(x^2 + y^2)\right].$$

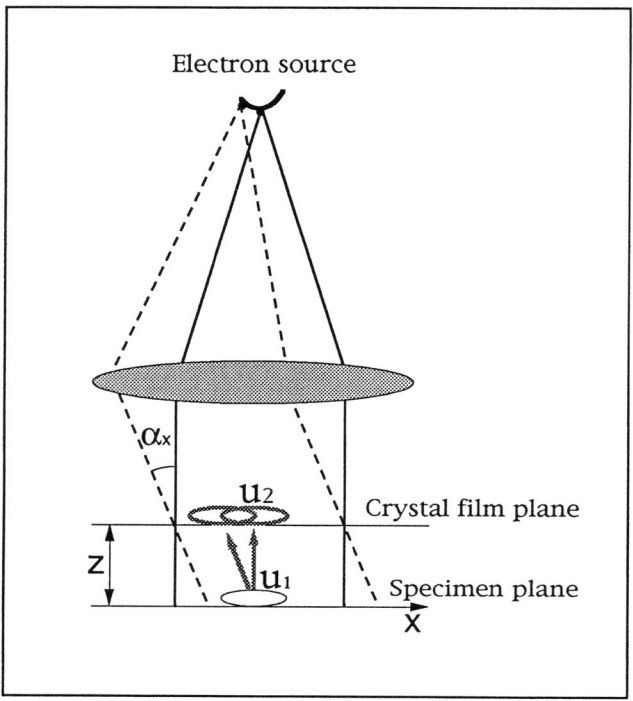

Figure 4. Geometry showing blur effect by finite source size.

The location of the defocused image depends on the illumination angle. The defocused object wave for an arbitrary collimated illumination with the incident angle (α_x, α_y) can be expressed by

$$u_2(x,y;\alpha_x,\alpha_y) = f(\alpha_x,\alpha_y)u_1(x-\alpha_x z, y-\alpha_y z) * g(x-\alpha_x z, y-\alpha_y z). \tag{3}$$

The object wave produced by each collimated beam is recorded in its own hologram, and all the holograms produced by the various illuminating beams are incoherently superposed together in the hologram plane. When a coherent laser beam is illuminated onto the superposed holograms, all these object waves are reconstructed and coherently superposed as

$$\begin{aligned} u_2(x,y) &= \iint f(\alpha_x,\alpha_y)u_1(x-\alpha_x z, y-\alpha_y z) * g(x-\alpha_x z, y-\alpha_y z) d\alpha_x d\alpha_y \\ &= \frac{1}{z^2} f\left(\frac{x}{z},\frac{y}{z}\right) * [u_1(x,y) * g(x,y)]. \end{aligned} \tag{4}$$

In reconstruction of an off-axis hologram, the hologram is first Fourier transformed using a lens to select one of the two conjugate diffraction waves in the Fourier plane. The Fourier transform of Equation 4 is then given by

$$U_2(f_x,f_y) = F(zf_x,zf_y) \cdot U_1(f_x,f_y) \cdot G(f_x,f_y), \qquad (5)$$

where

$$G(f_x,f_y) = \exp[-i\pi\lambda z(f_x^2 + f_y^2)].$$

Since the image plane of the object, not the hologram, is focused on the observation plane with the lens, a defocusing component which is exactly conjugate with $G(f_x,f_y)$ has been introduced in the diffraction wave. This focus condition means that the component $G(f_x,f_y)$ in Equation 5 has been canceled from the final reconstructed object wave. The Fourier transform of the final reconstructed object wave can therefore be expressed by

$$U_2(f_x,f_y) = F(zf_x,zf_y) \cdot U_1(f_x,f_y). \qquad (6)$$

The equation describes the frequency property of the holographic imaging system. The transfer function $F(zf_x,zf_y)$ of the system is characterized using the illumination conditions and the defocusing amount. In order to estimate the cut-off frequency of the transfer function, the illuminated beam is roughly described with rectangular functions as

$$f(\alpha_x,\alpha_y) = \mathrm{rect}\left(\frac{\alpha_x}{s_x}\right)\mathrm{rect}\left(\frac{\alpha_y}{s_y}\right), \qquad (7)$$

where s_x and s_y denote the spreads of the incident angles of the illuminated beam in the x and y directions. The transfer function $F(zf_x,zf_y)$ becomes

$$F(f_x,f_y) = \mathrm{sinc}(zs_x f_x)\mathrm{sinc}(zs_y f_y). \qquad (8)$$

The transfer function acts as a low-pass filter that limits the resolution of the reconstructed image. Accordingly, the cut-off frequency in the x direction for example is given by

$$f_{x-cutoff} = \frac{1}{zs_x}. \qquad (9)$$

It can be concluded from Equations 6 and 9 that any coherent or incoherent electron sources can be used to form an off-axis hologram. Using a coherent source, such as an FEG that can produce a very small and bright source, high resolution images can be obtained. Using an incoherent electron source, such as point filaments or LaB_6 tips that are well equipped with normal TEMs and usually have a large solid angle for a sufficiently bright illumination, holographic observations can also be done in place of a decreased resolution. If $s=10^{-4}$ radian and $z=10\,\mu m$, for example, the object structures larger than 1 nm can be observed using amplitude-division holography, whereas nothing can be observed using wavefront-division holography.

In the case of wavefront-division holography, the sensitivity (or fringe contrast) instead of the resolution is strongly depending on the coherence of the illumination. If an incoherent illumination is used, the visibility of fringes becomes to zero. So no fringes can be formed and object wave cannot be recorded and reconstructed.

5. EXPERIMENTAL RESULTS

A Hitachi HF-2000 microscope equipped with an FEG is used in the experiments to show the image performance under incoherent illuminations. A biprism is also installed and used to check if the illumination is coherent or not. A gold crystal thin film which is frequently used for checking the specifications of a TEM is used here to split the incident electron beam. A very convergent and bright beam was created. The spatial coherence of the beam was greatly reduced so that no fringes could be formed with a biprism under these illuminating conditions, whereas the lattice fringes of the gold crystal thin film are still clearly observed.

An example of such a hologram and its reconstructed phase image are shown in Figure 5 where gold lattice fringes of 0.2nm spacing are modulated by a defocused image of a polystyrene latex particle of 120 nm in diameter. The latex particles are supported by a thin carbon film. The gap between the specimen film and the gold crystal film was about 10 μm.

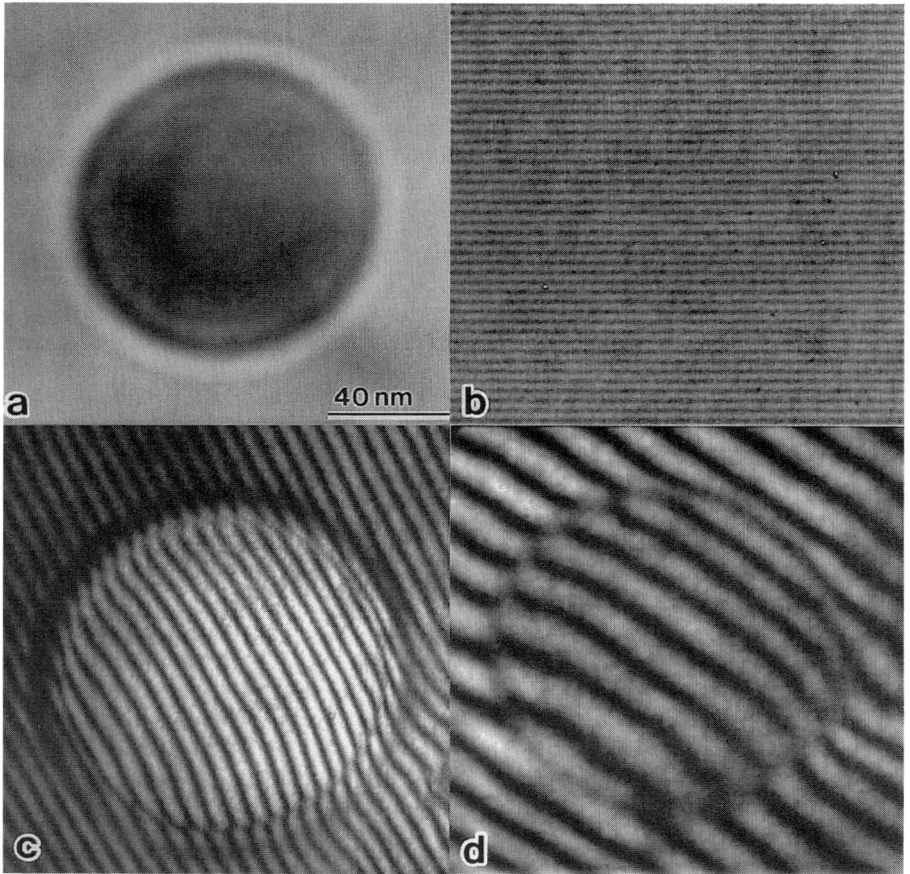

Figure 5. Experimental results showing (a) hologram of a latex sphere, (b) lattice fringes in the hologram, (c) phase image before defocus correction, and (d) after defocus correction.

The observation results of fine gold particles are shown in Figure 6. The gold particles were epitaxially grown on a molybdenite crystal substrate under a high-vacuum condition. The higher order diffraction beams from the molybdenite crystal and the gold particles have been cut off using a small objective aperture.

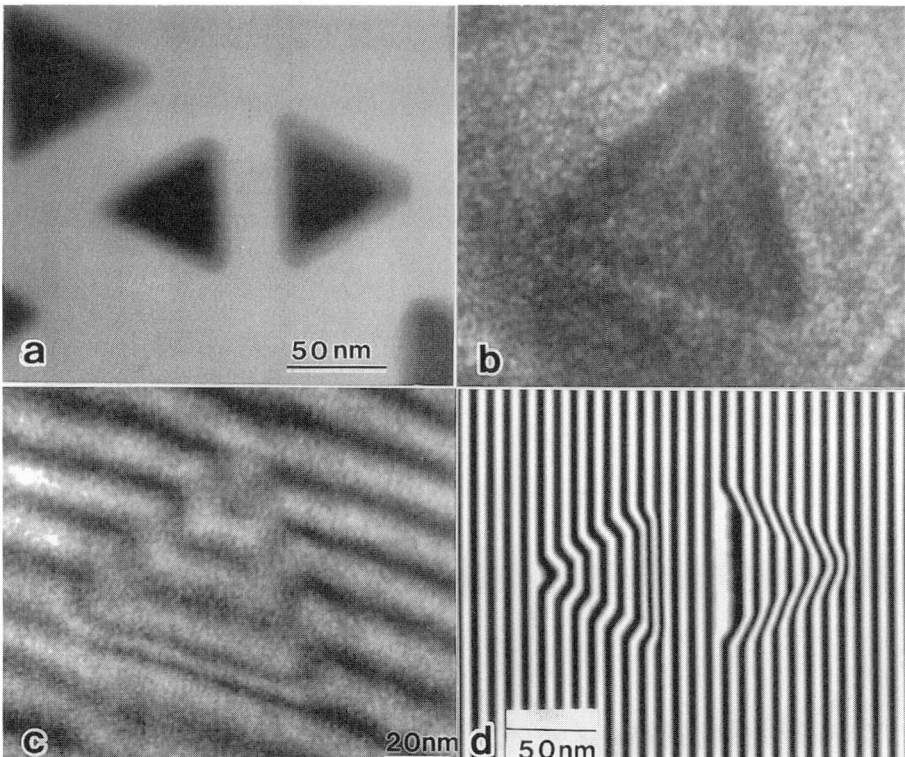

Figure 6. Experimental results showing (a) bright-field TEM image of gold platelets grown an molybdenite substrate, (b) optically reconstructed amplitude image of a platelet, (c) optically reconstructed phase image, and (d) numerically reconstructed phase image of two platelets.

From Equation 6 we can say that the holography is particularly useful for observing long-range objects such as electric or magnetic fields for which high resolution performance is usually not strongly required. In order to show the usefulness, a charged tip is used as the specimen. An equivalent electron-optical arrangement that has been proposed by Matteucci et al.[4] is used here to get a broad area of view. A micro tip hanging on a conducting wire is placed in the select-area-aperture plane, instead of the normal specimen plane, of the TEM and externally charged up. The beam-splitting gold crystal thin film is still placed in the specimen holder as before. The objective lens is adjusted to make the image of the crystal film in a plane over that of the micro tip. Consequently, almost the same holograms as those created before could be formed, except that the magnification of the specimen is greatly reduced. The optically reconstructed contour phase image is shown in Figure 7 which shows the contour electric-field distribution around the charged tip. Since the results are obtained with a very incoherent illumination, we believe that any other non-FEG source can be used to obtain such a phase image.

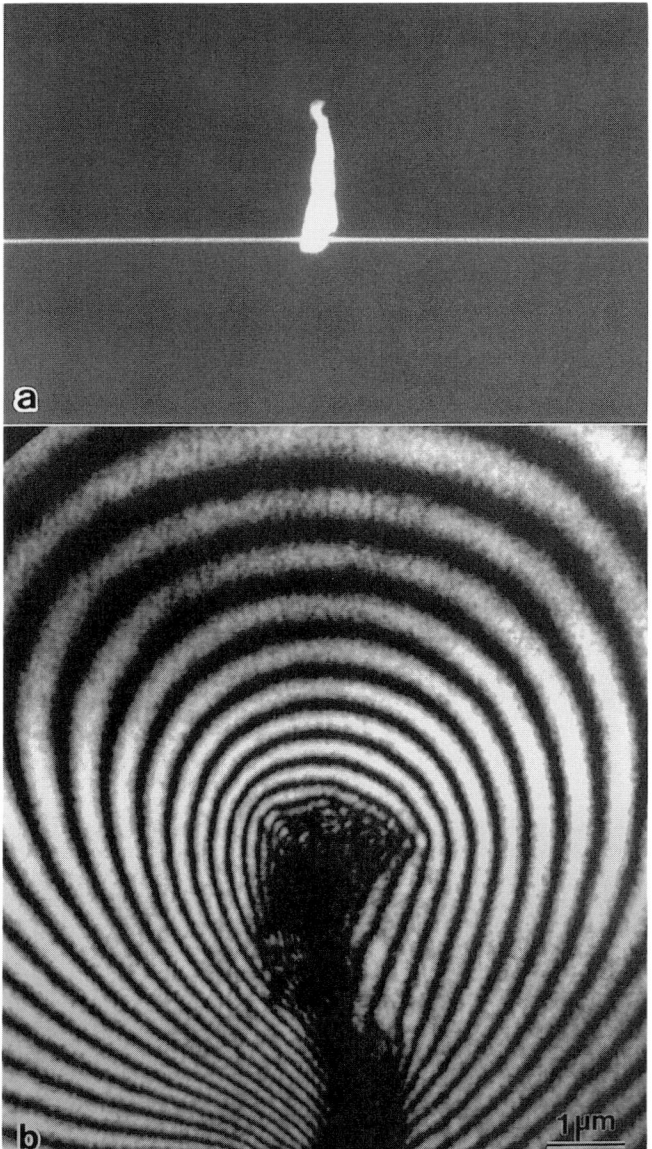

Figure 7. TEM image of a charged micro tip (a) and the contour phase map showing the electric-field distribution around the tip.

6. CONCLUSIONS

An amplitude-division electron holography has been presented to form off-axis electron holograms in a non-biprism TEM with a single crystal thin film. The theoretical description of the holographic imaging process shows that the spatial coherence of the illuminating electron beam does not reduce the fringe quality and the image sensitivity, but limits the resolution of the reconstructed images. Experimental results show that incoherent electron sources are also available to do holographic observations at low- or medium-resolution. [10]

7. REFERENCES

1. G. Möllenstedt and H. Düker, Z. Physik, 145 (1956) 377.
2. A. Tonomura, Electron Holography, Springer-Verlag, Berlin Heidelberg, 1993.
3. G. Pozzi, Optik, 66 (1983) 91.
4. G. Matteucci, G.F. Missiroli, and G. Pozzi, Ultramicroscopy, 8 (1982) 403.
5. Q. Ru, N. Osakabe, J. Endo, and A. Tonomura, Ultramicroscopy, 53 (1994) 1.
6. J. Endo, T. Matsuda, and A. Tonomura, Jpn. J. Appl. Phys., 18 (1979) 2291.
7. M. Takeda and Q. Ru, Appl. Opt., 24 (1985) 3068.
8. Q. Ru, T. Matsuda, A. Fukuhara, and A. Tonomura, J. Opt. Soc. Am., A8 (1991) 1739.
9. J.W. Goodman, Introduction to Fourier Optics, McGraw-Hill Book Company, New York, 1968.
10. Q. Ru, J. Appl. Phys., to be published in February 15, 1995 edition.

The First International Workshop on Electron Holography
- Concluding Remarks -

David C. Joy

EM Facility, University of Tennessee, Knoxville, TN 37996, USA
and
Oak Ridge National Laboratory, Oak Ridge, TN 37831, USA

As Professor Hannes Lichte said in his opening lecture, "this First International Workshop represents the end of the beginning for Electron Holography". Although it is now forty-five years or so since Dennis Gabor first described this technique, for much of this period electron holography - as Prof. T. Mulvey reminded us in his historical overview[1] - has remained the province of a few dedicated specialists in electron-optics and quantum physics. Now, with this meeting, holography performed in transmission, and scanning transmission, electron microscopes emerges into the mainstream as a powerful, quantitative, imaging technique with demonstrated applications covering the spectrum from vortices in superconductors to molecular biology. Major breakthroughs in experimental technique - such as the real time reconstruction of holograms using LCD displays, high speed phase shifting methods, and coherent diffraction techniques - have been matched with innovative new approaches to problems such as the imaging of electric and magnetic fields. Consequently, as commercial cold field emission gun TEMs, combined with high performance digital image recording systems and sophisticated computer programs for analysis, become more widely available, electron holography will increasingly represent the leading edge of microscopy technique.

Several, separate, events led to the idea of this International Workshop. One source of impetus was the desire to disseminate the results of the work of Tonomura Electron Wavefront Project of the Exploratory Research for Advanced Technology (ERATO) under Research Development Corporation of Japan (JRDC) at the completion of its five-year period. At the same time the Oak Ridge National Laboratory - University of Tennessee (ORNL-UT) group were keen to celebrate their three year Laboratory Directed Research and Development (LDRD) program on electron holography then nearing completion at ORNL. Discussions between Dr. Akira Tonomura of the ERATO project, Dr. Larry F. Allard of ORNL, and Professor Giulio Pozzi of the University of Bologna finally led to agreement on the idea of a meeting in Knoxville, Tennessee which would serve both to give a public account of the accomplishments of the two programs, and to act as a forum to discuss the current state-of-the-art in this rapidly developing field. The difficult job of organizing this International Workshop across three continents was only possible because of the efforts of Dr. Yoshimasa A. Ono of the ERATO project and Dr. Allard. Their drive, hard work, focus, and management skills, aided by the tentacles of the INTERNET, molded the scientific content of the Workshop program and ensured the smooth running meeting that we have all enjoyed.

Financially this Workshop was made possible by the generous help of the JRDC which underwrote much of the expenses of travel and hotel accommodation for invited speakers, and of workshop rooms and audio-visual equipment. JRDC also covered the costs of publishing these Workshop Proceedings, which will be distributed to all attendees. The organizers are grateful to Mr. Genya Chiba of JRDC, and his staff, for their invaluable support and cooperation. Substantial financial support was also provided by the LDRD program and the Metals and Ceramics Division of ORNL. Mr. Yasuo Ren of Nissei Sangyo America, Ltd. provided significant financial support for local costs of the Workshop, and administrative support for the multitude of hotel and catering arrangements that had to be made; and the Advanced Research Laboratory, Hitachi, Ltd. provided a grant to encourage student attendance at the meeting. The University of Tennessee and the Science Alliance provided financial help to students and took care of a variety of other necessary administrative functions. Ms. Rebecca Jones-Shupe and Ms. Billie Russell, of the High Temperature Materials Laboratory at ORNL, and Ms. Elisa Carr of Nissei Sangyo America, Ltd. provided a prodigious level of secretarial support over a period of many months.

Finally I, and the other organizers, thank all of the invited speakers and all of the participants. Without your scientific contributions - in the form of talks, or posters, or questions, and your enthusiasm and excitement this Workshop would have remained just another 'fringe' event.

Reference

1. T.M. Mulvey, in Advances in Imaging and Electron Physics, Vol. 91, P.W. Hawks (ed.), Academic Press, San Diego, 1995, p.259, "Gabor's pessimistic 1942 view of electron microscopy and how he stumbled on the Noble prize."

List of participants

Ade, George
Physikalisch-Technische Bundesanstalt
Lab 6.12
Bundesallee 100
D-38116 Braunschweig, Germany
　Phone: +49-531-5926121
　Fax: +49-531-5929292

Ailey, Shawn
Dept. of Material Science and Engineering
North Carolina State University
229 Riddick Hall
Raleigh, NC 27606, USA
　Phone: +1-919-515-6619
　Fax: +1-919-515-7724
　E-mail: ksailey@mte.ncsu.edu

Alexander, Kathi
Metals and Ceramics Division
Oak Ridge National Laboratory
PO Box 2008, MS 6376
Oak Ridge, TN 37831, USA
　Phone: +1-615-574-0631
　Fax: +1-615-574-0641
　E-mail: alexanderkb@ornl.gov

Allard, Larry
High Temperature Materials Lab.
Metals and Ceramics Division
Oak Ridge National Laboratory
Bldg. 4515, MS 6064
PO Box 2008
Oak Ridge, TN 37831, USA
　Phone: +1-615-574-4981
　Fax: +1-615-574-4913
　E-mail: allardlfjr@ma160.ms.ornl.gov

Allen, Charles W.
MSD 212/E211
Argonne National Laboratory
9700 S. Cass Ave.
Argonne, IL 60439, USA
　Phone: +1-708-252-4159
　Fax: +1-708-252-4798
　E-mail: allen@anlemc.bitnet

Anderson, Ian M.
Metals and Ceramics Division
Oak Ridge National Laboratory
PO Box 2008, MS 6376
Oak Ridge, TN 37831, USA
　Phone: +1-615-574-0632
　Fax: +1-615-574-0641
　E-mail: ia1@ornl.gov

Aoyama, Kazuhiro
Tonomura Electron Wavefront Project
ERATO, JRDC
c/o Advanced Res. Lab., Hitachi, Ltd.
Hatoyama, Saitama 350-03, Japan
(Present Address)
Max-Planck-Institut für Biochemie
Abteiung Molekulare Strukturbiologie
82152 Martinsried bei München
Germany
　Phone: +49-89-8578-2653
　Fax: +49-89-8578-2641
　E-mail: aoyama@vms.biochem.mpg.de

Asai, Shojiro
Advanced Research Laboratory
Hitachi, Ltd.
Hatoyama, Saitama 350-03, Japan
　Phone: +81-492-96-6111
　Fax: +81-492-96-6004
　E-mail: asai@harl.hitachi.co.jp

Ball, James B.
Oak Ridge National Laboratory
Bldg. 4500-N, MS 6182
PO Box 2008
Oak Ridge, TN 37831, USA
　Phone: +1-615-574-4650
　Fax: +1-615-291-4978

Beeli, Conradin
Electron Microscopy
Swiss Federal Institute of Technology
EPFL-I2M Bat. MXC
Lausanne, Ch-1015, Switzerland
　Phone: +41-21-693-4437
　Fax: +41-21-693-4401
　E-mail: cbeeli@i2msg3.epfl.ch

Bentley, Jim
Metals and Ceramics Division
Oak Ridge National Laboratory
Bldg. 5500, MS 6376
PO Box 2008
Oak Ride, TN 37831, USA
　Phone: +1-615-574-5067
　Fax: +1-615-574-0641
　E-mail: bentleyj@ornl.gov.

Berta, Yolande
School of Materials Science and
Engineering
Georgia Institute of Technology
778 Atlantic Drive
Atlanta, GA 30332-0245, USA
   Phone: +1-404-894-2545
   Fax: +1-404-853-9140

Bigelow, Wilbur C.
Dept. of Materials Science and Engineering
University of Michigan
3062 H.H. Dow Building
2300 Hayward Street
Ann Arbor, MI 48109-2136, USA
   Phone: +1-313-764-3321
   Fax: +1-313-763-4788
   E-mail: bigelow@umich.edu

Blackson, John H.
Analytical Sciences
The Dow Chemical Company
1897 Building
Midland, MI 48667, USA
   Phone: +1-517-636-6316
   Fax: +1-517-638-6443

Blass, William E.
Department of Physics and Astronomy
University of Tennessee
401 Nielsen Physics, Cir. Dr.
Knoxville, TN 37996-1200, USA
   Phone: +1-615-974-3342
   Fax: +1-615-975-7843
   E-mail: blass@utkux1.utk.edu

Bonevich, John E.
National Center for Electron Microscopy
Materials Science Division
Lawrence Berkeley Laboratory
MS-72-150
1 Cyclotron Road
Berkeley, CA 94720, USA
   Phone: +1-510-486-4716
   Fax: +1-510-486-5888
   E-mail: jebonevich@lbl.gov

Carim, Altaf H.
Dept. of Materials Science and Engineering
Pennsylvania State University
118 Steidle Building
University Park, PA 16802, USA
   Phone: +1-814-863-4296
   Fax: +1-814-865-0016
   E-mail: carim@ems.psu.edu

Carr, Elisa
Nissei Sangyo America, Ltd.
25 West Watkins Mill Road
Gaithersburg, MD 20878, USA
   Phone: +1-301-840-1650
   Fax: +1-301-963-0808

Chen, Jun
Tonomura Electron Wavefront Project
ERATO, JRDC
c/o Advanced Res. Lab., Hitachi, Ltd.
Hatoyama, Saitama 350-03, Japan
(Present Address)
Yamamoto Quantum Fluctuation Project
ERATO, JRDC
c/o NTT Musashino R&D Center
3-9-11 Midori-cho, Musashino
Tokyo 180, Japan
   Phone: +81-422-36-1894
   Fax: +81-422-36-1867

Chiba, Genya
Research Development Corporation of
Japan (JRDC)
4-1-8 Hon-cho
Kawaguchi, Saitama 332, Japan
   Phone: +81-48-226-5600
   Fax: +81-48-226-5651

Cunningham, Brian
IBM East Fishkill
Route 52
Hopewell Junction, NY 12533, USA
   Phone: +1-914-894-2164
   Fax: +1-914-892-6256

Datye, Abhaya K.
UNM/NSF Center for Micro-Engineered
Ceramics
University of New Mexico
Farris Engineering Center #203
Albuquerque, NM 87131-6041, USA
   Phone: +1-505-277-0477
   Fax: +1-505-277-1024
   E-mail: datye@unm.edu

Dravid, Vinayak
Dept. of Materials Science and Engineering
Northwestern University
2225 N. Campus Dr.

Evanston, Illinois 60208-3108, USA
  Phone: +1-708-467-1363
  Fax: +1-708-491-7820
  E-mail: v-dravid@nwu.edu

Endo, Junji
Tonomura Electron Wavefront Project
ERATO, JRDC
c/o Advanced Res. Lab., Hitachi, Ltd.
Hatoyama, Saitama 350-03, Japan
(Present Address)
Advanced Research Laboratory
Hitachi, Ltd.
Hatoyama, Saitama 350-03, Japan
  Phone: +81-492-96-6111
  Fax: +91-492-96-6006

Engel, Alan K.
ERATO Overseas Representative
c/o ISTA, Inc.
950 Conestoga Road
Rosemont, PA 19010, USA
  Phone: +1-610-527-4538
  Fax: +1-610-527-2041
  E-mail: 71420.3126@compuserve.com

Evans, Neal D.
Metals and Ceramics Division
Oak Ridge National Laboratory
Bldg. 5500, MS 6376
PO Box 2008
Oak Ridge, TN 37831, USA
  Phone: +1-615-576-4427
  Fax: +1-615-574-0641

Fink, Hans-Werner
IBM Research Division
Zurich Research Laboratory
Forschungslaboratorium Zürich
Säumerstrasse 4
8803 Rüschlikon, Switzerland
  Phone: +41-1-724-84-11
  Fax: +41-1-724-07-24

Frost, Bernhard G.
Department of Physics
University of Tennessee
Knoxville, Tennessee 37996, USA
(Present Address)
High Temperature Materials Laboratory
Metals and Ceramics Division
Oak Ridge National Laboratory
Bldg. 4515, MS 6064
PO Box 2008
Oak Ridge, TN 37831, USA
  Phone: +1-615-574-0811
  Fax: +1-615-574-4913

Gajdardziska-Josifovska, Marija
Department of Physics
University of Wisconsin-Milwaukee
Physics Building, Room 484
1900 E. Kenwood Blvd.
PO Box 413
Milwaukee, WI 53201, USA
  Phone: +1-414-229-4965
  Fax: +1-414-229-5589
  E-mail: mgj@csd4.csd.uwm.edu

Germinario, Louis T.
Eastman Chemical Co.
ECCR Labs.
PO Box 1972
Kingsport TN 37662, USA
  Phone: +1-615-229-4047
  Fax: +1-615-229-4558

Gonzalez, Roland E.
Center for Applied Energy
Analytical Services
3572 Iron Works Pike
Lexington, KY 40511, USA
  Phone: +1-606-257-0266
  Fax: +1-606-257-0302
  E-mail: gonzalez@caer.uky.edu

Griffith, Doug
Nissei Sangyo America, Ltd.
25 West Watkins Mill Road
Gaithersburg, MD 20878, USA
  Phone: +1-301-840-1650
  Fax: +1-301-963-0808

Herring, Rodney A.
Tonomura Electron Wavefront Project
ERATO, JRDC
c/o Advanced Res. Lab., Hitachi, Ltd.
Hatoyama, Saitama 350-03, Japan
(Present Address)
604 Aldershot Rd.
Baltimore, MD 21229, USA
  Phone: +1-410-747-1768
  E-mail: RodneyHerr@aol.com

Hirayama, Tsukasa
Tonomura Electron Wavefront Project
ERATO, JRDC
c/o Advanced Res. Lab., Hitachi, Ltd.
Hatoyama, Saitama 350-03, Japan
(Present Address)
Research and Development Laboratory
Japan Fine Ceramics Center
2-4-1 Mutsuno, Atsuta-ku
Nagoya, Aichi-ken 456, Japan
Phone: +81-52-871-3500
Fax: +81-52-871-3599

Hirche, Robert
2239 Chesterfield
Maryville, TN 37801, USA
(ICMAS, Inc.)
Phone: +1-615-984-8683

Hovington, Pierre
Universite de Sherbrooke
Etudiant-Societe de Microscopie Canada
2500 Boul. Université
J1k 2RI Sherbrooke, Canada

Ishizuka, Kazuo
Tonomura Electron Wavefront Project
ERATO, JRDC
c/o Faculty of Engineering, Toyo Univ.
2100 Nakanodai, Kujirai
Kawagoe, Saitama 350, Japan
(Present Address)
14-48 Matsukaze-dai
Higashi-matsuyama, Saitama 355, Japan
Phone and Fax: +81-493-35-3913

Jayaram, Raman
Metals and Ceramics Division
Oak Ridge National Laboratory
PO Box 2008
Oak Ridge, TN 37831, USA
Phone: +1-615-576-7738
Fax: +1-615-574-0641
E-mail: jyz@ornl.gov

Jesson, David E.
Oak Ridge National Laboratory
PO Box 2008
Oak Ridge, TN 37831, USA

Joy, David C.
Director, EM Facility
University of Tennessee
F241 Walters Life Sciences Building
Knoxville, TN 37996-0810, USA
Phone: +1-615-974-3638
Fax: +1-615-974-3642
E-mail: joy@utkvx.bitnet

Kersker, Michael M.
JEOL USA, Inc.
11 Dearborn Road
Peabody, MA 01960, USA
Phone: +1-508-535-5900
Fax: +1-508-536-2205

Lai, Guanming
Tonomura Electron Wavefront Project
ERATO, JRDC
c/o Advanced Res. Lab., Hitachi, Ltd.
Hatoyama, Saitama 350-03, Japan
(Present Address)
Department of Electronic Engineering
Faculty of Engineering
Shizuoka University
3-5-1 Johoku
Hamamatsu, Shizuoka 432, Japan
Phone and Fax: +81-53-478-1111
E-mail: lai@el.shizuoka.ac.jp

Lehmann, Michael
Institut für Angewandte Physik
Universität Tübingen
Auf der Morgenstelle 12
D-72076 Tübingen, Germany
Phone: +49-7071-294052
Fax: +49-7071-295400
E-mail: Lehmann@castor.tat.physik.uni-tuebingen.de

Libera, Matthew
Dept. of Materials Science and Engineering
Stevens Institute of Technology
Hoboken, NJ 07030, USA
Phone: +1-201-216-5259
Fax: +1-201-216-8306

Lichte, Hannes
Institute für Angewandte Physik
Universität Tübingen
Auf der Morgenstelle 12
D 72076 Tübingen, Germany
Phone: +49-7071-29-2428
Fax: +49-7071-29-5400
(Present Address)
Institut für Angewandte Physik

Technische Universität Dresden
Zellescher Weg 16
D 01062 Dresden, Germany
   Phone: +49-351-463-6050
   Fax: +49-351-463-3199
   E-mail: hl@padns1.phy.tu-dresden.de

Lin, Xiwei
Dept. of Materials Science and Engineering
Northwestern University
2225 N. Campus Drive
Evanston, IL 60208, USA
   Fax: +1-708-491-7820

Mahan, Stephen
Department of Physics and Astronomy
University of Tennessee
401 Nielsen Physics Bldg.
Knoxville, TN 37996, USA
   Phone: +1-615-974-7847
   Fax: +1-615-974-7843
   E-mail: mahan@aurora.phys.utk.edu

Maher, Dennis M.
Dept. of Material Science and Engineering
North Carolina State University
2140 Burlington Engineering Laboratory
Raleigh, NC 27695-7916, USA
   Phone: +1-919-515-7149
   Fax: +1-919-515-7724
   E-mail: maher@mat.mte.ncsu.edu

Mancuso, James F.
Advanced Microscopy Techniques
Box 661, South Prospect Street
Rowley, MA 01969, USA
   Phone: +1-508-948-5507
   Fax: +1-508-948-7376

Mankos, Marian
Department of Physics and Astronomy
Arizona State University
Box 871504
Tempe, AZ 85287-1504, USA
   Phone: +1-602-965-3561
   Fax: +1-602-965-7954
(Present Address)
IBM Corporation
Thomas J. Watson Research Center
PO Box 218
Yorktown Heights, NY 10598, USA

Matsumoto, Takao
Tonomura Electron Wavefront Project
ERATO, JRDC
c/o Advanced Res. Lab., Hitachi, Ltd.
Hatoyama, Saitama 350-03, Japan
(Present Address)
Advanced Research Laboratory
Hitachi, Ltd.
Hatoyama, Saitama 350-03, Japan
   Phone: +81-492-96-6111
   Fax: +81-492-96-6006
   E-mail: matsumoto@harl.hitachi.co.jp

Matteucci, Giorgio
Department of Physics
University of Bologna
Via Irnerio 46
I-40126 Bologna, Italy
   Phone: +39-51-630-5145
   Fax: +39-51-630-5153
   E-mail: matteucci@gpxbof.df.unibo.it

McCartney, Martha R.
Center for Solid State Science
Arizona State University
Tempe, AZ 85287, USA
   Phone: +1-602-965-4558
   Fax: +1-602-965-9004
   E-mail: mccartney@csss.la.asu.edu

McGibbon, Alistair
Oak Ridge National Laboratory
PO Box 2008
Oak Ridge, TN 37831, USA

McKernan, Stuart
High Resolution Microscopy Center
Department of Chemical Engineering and
Material Science
University of Minnesota
100 Union St., SE
Minneapolis, MN 55455, USA
   Phone: +1-612-625-8508
   Fax: +1-612-626-7530
   E-mail: stuartm@staff.tc.umn.edu

Mertens, Bas M.
Department of Physics
Delft University of Technology
Lorentzweg 1
Delft, 2628 CJ, The Netherlands

Mohan, A.
1611 Laurel Ave. #1218
Knoxville, TN 377916
   Phone: +1-615-546-5316
   E-mail: amohan@utkvx.utk.edu

More, Karren L.
Metals and Ceramics Division
Oak Ridge National Laboratory
Bldg. 4515, MS 6064
PO Box 2008
Oak Ridge, TN 37831, USA
   Phone: +1-615-574-7781
   Fax: +1-615-574-4913

Mulvey, Tom
Electronic Engineering and Applied Physics
Aston University
Aston Triangle
Birmingham B4 7ET, United Kingdom
   Phone: +44-21-359-3611 ext. 5269
   Fax: +44-21-359-0156
   E-mail: t.mulvey@aston.ac.uk

Myers, Alline F.
Dept. of Material Science and Engineering
North Carolina State University
PO Box 7907
Raleigh, NC 27695, USA
   Phone: +1-919-515-7148
   Fax: +1-919-515-7724
   E-mail: myers@mte.ncsu.edu

Nolan, Ted
High Temperature Materials Laboratory
Metals and Ceramics Division
Bldg. 4515, MS 6064
Oak Ridge National Laboratory
PO Box 2008
Oak Ridge, TN 37831, USA
   Phone: +1-615-574-0811
   Fax: +1-615-574-4913

Ono, Yoshimasa A.
Tonomura Electron Wavefront Project
ERATO, JRDC
c/o Advanced Res. Lab., Hitachi, Ltd.
Hatoyama, Saitama 350-03, Japan
(Present Address)
Advanced Research Laboratory
Hitachi, Ltd.
Hatoyama, Saitama 350-03, Japan
   Phone: +81-492-96-6111
   Fax: +81-492-96-6006
   E-mail: yaono@harl.hitachi.co.jp

Op de Beeck, Marc
EMAT
University of Antwerp (RUCA)
Groenenborgerlaan, 171
B 2020 Antwerpen, BELGIUM
   Phone: +32-3-218-02-61
   Fax: +32-3-218-02-57
   E-mail: modb@ruca.ua.ac.be

Ott, John A.
Dept. of Materials Science and Engineering
Stevens Institute of Technology
Castle Point on the Hudson
Hoboken, NJ 07030, USA
   Phone: +1-201-216-8312
   Fax: +1-201-216-8306

Pennycook, Stephen J.
Solid State Division
Oak Ridge National Laboratory
PO Box 2008
Oak Ridge, TN 37831-6030, USA
   Phone: +1-615-574-5507
   Fax: +1-615-574-4143
   E-mail: pyk@solid.ssd.ornl.gov

Pozzi, Giulio
Department of Physics
University of Bologna
Via Irnerio, 46
I-40126 Bologna, Italy
   Phone: +39-51-630-5146
   Fax: +39-51-630-5153
   E-mail: pozzi@gpxbof.df.unibo.it

Ravikumar, V.
Dept. of Materials Science and Engineering
Northwestern University
2225 N. Campus Drive
Evanston, IL 60208, USA
   Phone: +1-708-491-7798
   Fax: +1-708-491-7820
   E-mail: ravi@casbah.acns.nwu.edu

Ren, Yasuo
Nissei Sangyo America, Ltd.
755 Ravendale Drive
Mountain View, CA 94943, USA
   Phone: +1-415-961-0461
   Fax: +1-415-961-7259

Ren, Shelly
Dept. of Materials Science and Engineering
University of Tennessee
434 Dougherty Engineering Bldg.
Knoxville, TN 37996, USA
  Phone: +1-615-574-4343
  E-mail: xlwang@utkvx.utk.edu

Rice, Philip M.
Metals and Ceramics Division
Bldg. 5500, MS 6376
Oak Ridge National Laboratory
PO Box 2008
Oak Ridge, TN 37831-6376, USA
  Phone: +1-615-574-2480
  E-mail: icu@ornl.gov

Richardson, Wayne H.
Yamamoto Quantum Fluctuation Project
ERATO, JRDC
Edward L. Ginzton Laboratory
Stanford University
Stanford, CA 94305-4085, USA
  Phone: +1-415-725-2161
  Fax: +1-415-725-4115
  E-mail: wayne@sierra.stanford.edu

Rodrigues, Richard P.
Department of Materials Science and Engineering
Northwestern University
2225 N. Campus Drive
Evanston, IL 60208, USA
  Phone: +1-708-491-7798
  Fax: +1-708-491-7820
  E-mail: richard@casbah.acns.nwu.edu

Ru, Qingxin
Tonomura Electron Wavefront Project
ERATO, JRDC
c/o Advanced Res. Lab., Hitachi, Ltd.
Hatoyama, Saitama 350-03, Japan
(Present Address)
Takayanagi Particle Surface Project
ERATO, JRDC
c/o JEOL Ltd.
Musashino 3-1-2, Akishima-shi
Tokyo 196, Japan
  Phone: +81-425-45-7175
  Fax: +81-425-45-7231

Rudee, M. Lea
Dept. of Electric and Computer Engineering
University of California at San Diego
9500 Gilman Drive
La Jolla, CA 92093-0407, USA
  Phone: +1-619-534-4575
  Fax: +1-619-534-4771
  E-mail: rudee@ucsd.edu

Sanguedolce, Joseph A.
Nissei Sangyo America, Ltd.
25 West Watkins Mill Road
Gaithersburg, MD 20878, USA
  Phone: +1-301-840-1650
  Fax: +1-301-963-0808

Scheerschmidt, Kurt Erich
Max Planck Institute of Microstructure Physics
Weinberg 2, D-06120 Halle/Saale
Germany
  Phone: +49-345-5582-910
  Fax: +49-345-5511-223
  E-mail: schee@secundus.mpi-msp-halle.mpg.de

Sewell, Peter B.
LAB-6
4890 Opeongo Road
Box 259, RR#3
Woodlawn, Ontario K0A 3M0, Canada
  Phone: +1-613-832-1090
  Fax: +1-613-832-3735

Shupe-Jones, Rebecca
High Temperature Materials Laboratory
Metals and Ceramics Division
Bldg. 4515, MS 6064
Oak Ridge National Laboratory
PO Box 2008
Oak Ridge, TN 37831, USA
  Phone: +1-615-574-0811
  Fax: +1-615-574-4913

Smith, David J.
Center for Solid State Science
Arizona State University
Tempe, AZ 85287-1740, USA
  Phone: +1-602-965-4540
  Fax: +1-602-965-9004
  E-mail: smithd@csss.la.asu.edu

Smith, David R.
Department of Physics
University of California at San Diego
9500 Gilman Drive
La Jolla, CA 92093-0319, USA
   Phone: +1-619-534-3719
   Fax: +1-619-534-6301
   E-mail: drs@ucsd.edu

Spence, John C.H.
Department of Physics and Astronomy
Arizona State University
Tempe, Arizona 85287-1504, USA
   Phone: +1-602-965-6486
   Fax: +1-602-965-7954
   E-mail: spence@phyast.la.asu.edu

Steeds, John W.
H.H. Wills Physics Laboratory
University of Bristol
Royal Fort, Tyndall Avenue
Bristol BS8 1TL, United Kingdom
   Phone: +44-272-288730
   Fax: +44-272-255624

Sung, Changmo
University of Massachusetts
0318C One University Avenue
City of Lowell, MA 01854, USA
   Phone: +1-508-934-3650
   Fax: +1-508-970-2435

Tanji, Takayoshi
Tonomura Electron Wavefront Project
ERATO, JRDC
c/o Advanced Res. Lab., Hitachi, Ltd.
Hatoyama, Saitama 350-03, Japan
(Present Address)
Advanced Research Laboratory
Hitachi, Ltd.
Hatoyama, Saitama 350-03
Japan
   Phone: +81-492-96-6111
   Fax: +81-492-96-6006

Tonomura, Akira
Tonomura Electron Wavefront Project
ERATO, JRDC and
Advanced Research Laboratory
Hitachi, Ltd.
Hatoyama, Saitama 350-03, Japan
(Present Address)
Advanced Research Laboratory

Hitachi, Ltd.
Hatoyama, Saitama 350-03, Japan
   Phone: +81-492-96-6111
   Fax: +81-492-96-6006
   E-mail: tonomura@harl.hitachi.co.jp

Tsung, Lancy
Philips Electronic
85 McKee Drive
Mahwah, NJ 07430, USA
   Phone: +1-201-529-6160
   Fax: +1-201-529-2252

Turner, Shirley
Microanalysis Group
Bldg. 222, Rm A113
National Institute of Standards and
Technology
Gaithersburg, MD 20899, USA
   Phone: +1-301-975-3923
   Fax: +1-301-216-1134

Van Dyck, Dirk
EMAT
University of Antwerp (RUCA)
Groenenborgerlaan 171
B-2020 Antwerpen, Belgium
   Phone: +32-3-218-02-58
   Fax: +32-3-218-02-57
   E-mail: dvd@ruca.ua.ac.be

Venables, David
Department of Material Science and
Engineering
North Caroline State University
Raleigh, NC 27695-7907, USA
   Phone: +1-919-515-7622
   Fax: +1-919-515-7724
   E-mail: venables@mat.mtu.ncsu.edu

Völkl, Edgar
High Temperature Materials Laboratory
Metals and Ceramics Division
Bldg. 4515, MS 6064
Oak Ridge National Laboratory
PO Box 2008
1 Bethel Valley Road
Oak Ridge, TN 37831-6064, USA
   Phone: +1-615-574-8181
   Fax: +1-615-574-4913
   E-mail: voelkle@ornl.gov

Wang, Young-Chung
Dept. of Materials Science and Engineering
Stevens Institute of Technology
PO Box S-1660
Castle Point on the Hudson
Hoboken, NJ 07030, USA
   Phone: +1-201-216-5270
   Fax: +1-201-216-8306
   E-mail: y1wang@vaxc.stevens-tech.edu

Wittig, James E.
Dept. of Material Science and Engineering
Vanderbilt University
Box 1593, Station B
Nashville, TN 37235, USA
   Phone: +1-615-343-6028
   Fax: +1-615-343-8645

Yoda, Yasunori
Nissei Sangyo America, Ltd.
25 West Watkins Mill Road
Gaithersburg, MD 20878, USA
(Present Address)
Scientific Instruments International
First Department
Nissei Sangyo Co., Ltd.
24-14, Nishi-shimbashi, 1-Chome
Minato-ku, Tokyo 105, Japan
   Phone: +81-3-3504-7295
   Fax: +81-3-3504-7302

Zhang, Xhu
Department of Physics and Astronomy
Arizona State University
Tempe, AZ 85287, USA
   Phone: +1-602-965-6486
   Fax: +1-602-965-7954

Zhang, Xiao
Polymer & Inorganic Systems Laboratory
General Electric Company
Bldg. K-1, Room CE-111
PO Box 8
Schenectady, NY 12301, USA
   Phone: +1-518-387-6709
   Fax: +1-518-5595

AUTHOR INDEX

Ade, G., 33
Allard, L.F., 103, 169, 199, 219
Aoyama, K., 239
Bonevich, J.E., 125, 135
Cavalcoli, D., 159
Chen, J., 81, 145
Coene, W., 307
Cowley, J.M., 329
Datye, A.K., 199
de Haan, P., 329
de Ruijter, W.J., 181
Dravid, V.P., 209
Fink, H.-W., 257
Frost, B.G., 103, 169, 189
Fukuhara, A., 93
Gajdardziska-Josifovska, M., 181
Harada, K., 125, 135
Herring, R.A., 287
Hirayama, T., 81, 93, 145
Hull, R., 189
Ishizuka, K., 45, 81, 93, 145, 317
Joy, D.C., 169, 355
Kalakkad, D.S., 199
Kambersky, V., 329
Kasai, H., 125, 135
Knoll, F., 117
Kreuzer, H.J., 257
Lai, G., 81, 93, 239
Lehmann, M., 69
Libera, M., 231
Lichte, H., 11, 69
Lin, X., 209
Mankos, M., 329
Matsuda, T., 125, 135

Matsumoto, T., 249
Matteucci, G., 159, 329
McCartney, M.R., 181, 189, 329
Midgley, P.A., 277
Muccini, M., 159
Op de Beeck, M., 297, 307
Ott, J., 231
Pozzi, G., 125, 135, 287
Qian, W., 267
Ravikumar, V., 209
Rodrigues, R.P., 209
Ru, Q., 55, 145, 239, 343
Ruoff, R.S., 219
Scheerschmidt, K., 117
Scheinfein, M.R., 189, 329
Schmidt, H., 257
Smith, D.J., 181, 189
Spellward, P., 277
Steeds, J.W., 277
Spence, J.C.H., 267
Subramoney, S., 219
Tanji, T., 45, 81, 93, 145, 249
Tonomura, A., 1, 81, 93, 125, 135, 145, 249
Van Dyck, D., 297, 307
Vincent, R., 277
Völkl, E., 103, 169, 189, 199, 219
Wang, Y.C., 231
Weiss, J.K., 181
Wilkox, N., 209
Yang, Z., 329
Yoshida, T., 125, 135
Zhang, X., 267